Restoring the Balance

WHAT WOLVES TELL US ABOUT OUR RELATIONSHIP WITH NATURE

John A. Vucetich

 JOHNS HOPKINS UNIVERSITY PRESS | *Baltimore*

Johns Hopkins Paperback edition, 2024
9 8 7 6 5 4 3 2 1

Johns Hopkins University Press
2715 North Charles Street
Baltimore, Maryland 21218
www.press.jhu.edu

The Library of Congress has catalogued the hardcover edition of this book as follows:

Names: Vucetich, John A., author.
Title: Restoring the balance : what wolves tell us about our relationship with nature / John A. Vucetich.
Description: Baltimore : Johns Hopkins University Press, 2021. | Includes bibliographical references and index.
Identifiers: LCCN 2020045458 | ISBN 9781421441559 (hardcover) | ISBN 9781421441566 (ebook)
Subjects: LCSH: Wolves—Behavior—Michigan—Isle Royale. | Human-animal relationships—Michigan—Isle Royale. | Nature—Effect of human beings on—Michigan—Isle Royale. | Isle Royale (Mich.)—Environmental conditions.
Classification: LCC QL737.C22 V83 2021 | DDC 599.77309774/997—dc23
LC record available at https://lccn.loc.gov/2020045458

A catalog record for this book is available from the British Library.

ISBN 978-1-4214-4908-1 (paperback)

Special discounts are available for bulk purchases of this book. For more information, please contact Special Sales at specialsales@jh.edu.

For Leah

CONTENTS

Foreword, by David W. Macdonald vii

Preface xvii

Acknowledgments xxiii

1. Why Wolves? 3

2. Thoughts of a Moose 33

3. Beginnings 67

4. Balance of Nature 93

5. Exogenous Forces 133

6. The Old Gray Guy 160

7. The Unraveling 194

8. Sense of Place 233

9. All Natural 261

10. Restoring the Balance 286

Coda 324

Notes 335

References 361

Index 379

WHY WOLVES?

The kaleidoscope of captivating facts about wolves is certainly induce-
ment enough to study, publish, and read about them, but what of the
big questions? Why do wolves capture our intellect, excite our ardor so
powerfully? As a 10-year-old bewitched by Jack London's novels, I knew
the answer—I wanted, really wanted, to be a wolf. In truth, it seems to
me that all insightful biologists feel what it's like inside their subjects'
skin. In this book, John Vucetich asks you to imagine yourself as a
young wolf, dreaming of attempting to kill your first moose, 10 times
your size, using only your teeth. He asks the big question (bravely, for
a hard-nosed quantitative biologist in a profession neurotic about anthropomorphism), *What is it like to be a wolf?* He thinks, as do I, that this
is a more sensible question than you might suspect, in part because it
turns out there's so much similarity between us and them.

THE AUTHOR

Which is the more interesting, the author or his subject? For me, it's
hard to say in this case, because John Vucetich is—and I am proud to
say it—my friend; as I read his book, enthralled by the topic, I do so
with the added benefit of hearing his voice in my mind. So let us start
with John himself, in case I may offer some insight into the man behind
the narrative. His simple statement, "*I am comfortable in the wilderness.
I like science, math, and philosophy. I enjoy a modicum of competency in
each,*" opens a small door into the much larger space of his inner world.

There was a generation whose hearts pumped to the rhythm of the
backwoods, while their brains dissected the natural world with a scientific scalpel. I revere the legacies of Niko Tinbergen and Hans Kruuk,

but think too of the Schallers, Terborgs, Grants: naturalist scientists whose estimable scholarship reflected, and depended on, their mastery of fieldcraft. I fear this rugged, eclectic skillset is becoming rare, but clearly, John is their descendant. The lessons that lie ahead for you in this book, reader, have been distilled from years during which John's boots were muddy, and his entire office consisted of a laptop *"balanced on the handlebar of a snowmobile."* But the cocktail that is *Restoring the Balance* would be flat were it not for the fizz of scholarship, the cerebral thought, that bubbles through every sentence. This active intellect that readers will savor in John's scholarship, he conspicuously appreciates in others. In pondering the deceptively trivial matter of whether cats scratch because they itch (the answer turns out to have implications that reverberate from tiny ticks to the vast action of forest regeneration), he introduces us to *"the elite cadre of scientists who study this sort of thing—few in ranks, distinguished in ambition"* and *"with imaginations big enough to see what others had overlooked."* This becoming admiration of intellect blossoms into a reverence for great scholars whose work has shaped modern ecological science and is woven with erudition through this narrative—we are reminded of the *"grandest insight"* of Charles Elton, *"the Father of Animal Ecology,"* who brilliantly, and perhaps stunningly, asserted that what we imagine to be an inherent balance of untouched nature *"does not exist, and perhaps never has existed."* (Having occupied an office next to Elton in the 1980s, I remember a liberating moment when, burdened by my own relentless backlog of uncompleted tasks, Elton remarked to me, "I'm a bit behind with my work—just writing up some notes from my 1919 Spitzbergen Expedition.")

John leads us through a history of crucial ideas that recounts Elton's subsequent skirmish with Alexander Nicholson; the *"Father of Ecosystem Science,"* Eugene Odum, and his fluxes of energy; Volterra's equations; Gause's protozoans; Huffaker's oranges; and Utida's coexisting beetles and wasps. But then—and this is what matters if you are to understand John and the book he has written—he moves on to consider logicians who, as he puts it, *"endure soul-scarring education* to be worthy of . . . *abstracting the logic of specific, detail-filled arguments into generic expressions stripped of distracting minutiae.*" Then, at a gallop, John's off, clearing the hedges of logical fallacies, such as slippery slope, red herring, straw man, and cherry picking. He really hits his stride through *De Natura Deorum*, where, 2,000 years before Elton and Nicholson (and John specially enjoys juxtaposing the time it has taken people to think things through), Cicero attributes the balance of nature to mutual relationships between species that were all, in Cicero's time, empirical *"evidence for the wisdom and benevolence of the Creator."* John inimitably matches breadth of biological knowledge with depth of historical insight: how could Cicero's ecological God be benevolent and all powerful but still allow the world to be filled with so much evil—predation included? Plotinus (204–270 CE) answers that predation is, in fact, a benevolence: the goodness of abundant life more than justifies the death needed for sustaining other life. John, the mathematical biologist, arrives at a different answer (though always deferring to Darwin, *"In ecology, the ultimate explanations must be approved by Darwin"*). Nonetheless, his modernity builds solidly on awareness of centuries, millennia, of thought.

And there you have it—John is a real field man, a dauntingly quantitative biologist, and a dedicated student of logic: the coalescence of this whole emerges as a leading conservation ethicist. The adage would have us believe that a jack of so many trades can be master of none; the reader will need to discover whether John defies the aphorism. So much for the man—what then, as you prepare to take the plunge, of the book?

THE BOOK

What sort of book is this? Well, there are many sorts of book that it is not. It is not a textbook—while you cannot readily look up fecundity, fetuses, and foxes in indexed alphabetical completeness, you could find almost every nugget worth knowing about these things if you panned for gold in this text. Neither is it a natural history book (although, gosh, the natural history is top notch). Nor is it an autobiography (although, as my opening hints, you do learn a lot about the author). It is not just a science book entirely about empiricism: *"Empiricists yearn to document that variation. Why the yearning? They don't even know why; it's their instinct."* Nor yet is it one entirely about theory. In fact, *Restoring the Balance* is a much more remarkable book than can be defined by any of these familiar types.

Before turning from the genre of the book to the detail of its topic, have in mind a powerful message at the interface of natural processes, fieldwork, and scholarship, and while John does not thump this tub explicitly, it is a subliminal drumbeat to every chapter. There is, as evolutionary theorists will know, a modern corpus of thinking named Pace of Life theory—how individual animals trade off the costs and rewards of living fast and dying young (*"reproduction is inviolably traded against longevity"*). Spend your limited energy on reproduction and that energy is unavailable for maintaining your soma. The book is redolent of parallel trade-offs between working fast and finishing too quickly. John Vucetich himself has spent 25 years plodding around, and giddily flying above, Isle Royale (performing hundreds of moose postmortem examinations), but the legacy that he has inherited from Durward Allen and Rolf Peterson (which they inherited from Adolph Murie's 1935 moose monograph) has accumulated over 60 years of data. In the first year of fieldwork on Isle Royale, rookie graduate student Dave Mech (friend and hero of my early career), aspiring, as John puts it, *"to acquire more knowledge about wolves than any other human ever had,"* clocked not a single sighting of a wolf: zero data. The questions tackled in this book do not lend themselves to the quick academic kill or three-year grant cycle. You might think that devoting over half a century (1958–2021) to researching wolf-moose dynamics constitutes working at a leisurely pace, but it's only 15 wolf generations. Young researchers, choose your subjects carefully: predator-prey aficionados will remember (and if you don't, John will remind you) that Gause's classical predator-prey experiments lasted only six days—but those six days of research saw the passage of more paramecium generations than 60 years of wolf research.

The impacts of disease, inbreeding, and climate change have required decades to puzzle out. To be clear, only part of the dogged pace of research can be attributed to tracking a moving target and new factors entering the concoction—much of this pace is the fruit of people working hard and intelligently, and as fast as their snowshoes would permit, on a natural system that cannot be understood quickly. None of the big questions about whether, how, and why wolf and moose populations ebb and flow could ever be answered by simply knowing the number of wolves at a single point in time. As John says, *"No pulse can be known by hearing just one beat."*

So, to whet your appetite, what am I to tell you of the (moose) meat of the book? Let me confess that in preparing for this task, I took 41 pages of notes, so squeezing that into the allotted few pages is going to require a diamond of compression. First, some fascinating facts. A fortunate wolf will be present for the death of about 300 moose. A heavily infested moose routinely hosts about seven ticks on every square inch of its hide. A wolf has about 50 times the number of olfactory receptors in its nose than we do. The rate that wolves acquire food during the winter fluctuates greatly from year to year, with food being more than twice as plentiful in the good years as compared to the bad years.

Beyond these kinds of intriguing insights, what does this book teach us about wolves? First, through 60 years of exhilarating data, it shares the remarkable, Machiavellian complexity of individual lives, even when viewed from outside their skins. The trajectory of a population often drives the lives of its individuals. But sometimes, the life of an individual can drive the trajectory of a population and even entire food webs (as you read on, remember, these wolves eat moose, and the moose eat trees and are also eaten by ticks). With socializing occupying about eight hours of each wolf's day, wolves and their packs came and went. At times their social order fissured: two packs became three and three packs became five. Once, in 1980, the population *"swelled to 50 sets of paws."* Pups were born, some timid, some assertive: personalities matter, a lot. So too do circumstances, with the waning or prospering of moose, so generations of pups were privileged or rationed by their adults, their lives set to be longer or shorter. In lean times, one pack (the Gang of Four), triumphing by strength of character, monopolized the entire island. In another lean year, half the wolves died. Over the years, John recounts stories of trespassers killed, mates pursued, harrowing hunts, packs founded, and one remarkable alpha, the Old Gray Guy, who built a dynasty with enough power and incest to rival any medieval royal house.

The narrative is replete with the stories of individual wolves like the Old Gray Guy, and others like Cinderella and the appropriately named Romeo. Yet we are urged to remember: *"The fate of every individual depends intimately on the course of the whole. The course of the whole is sometimes nudged by the life of an individual."* Remember, too, *"the inseparability between individuals and the collectives to which they belong."*

If only there was time for me to whet your appetites further with John's insights into how individual wolves interact exponentially to create the affairs of populations and food webs! But many pages of delight lie ahead for you. Therein you will find remarkable science (why John is different is that he is at least as good as his predecessors at getting inside his subject's skin in the field, but he is also world-class at math and theory). Rather little of Isle Royale's contribution to science is about hypothesis testing, *"an over-credited, human-designed incubator of scientific knowledge."* More of it is about John's exhortation to *"clear your mind, open your eyes, and let nature show you something."* What is the main insight of that science? Holism. Consider, for example, how climate change makes for warm years, warm years make for more ticks, ticks weaken moose, wolves select weakened moose, moose abundance falls, and the forest thrives. *"The skin-pricks of such a small creature just might resound throughout the forest."* John calls holism a rhapsodic realization (one owed to plant ecologist Frederic Clements) and recommends you consider your relationship with your kidney: it can no more survive without you than a trillium flower on the forest floor can survive outside its ecosystem.

Of the myriad questions John Vucetich is clever enough to pose, one particularly catches my eye. Why, if only five or six of a pack of 15 hunting wolves expend most of the labor, and shoulder most of the risk, don't they split into two smaller packs sharing more evenly the investment in hunting and the return on moose meals? The question is notable because it is the same question some of us have posed for species (such as badgers) precisely because, unlike wolves, they appear not to cooperate. The short answer to the badger version of the question is that the richness of a single sharable food patch (likely to be an area of pasture squirming with nutritious earthworms) determines the size of social group that can feed together, and that group can't fission its territory and still secure sufficient food patches to sustain even one or two group members. The answer for wolves turns out to be less about cooperative hunting and much more about scavenging ravens.

THE PURPOSE

Being interesting would be enough, but *Restoring the Balance* is more than interesting—it matters, and it matters most when John dons his logician's hat and crosses the Styx from science to policy and starts the arduous climb to value and purpose. His immediate mission on the far side is to evaluate, and inform, discussion on whether wolves should be reintroduced to Isle Royale. Why does this question arise? Because the forest is on its last legs. Why? Because as a result of the accumulated misfortunes of the wolf population (and it will be critical to appreciate the causes of those misfortunes), moose abundance began rising, and each year from 2012 moose have eaten more than the year before. Was this natural? The processes linking predator and prey are natural, but the factors causing this particular expression of these processes were not.

John's sortie into history knocks further nails in the coffin of naturalness, if ever that was to be our yardstick for what is right. That history enmeshes the impact of smallpox brought by colonists on the Ojibwe people, Benjamin Franklin's negotiations with the British to draw the US border, copper mining (Paleolithic and present), and 1920s golf resorts. The narrative's laser highlights personalities and politics, journalists and rivalries, arriving at, for no particularly scientific reason, congressional authorization of Isle Royale as a national park in 1931 amidst a *"rising tide of wilderness recreation inspired by fear that Americans were losing their 'manly' character."* Naturalness is exposed here as too slippery a notion to dictate policy.

So, what logic guides us? With elegantly soft-spoken scholarship, John concludes that wolves and moose (and beyond) are not only alive, they have lives—lives that give them interests (in the sense that humans have interests in surviving, achieving pleasure, and avoiding pain). Extending the Golden Rule (the principle of ethical consistency), those who wish their interests to be treated with fair concern are obliged to treat others with those interests in the same way. If this sounds worryingly hippieish, it isn't: this is a scholarly book by a clinically intellectual author, steering clear of both anthropocentrism and misanthropy. You might not be impressed that the Ojibwe who once inhabited Isle Royale believed the wolf to be their brother—not metaphorically but literally. You should be impressed that exactly the same reality is demonstrated by the DNA-based tree of life, which proves all animals have a common ancestor.

This book is about understanding what counts as a good relationship with nature (and to help you decide, calls on the virtues of equality, equity, need, and entitlement—virtues that logic dictates are not the prerogative of humans but of all those who, like us, have interests). But when it comes to interests, whose do we rank highest? Wolves, after all, kill moose, and moose have an interest not to be killed. Would the reintroduction of wolves to Isle Royale, even if good for individual wolves and populations of moose, be fair to individual moose? Recognizing a moral dilemma is not the same as solving it, but it's a start. Furthermore, John Vucetich will not let you off lightly—he dismisses most of the arguments as *"excuses for absolution from failing to give tragedy due attention."* Ultimately, this is neither textbook nor wolf book: it is a paean to reason.

But where does reason, culpability in hand, lead us? History repeats itself—although we are luckier this time because we have John Vucetich: once again, the question is whether to (re)introduce wolves. John's voice, never raised but always clear, has been heard, if not always heeded, in the highest circles.

I won't spoil the suspense by telling you what he advised or what has happened, but I will tell you that the maxim by which John concludes we might think our way toward coexistence is akin to what I have elsewhere called "respectful engagement with nature." Insomuch as the principles of restorative justice apply to the relationship between wolves and humans, we, descendants of the transgressors, have an obligation to right the past wrongs.

This notion, illuminated with the twin halogens of reason, leads John to offer this positive aphorism: *"Live, let live, and right past wrongs."* The fact that I think it's tricky, when deciding how to balance looking forward and backward, to know when restitution gives way to letting bygones be bygones, should be no impediment to appreciating that when it comes to infringing on the well-being of other species, the report card on humanity reads "must try harder."

Why does it matter? Because, as John observes, *"the wolves of Isle Royale share something very important with a large portion of biodiversity."* It is that, when the wolves vanish, the human enterprise will roll on unperturbed. None of the monetized, utilitarian arguments for conserving biodiversity will cause humanity to miss a beat. The utility of carnivores is insufficient to motivate us to coexist with them. John leads us toward *"a streetwise reason"* for preventing biodiversity loss that goes far beyond Isle Royale. That reason, and the purpose for conservation that must deliver it, will surely be holistically transdisciplinary and tightly laced with ethics—and it will owe a lot to what John Vucetich learned from the wolves of Isle Royale.

David W. Macdonald, CBE, FRSE, DSc
Lady Margaret Hall, Oxford

I have been studying wolves and moose on a wilderness island for more than a quarter century. I have written more than 100 scientific papers stemming from that experience, and I testified on behalf of wolves to the US Congress. The lessons that have meant the most to me are not conveyed in those papers or that testimony. They are conveyed in this book.

I am comfortable in the wilderness. I like science, math, and philosophy. I enjoy a modicum of competency in each. The lessons that have been most important to me—the ones that compose this book—are cocktails of physical experience, tested knowledge, and cerebral thought. *Restoring the Balance* is a concoction. Its first ingredient is my field notes, which chronicle the day-to-day happenings of a scientist (me) and the wolves and moose that I follow. The other ingredients are the science and math that explain the chronicling, and finally the environmental philosophy, that has helped me attach meaning to it all. These insights, gifted from the wilderness, depend on all three ingredients. Just as a margarita depends on lime juice, triple sec, and tequila, no two out of three will do.

This is not a wolf book. It's disqualified from that designation by the prominence of introspection. Nor is it a science book. There's too much triumph and heartbreak in it for that. And it's not a philosophy book. I don't care how many angels can fit on the head of a pin or whether "this sentence is false" might be true. The philosophical parts are simply recounting thoughts that have helped me understand the purpose of my relationship with others—especially nonhuman others. I have no admonishments, just considerations to share, some of which are likely to be new to you.

The book begins with a portrait of wolves. If you know that you like wolves, then Chapter 1 is candy. The nubs of Chapter 1 are excerpted from my field notes. There is blood-soaked snow, an airplane named *Flagship*, a terror-stricken calf saved by her mother, and a wolf dream. Not a dream about a wolf, but a wolf's dream. I do not take for granted that you like wolves. If you don't know whether you like them, Chapter 1 explains what it is about wolves that has held my attention for so many years. It might grab your attention too. At least you'll know what all the fuss is about.

My habit of following wolves and moose extends beyond their tenure among the living. In fact, my understanding of each moose's life usually begins with an at-the-site-of-death necropsy. I'll bring you along for one. That'll put us face-to-capitulum with a tick. Well, not one tick. Hundreds, actually. I'll tell you what that's all about. The ticks will make us think about parasites, which will lead us to consider bot flies, who are, we'll say, "hosted" by mice that live on Isle Royale. These mice brokered my marriage with Leah, and she ends up being the reason we've spent so much time at a mud lick. Those are places where moose come to slurp mud. There will be time to tell you why they do that. I mention it here because the mud lick is where I was presented with the lesson that's meant the most to me. That's all in Chapter 2. You might think I'd save the most important part for later in the book. Nope.

The remote island where I work is Isle Royale. It is situated in the northwest corner of Lake Superior—the largest freshwater lake in the world. Research on Isle Royale's wolves and moose began in 1958, making this the longest-running study of any kind of predator-prey system in the world. I am writing this sentence at base camp during the 62nd winter study. A winter study older than astronauts, Barbie dolls, and Super Bowls. The research has been led, over the decades, by just three scientists. I am the third. Chapter 3 explains how the first leader, Durward Allen, and his student, Dave Mech, pioneered the research in the late 1950s and early 1960s.

Chapter 3 leaves us with the revelation from Isle Royale that resonated more for Allen than any other. This lesson taps the vein of an ancient and tenacious belief. It has been lauded by some and vilified by others. The discord has endured for centuries and reflects a tripartite tension among theology, science, and environmental consciousness. This antagonizing idea is "balance of nature," and it has led to disputes over how to understand nature's essence.

What does this idea really mean?

If the idea of "balance of nature" is true, what does it demand of us?

And, at the heart of it all, is it true?

Allen had philosophical answers for the first two questions, and Isle Royale provided him with a scientific answer for the third. In Chapter 3, I relate that answer as Allen and Mech saw it. It follows this book's recipe—action and thought. Their views will draw our desire to know how others have responded to those questions. That's what Chapter 4 is for. It is cerebral, but it features a shipwreck, a labyrinth coursing with eight-legged Minotaurs, and the discovery of an exotic lifeform that permeates the air we breathe and the ground on which we walk.

In 1974, Durward Allen passed the research reins to Rolf Peterson. Chapter 5 tells how Rolf Peterson's obsession with snow—simple, yet carefully cultivated—led to unanticipated insight: wolf and moose populations are perpetually pulled toward balance by predation and repeatedly knocked from balance by perturbations, such as the occasional severe winter. Chapter 5 concludes with Rolf Peterson's pensive conviction that Isle Royale had become a "broken balance" after the wolf population was hammered down by a disease that humans had inadvertently brought to the island.

With wolves laid low by disease, the moose abundance soared. Moose chewed the forest, day and night, week after week, until there was no more to chew. Starvation conspired with the winter of 1996, the most severe on record, and moose abundance was cut in half. The wake was a mixture of shock and relief, like what follows a stunning financial market correction. But the market kept correcting—moose abundance sank even further over the next several years, and we didn't know why. Over the course of a decade, the moose population fell from its all-time high of 2,400 individuals to its all-time low of 400. I arranged the evidence and assembled the most likely explanation. Predation was potent, though wolf abundance was not high. Ticks had scourged moose for several consecutive years. The ticks themselves were likely favored by a spate of warm years. Running that sequence in reverse produced an explanation: warmer years favored ticks, too many ticks weakened the moose (predisposing them to wolf predation), and the moose population sank. I was honest, but timorous, about that being merely a best guess. Then, a pattern that had been festering over the decades presented itself all at once. This is what Chapter 6 is about—the moose collapse, the red herring, and a sad secret whispered by the Old Gray Guy.

The life of a population unfolds over generations. Nested within, the life of an individual unfolds day to day, season to season, and year to year. Furthermore, the fate of every individual depends on the course of the whole, and the course of the whole is sometimes nudged by the life of an individual. Chapter 7 tells the story of the wolf population's final collapse through the eyes of two individual wolves.

The wolf collapse was anticipated, by several years, based on the evidence that I had assembled with my colleagues. I alerted the National Park Service to the situation and inquired as to what they would like me to say when journalists asked me how they were planning to respond. I expected to be told how they would begin a decision-making process. Instead, I was informed that NPS's wilderness policy was clear. There would be no intervention. I was also left with a clear impression that there would be consequences if I decided to speak out.

A hodgepodge of reasons for letting wolves disappear from Isle Royale percolated over the next few years from individuals who worked for the NPS. Some of the reasons depended on misleading characterizations of the science. Some claimed that there was no need for intervention because the wolves were doing fine. Some acknowledged that the wolves were not fine but argued there was no need to conserve wolf predation; instead, we should hunt moose. At least one NPS employee thought wolves should be allowed to go extinct to prove to the world that they are "not too big to fail." (Seriously, that was a reason.) A prominent voice indicated that the history of Isle Royale itself tells us to refrain from conserving wolf predation. Chapter 8 is about these reasons.

In fact, more justifications for nonaction were offered than one chapter can contain. Chapter 9 explores a particularly potent pair of related reasons to do nothing. The first is that the wolves of Isle Royale and the moose upon which they depend are unnatural, while extinction on an island is natural; therefore, the wolves do not deserve to be protected from extinction. The second reason is that Isle Royale is a wilderness where nature should be allowed to take its course, even if that means extinction. That humans contributed to the extinction—by disease, drownings, and climate change—matters not because only an addict would prescribe additional interference as a cure for an obsession with meddling in the affairs of nature. These reasons are powerful for being insidiously polymorphous, adapting themselves to misguide a wide range of relationships between humans and nature.

In what appears to have been a late reversal, the NPS ultimately decided to restore wolf predation on Isle Royale (Mlott 2016). But was it the right thing to do? If so, how do we know? It's important to ask because the Isle Royale case is emblematic of many instances where the biodiversity of a protected area is, or will soon be, lost or harmed by inadvertent and indirect effects of climate change.

The wolves of Isle Royale share something very significant with a large portion of biodiversity: if Isle Royale wolves vanish, humans will get along fine without them. Consequently, the Isle Royale case is crucial far beyond the confines of the island because responses to the case imply much about what our judgments will be for protecting and restoring that large share of biodiversity that is not especially valuable to humans. Frankly, the questions stemming from the Isle Royale case are more important than the Isle Royale case itself.

The go-to reason to restore any predator, anywhere, is that the health of ecosystems inhabited by large herbivores, like moose, depends on the presence of predators, like wolves. The explanation has merit, but it is inadequate. Chapter 10 is a rumination on that reason, its limitations, and the possibility that the loss of biodiversity is fueled by failing to have forged a streetwise argument for preventing its loss. Chapter 10 also describes the dragons that this redeeming reason would have to slay and what's at stake if we fail to ennoble so capable a rationale.

In the Coda, I describe the NPS's effort to implement their decision to restore wolves on Isle Royale and the status of wolves and moose up to March 2020.

This book was made possible because of the people to whom I belong and their belief in me.

Leah, my wife, believes in me more than anyone. She has invested much in her belief, including my various absences, from hours on the computer screen to weeks away from home and in the field. She contributes much to the success of wolf-moose research on Isle Royale, from managing the lab to organizing logistics and personnel. On countless occasions she's helped me through various phases and stages of this book, which would not exist without her love and support.

Rolf Peterson, my mentor and partner in Isle Royale research, has believed in me for a very long time. I'm forever indebted to him for opportunities that shaped the course of my life, enduring friendship, and the joy of working hard together.

Perhaps only my mom and dad have believed in me longer than Rolf. My mother deserves much credit for igniting my curiosity, and she would have been proud of this book. I'd never have made it far enough in life to write a book were it not for my father. Emblematic memories of what he provided include what must, at times, have seemed a tiresome schedule of scouting events, which nurtured my love of the outdoors, and his help with my morning paper route to pay for an education at a private high school.

Andrew J. Storer, dean of Michigan Technological University's College of Forest Resources and Environmental Science, and his predecessor, Terry L. Sharik, consistently express their belief in me and have demonstrated that belief with valuable support, often at moments when I've been in most need.

This book is, in part and indirectly, about my personal journey in being an academic engaged with the world. Many people contributed to that journey, including, but certainly not limited to, Joseph Bump, Scott Carter, George Desort, Camilla Fox, Jill Fritz, Brett Hartl, Ron Kagan, Joe Kaplan, Wayne Pacelle, Paul Paquet, Nicole Paquette, Dave Parsons, Candy Peterson, Mike Phillips, Tim Presso, Doug Smith, Adrian Treves, Tom Waite, and Senator Gary Peters. David Macdonald and Dawn Burnham have been many good things to me, from dear friends to valued colleagues; especially for inspiring me to reach high, thanks. I owe tremendous intellectual debt to Jeremy Bruskotter and Michael Paul Nelson. Without their intellect, I'd still be swimming in a much smaller pool of knowledge and insight; with their friendship, I've become rich. Thanks fellas.

Research on the wolves and moose of Isle Royale has been successful and enduring because of the skill, commitment, and good cheer of the pilots who've served. I'm grateful to Don E. Glaser for showing me how to observe nature from an airplane with the skill of an expert and the wonder-like joy of a child. Don E. Murray (grandson of the Don Murray featured in Chapter 3) is near the beginning of what I hope becomes a tenure on Isle Royale as long as Glaser's and is becoming as good a friend and compatriot.

Wolf-moose research on Isle Royale has benefited immeasurably, year in and year out, from many hundreds of benefactors and volunteers. Thanks to all the college interns who have worked with me over the years. Thanks to Ken Vrana for administering our team of citizen-scientist volunteers. That effort has been key to the success of the research and has brought tremendous joy and a taste of science to so many. I cannot resist mentioning a few volunteers by name: Karen Bacula, David Beck, Bob Bollinger, Jim Clink, Clay Ecklund, Ron Eckoff, Erik Freeman, Larry Fuerst, Michael George, Velda Hammerbacher, Hal Hanson, Jeff Holden, Dick Murray, Loreen Niewenhuis, Joe Olenik, Tim Pacey, Ron Porritt, David Rolfes, Thomas Rutti, Wayne Shannon, Barrett Warming, and John Warming. Thanks for all the miles you've hiked in the wilderness searching for clues about the lives of moose and wolves.

I'm grateful to Sarah Hoy for picking up slack in our work, which created time I needed to work on this project.

Research on the wolves and moose of Isle Royale National Park has been supported, financially and institutionally, by the College of Forest Resources and Environmental Science of Michigan Technological University, the US National Park Service, the US National Science Foundation, McIntire-Stennis grants, the Robert Bateman Endowment at the Michigan Tech Fund, the James L. Bigley Revocable Trust, and the Detroit Zoological Society. Much financial support has also come from hundreds of private citizens making large and small donations. For all that support, I am indebted.

Thanks go to Matthew Monte and his staff at Monte Consulting for preparing the illustrations in this book.

I am especially grateful to the people of Johns Hopkins University Press who made my experience with this book a dream. Esther P. Rodriguez made the incorporation of illustrations a breeze. Carrie Love's masterful copyediting has resulted in subtle and valuable improvements. I owe much to Tiffany Gasbarrini for believing in the project from the beginning and for leading me to improvements in writing for which I will be forever grateful.

RESTORING THE BALANCE

SOME PLACES ON ISLE ROYALE

1. Angleworm Lake
2. Bangsund Cabin
3. Beaver Lake
4. Big Siskiwit Swamp
5. Card Point
6. Chickenbone Lake
7. Chippewa Harbor
8. Cumberland Point
9. Daisy Farm Campground
10. Duncan Bay
11. Feldtmann Lake

12. Florence Bay
13. Grace Creek
14. Grace Creek Overlook
15. Grace Harbor
16. Grace Island
17. Harvey Lake
18. Hatchet Lake
19. Hay Bay
20. Hidden Lake
21. Houghton Point

22. Hugginin Cove
23. Lake Desor
24. Lake Eva
25. Lake LeSage
26. Lake Mason
27. Lake Richie
28. Lake Whittlesey
29. Lily Lake

30. Little Siskiwit River
31. Little Todd Harbor
32. Livermore Lake
33. Long Point
34. Malone Bay
35. McCargo Cove

36. Middle Point
37. Moskey Basin
38. Mount Franklin
39. Mount Siskiwit
40. Mud Lake
41. Pickett Bay

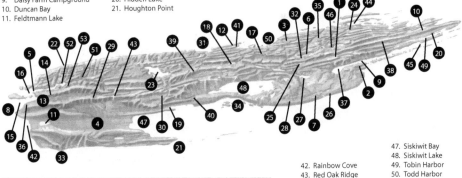

42. Rainbow Cove
43. Red Oak Ridge
44. Robinson Bay
45. Rock Harbor
46. Sargent Lake

47. Siskiwit Bay
48. Siskiwit Lake
49. Tobin Harbor
50. Todd Harbor
51. Washington Creek
52. Washington Harbor
53. Windigo

The Greenstone Ridge runs the length of Isle Royale, down the middle of its long axis.
The Minong Ridge runs parallel, just a couple miles north of the Greenstone Ridge.

Base map from the US National Park Service

1 |

Why Wolves?

February 18. Last night the wolves of Chippewa Harbor Pack crossed the Greenstone Ridge to hunt in remote corners of their territory. By morning they were traveling back toward the core of their territory.

Shortly after we caught up with them, not far from Little Todd Harbor, we watched them change course, abruptly and straight into the wind. We saw what they smelled—a cow moose and her calf, who had themselves been foraging. It didn't look good for the cow and calf right from the beginning. The calf was too far away from her mother, and they may have had different ideas about how to handle the situation. The wolves rushed in. The cow turned to face the wolves, expertly positioned between the wolves and her calf, but only for a second. The calf bolted. After a flash of confusion's hesitation, the cow pivoted and did the same. Had she not, the wolves would have rushed past the cow and bloodied the snow with her calf. The break in coordination between cow and calf put four or five wind-thrown trees lying in a crisscrossed mess between the cow and her tender love. The cow hurled herself over the partially fallen trunks that were nearly chest-high on a moose. She caught up with her frantic calf before the wolves did.

Then the chase was on, led by the least experienced of them all—the calf. The cow, capable of running faster, stayed immediately behind the calf, no matter what direction the terror-ridden mind of that calf decided to take. Every third or fourth step the cow snapped one of its rear hooves back toward the teeth of death. One solid knock to the head would rattle loose the life from, even, a hound of hell.

After a couple of minutes and perhaps a third of a mile, the pace slowed. By the third minute everyone was walking. The calf, the cow, and the wolves. The stakes were high for all, but not greater than the exhaustion they shared. Eventually they all stopped. Not a hair's width separated the cow and calf, and the wolves were just 20 feet away. The cow faced the wolves. A few minutes later the wolves walked away. By nightfall Chippewa Harbor Pack had pushed on another six miles or so, passing who-knows-how-many-more moose. Their stomachs remained empty.

Those are my field notes from the winter of 2012. The island is Isle Royale, a wilderness surrounded by the largest freshwater lake in the world. I make these observations from the *Flagship*, an airplane just large enough for a pilot and one observer. After the flight, questions hack their way through the recursive web of dendrites that is my consciousness. What is the life of a wolf like? What is it like to be a wolf? Those questions are too presumptuous. The first questions should penetrate down to the foundation: *Of all the millions of species on planet Earth, why wolves, why not some other? Why concern ourselves with an impossible and dangerously anthropomorphic question, what is it like to be a wolf?* The first question should be, *Why wolves?*

Long ago, no one can say when, we made wolves into something more than the fur and paws and teeth that they are. We turned wolves into symbols, representing all that many of us love and cherish about nature. The Mongolian people believe they were created through the union of a deer and a wolf. The Ojibwe people profess the wolf as a principal source of knowledge. Western civilization rose from the she-wolf who suckled Romulus and Remus.

But then there are werewolves and Little Red Riding Hood. For wolves' occasional and innate habit of killing cattle and sheep, we spin them into symbols of all that many fear and hate about nature. We transmogrified wolves into partisan, bipolar symbols of our existential struggle to understand our relationship with nature on the whole.

Wolves are not just another species contributing to the health of the ecosystems they inhabit. We also press wolves to serve as our symbols. Symbols are vital to our humanity, but they flatten rich living realities into simpler inanimate objects, bandied about to serve one story or another. A wolf is a living creature, with a perspective, memories of yesterday, an interest in how tomorrow turns out, joys and fears of its own, and a story to be told. Those realities create obligations for us to be concerned for their lives. The harm in symbolizing a wolf is the same that rises when a young person idolizes a parent or lover. It's unfair.

The antidote to having unfairly turned wolves into a symbol is to ask, "What is it like to be a wolf?" Answering this question would seem to require overcoming insurmountable differences between ourselves and the subject of inquiry. Imagine panting every time you're just a little bit too warm, killing prey 10 times your size with your teeth, learning more about the world through your nose than your eyes or ears. Go ahead, try to learn what your neighbors are up to today with your nose. The life of a wolf lies on the murky fringes of our imaginations. Think of losing one in four family members to death every year of your life—that's typical for a wolf. This comparison is warranted because a wolf is more than able to grieve a lost sibling or parent. Even your pet dog misses you when you've merely been gone on a weekend trip.[1]

Our similarities are no less striking. Wolves spend about a third of their lives walking, roughly a third of their time socializing with packmates, and the balance of their time sleeping. They prefer to avoid their neighbors, except when appropriating territory or courting a love interest. If you like your family, but not your neighbors, if you like napping after a big meal, or if you like long walks, then you can understand the essence of what it is like to be a wolf. These similarities provide all the toehold necessary for developing an honest relationship with them.

One more point of connection deserves attention. Meat. Wolves eat meat, as do most of us. Living through the flesh of other animals is not at all distinctive. Hundreds of thousands of species (fishes, birds, frogs, shrews, wasps)—millions of species, really, live by taking the lives of others. But wolves don't merely eat flesh; they take nourishment from the same kind of flesh that we humans desire most. We share a craving for the flesh of large herbivores, especially deer, elk, moose, reindeer, cows, and sheep.

So, wolves confuse us. We daydream about wolves in admiration of how we perceive ourselves—exquisitely capable of putting meat on the table, conspicuously social, well-exercised, and beautiful. But from deep within, we project on wolves gluttony, deception, wasteful killing, sexual predation, and the incarnation of evil. We seethe at God's audacity for making another in our image. We obsess over exterminating the competition.

Of course, there are deeper differences and more basic similarities. For now, it is enough to recall that we have a choice—a choice to be taken by our differences or by our similarities. When taken by the similarities, we can discover kinship or conjure competition. When taken by the differences, we might marvel at the diversity or boil with repulsion. The value in asking "What is it like to be a wolf?" is what it will teach us about ourselves.

What might I learn personally from asking that question? Well, the root of my surname is *Vuk*, a Croatian word for wolf. Perhaps unsurprisingly, I struggle to understand the extent of coincidence and limits of predetermination. But somewhere in between lies the comfort of purpose. For the past 25 years, my purpose has been to study the wolves and moose of Isle Royale.

I am third in a succession of scientists to have that purpose. Durward Allen started the research in 1958. He recounts the first 18 years of the wolf-moose project in *Wolves of Minong*.[2] It is right to begin as he did: "My story is of an island and the creatures that inhabit it. It is a blend of natural and human history." For a taste of that blend between their lives and mine:

February 19. Chippewa Harbor Pack covered only a few more miles through the night. By midmorning we found them traveling through thick, tangled cedars in a drainage just southwest of McCargo Cove. On at least two occasions in the past month, they tracked down and chased the cow and calf that live in this area. On both occasions the wolves failed to make a meal. They almost certainly know half a dozen sites or more where they can find another cow and calf. There must be something about this cow and calf. Maybe the mother is inexperienced or old, or maybe the calf is underdeveloped.

Even from the privileged vantage of the *Flagship*, it was nearly impossible to see the wolves through the dense cedars. Maybe once every third circle, I'd glimpse just one wolf as he or she passed from beneath one cedar to the next. After quite a few more circles, a moose ran out of the swamp and over an open ridge top. A moment later the wolves appeared on the same ridge top. The moose was long gone. The wolves laid down on the ridge and slept for several hours. Chippewa Harbor Pack has chased or tested, quite possibly, half a dozen moose or more in the past 36 hours. To kill an 800- or 900-pound moose with your teeth is no mean feat.

February 20. Chippewa Harbor Pack hunted throughout the night. It's been 10 days since they last killed a moose. In the past 30 days, they've eaten only one adult cow moose, one yearling cow, and one calf. That's probably not enough to sustain the pack. Someone will likely starve to death. Chippewa Harbor Pack entered the winter with six wolves, and it has been three weeks since I've seen that sixth wolf. He may have dispersed in search of an opportunity to mate. Or he may have starved. I do not know. If he starved, he may not be the last.

As morning came, they walked out of the forest and onto the frozen ice of Moskey Basin. They rested some, then reentered the forest to hunt. Within an hour, they found and chased a cow and her calf. Exhaustion was the only result. They came up out of the forest again to rest on a south-facing ridge. A few hours later they searched the thick stand of fir trees north of Moskey Basin, where they found and chased another cow and calf, this time injuring both. But as night fell, nobody had fed and nobody had died.

February 21. Fresh snow indulges our interest to find and follow the tracks that trace the faint profile of these wolves' lives. So, I was excited for the first measurable snow accumulation in a month, a meager few inches. Up to today, a month's worth of tracks—moose, otter, foxes, and wolves—had accumulated. Wolves were also walking in their old tracks, in old moose tracks, or weren't leaving any tracks at all as the snow became hard and crusty. The island is our teacher, but the lessons are easier to grasp when she occasionally erases the chalkboard.

I spent the day boiling flesh off the bones of wolf-killed moose, collected from necropsies conducted earlier in the field season. Pretty light duty this season, as there have been only four necropsies so far. When moose escape predation and wolves suffer food shortage, work at the bone pot is less demanding.

February 22. A couple of hours before nightfall, the clouds had cleared just enough to permit a flight. It was enough time to discover that Chippewa Harbor Pack had finally killed a moose—probably the calf they injured two nights ago. The wolves fell asleep with the content feeling that comes from having full bellies. The feeling is almost certain to be ephemeral.

Wolves live in packs with other wolves, and they kill prey larger than themselves. You probably knew that. Those facts are so basic that we could be done and move on to some lesser known aspect of wolves that would make you think, *Well, isn't that something! I didn't know that.* Doing so would be like telling the natural history of humans by noting that we have a conspicuously large prefrontal cortex and dexterous hands, and then moving on to some tidbit about our kidneys. We will profit by taking it slower and in more detail.

Wolves live in packs, typically led by an alpha pair. They are the parents. Most or all of the remaining wolves are offspring born earlier that year or in previous years. Packs sometimes include an aunt, a cousin, or a grandparent. Most wolves on Isle Royale live in packs with four to seven other wolves.[3]

Pace and detail in this story are important, but insufficient. We also need to narrow the alienating gap between wolves' experience and our understanding of that experience. Let's try again.

Packs are families with much in common with your own. At the core, packs are parents and offspring. Some alpha wolves are naïve parents raising their first pups with experiences as fresh and novel as the pups'. Other alphas are old pros on their third or fourth set of quintuplets(!).[4]

Pups are born each year around the end of April. They are mostly full-sized by about the following January, but they don't know much, and they aren't useful for getting much of anything done around the pack. A pup's role in a pack is a little like that of a six-year-old human in a family—and both cases are as they should be, insomuch as packs and families exist for their young children, not the other way around.

Some pups are timid toward others and shy about new things. Others are aggressive and adventurous. Pups develop personalities that are fodder for much of the socializing that occurs in a pack— socializing that consumes about eight hours of each day.[5]

Pups in successful packs expect to be fed, but in floundering packs they only hope. Even the best wolf parents routinely falter. Starvation is a common cause of death. Even in the rare wolf population where humans refrain from killing wolves, it's common for one in five wolves to die each year.

Wolves can be understood in some ways by knowing domestic dogs. After all, dogs are *Canis lupus familiaris*, a subspecies of wolf. If you live with a dog, you know she likes some members of your family more than others and listens to some better than others. She experiences sadness when you're gone for an unexpectedly long spell and happiness when you return. She is sad upon the anticipation of prolonged absence; she knows what follows the packing of a suitcase. When a beloved family member dies, dogs mourn. The cognitive abilities of dogs have atrophied in the generations since they descended from wolves; if your dog can understand a loss, rest assured a wolf can too. To lose a family member each year and call that "normal"—if that circumstance was wholly beyond our imagination, we'd have no responsibility to try to understand it. But we can, and we should, strain our mind's eye for the shadowy outlines.

The older offspring of a pack, typically one to four years of age, are preadults and adults. These older progenies are sometimes merely tolerated, sometimes helpful, sometimes capable but fallen on hard times, and sometimes just not very capable.

When unable to make it on their own, some young adult humans all but wear out their parents' welcome. This is also a common occurrence in a wolf family. Such a hanger-on lives only where the pack was yesterday, never where they are today. Close enough to know what she is missing, far enough to avoid violence, and often a half mile or so behind the pack. She follows the tracks of her family, sleeps alone, and scavenges from carcasses after her estranged family has finished and moved on. Humans are not the only ones who ostracize.

The only way to learn anything about anything in the world is through metaphor and simile. My life is an open book. Robert Burns's elegant insight is that love is "like a red, red rose." And, the penetrating observation of children's poet Denise Rodgers is a simile: your feet smell so bad, just like limburger cheese. Unlike poets, the overly science-minded only hack up measurements and calculations. They haven't discovered that learning blossoms in comparing the familiar (human families) to the unfamiliar (wolf families). The insight carried by a metaphor depends on the very uncertainty we are left feeling about the proximity between the two poles of a metaphor. When we are unquestioningly sure about a relationship, learning is stifled. Comparisons between the familiar and unfamiliar are not gratuitous or risk free. But they are essential.[6]

How deep is the comparison between human families and wolf packs?[7] Is the similarity literal or literary license? Do we see superficial differences concealing a common core? Or, are we covering two different phenomena in the same veneer for the warm, but unjustified, feeling of kinship with other living creatures?

Behavioral scientists have crafted Darwinian theories to explain subtle details of parenting, parent-offspring conflict, and the stability of families, such as how those relationships are affected by wealth or the presence of stepparents. These theories rise from the presupposition that kindness is motivated by selfishness and that a parent's sacrifice is not quite the concession that it seems. These acts of apparent charity half help the parent because a parent and offspring share half their genes. So, a child can expect half—no more, no less—the sacrifice that a mother would make for herself. The theories place genes, rather than whole creatures, at the center of life. There's much more, but that's the essence.

The theories seem to explain, for example, the age of weaning, temper tantrums, favored siblings, spontaneous abortions, and a parent's influence on their grown offspring's selection of a mate. These theories are informed as much by human families as by families of ants, herring gulls, fur seals, baboons, and woodpeckers. Ungrateful adolescents and other intrigues of family life existed long before humans. Families are not a uniquely human adaptation or a quintessentially human experience. Rather, humans are among the many creatures who are fundamentally shaped by some kind of family life. And, each of us is free (obligated, really) to ask whether the similarities and differences inspire kinship or alienation.

A pack sometimes includes an extranuclear member, like an uncle or a grandmother. A grandmother is probably a former alpha female who lost the post when she could no longer outcompete one of her daughters. But she is still healthy and still respected for her judgments on killing moose and defending territory—judgments that can be as important as brutish strength.

The uncle may have been traveling with his sibling when he became an alpha. He may be capable and helpful, just unable to start his own family. Nevertheless, the alpha wolf will infer the intentions and monitor the motives of his brother. Every member is continuously reevaluating an ever-shifting balance on a scale weighted by genetic overlap.

A wolf family gets its start when would-be parents successfully defend real estate from neighboring packs and learn how to kill moose on a regular basis. Like families you know, the roles are broadly overlapping but varied. In some packs the female is better at leading kills, in other families it's the male, and in some cases it's a coordinated effort.

In coming to know the wolves and moose of Isle Royale, we will continue encountering an idea that can cause discomfort but cannot be avoided: anthropomorphism. While we are still here in the beginning, we should, for a just a few moments, find the humanistic and scientific interface of anthropomorphism.

The word "anthropomorphism" is most commonly understood to be deprecatory and used to communicate wrongheaded comparisons between humans and nonhumans, as if anthropomorphism were a kind of alchemy whereby things take human forms by unholy transmutation. The word has been burdened with all the pejorative tendencies and undertones that we attach to alchemy. But what if anthropomorphism means turning a thing, not into a human, but into a person? If humans and people are one-in-the-same, then we've gone nowhere. But humans and people are not one-in-the-same.

Allow me to clarify. "Person" is rooted to a Greek word, *prosōpon*, referring to a mask worn in a theatrical play.[8] A person is a character on the stage of life. A person has stories that can be told and interests that

The word "person" comes from the Greek "*prosōpon*," which refers to a mask worn on stage. © www.gograph.com/lhfgraphics

can be cared for. "*Homo sapiens*" is a narrow and exclusive category in Linnean taxonomy. But "person" is not. "Person" is a moral category that includes beings who have needs and desires deserving of our concern.

Wolves, like people, are characters with stories to be told and are deserving recipients of compassion. So a wolf is, sensu strictiore, a person. There, I said it. Wolves are people. They are one of millions of different kinds of people living on the planet. There are roaring people, peeping people, feathered people, scaled people, cold-blooded people, and warm-blooded people. While many are powered by mitochondria, I think some people might be fueled by chloroplasts. Leastwise, I have met trees with stories and interests. And so have you.

Anthropomorphism is not a clever literary tool for poets like Shel Silverstein to make up stories about trees. When performed properly—that is, when based on genuine similarities—anthropomorphism is not alchemy. But anthropomorphism is not without consequence. As one's capacity to observe similarity increases, so does one's capacity for benevolence. For an idea stamped with "arcane" and "recondite" all over it, anthropomorphism is practically important for shaping everyone's relationship with nature.

Some refuse similarities between humans and wolves (and other kinds of people) as a way of understanding wolves (and other kinds of people), and they refuse a priori.[9] The refusal is prejudged to avoid adding to the weight of things one ought to care for. Overly harsh? Probably not on the account of the white heterosexual men who have been refusing for millennia to acknowledge basic similarities between themselves and humans of other races and genders. These refusers deny elements of others' personhood (cognitive capacities and possession of souls) to dissolve obligations that rise from personhood, obligations such as fair treatment. If we so easily refuse comparisons and deny similarities among our own species, we should be on guard for the same tendency in our relationship with other species.

Surrounded by the fog of a powerful metaphor or an arresting comparison, we are confronted with a choice: refuse the similarity, and remain indifferent until the similarity is shown without doubt to be warranted, or accept the similarity and attending call for compassion until or unless the similarity is unquestionably demonstrated to be fanciful. Each choice carries its own risk.[10] A useful guide through the fog is knowing that, while humans suffer many sins of excess, gratuitous caring is far from epidemic.

This view on anthropomorphism carries us to empathy and the value of understanding empathy as the result of the cold evolutionary pressure applied to those who survive among the fittest. Empathy's neurological basis was discovered in the 1990s and boasts a eureka moment with a legendary hue that reminds one of Isaac Newton, his discovery of gravity, and the apple that conveyed the insight. A team of researchers had been conducting laboratory experiments on the premotor cortex of macaques—a kind of monkey commonly used in laboratory research. Your premotor cortex is a narrow band of gray matter stretching from just above your left temple over the top of your head to just above your right temple. This network of neurons is just an inch or two wide, from front to back, and, like the rest of your brain's cortex, is just a few millimeters thick. This portion of a mammal's brain is where neural communications arise and pilot the muscles that allow us to walk, sit, lie down, and generally animate our bodies during our time on stage. A particular region of the premotor cortex, known as field 5, or F5, is devoted to all the various hand motions. F5 is the origin of many auspicious actions such as grasping and bringing food to our mouths.

The researchers had surgically implanted electrodes into the F5 region of some macaques. The electrodes led to a machine that issues audible static when neurons in F5 fired. The macaque was trained to perform certain actions with his hand and the researchers were trained to interpret the resulting static. One afternoon, during an unscripted moment between experimental protocols, a researcher was scurrying around the lab. The macaque, bored out of his prefrontal cortex, followed the researcher with his eyes. The researcher picked up some food, the static machine sounded, and he (the researcher) was struck in the head by Newton's apple.

The researcher had engaged some neurons in F5 that said, "Use your hand to pick up the food." The macaque observed his lab mate do so. In his attentive state of mind, the macaque also engaged neurons in the F5 region of his brain. Empathy, by its dictionary definition, is understood to be a vivid, knowledge-based imagination of another's perspective or situation. Like an iceberg, empathy is much more than what floats above the surface. The mindful macaque had experienced a neural representation of his lab mate's perspective using neurons otherwise devoted to representing his own self. The macaque's mind had blurred the boundary between self and other just by observing. He created a bridge from self to other through the eyes.

Some people (of the human kind) believe it impossible to sensibly empathize with other kinds of people. Accepting that belief would be a confession that the cognitive and imaginative capacities of a macaque exceed our own.[11]

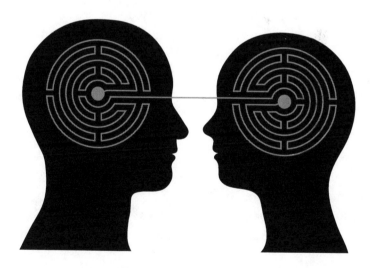

Empathy is, in a manner of speaking, a connection between minds.
© www.gograph.com/adriann

Let's return to the idea *wolves kill moose* and see it not as a basic fact, but let's try to know it as the essential truth that it is. Prepare to descend to the deepest recesses of your imagination. Unpracticed imaginauts may not get all that close.

The most powerful equipment at our disposal for this vivid imagining is dreams, where the real and imagined are not well distinguished. To get this right, some dreaming might be necessary. But the empathic exercise is not to dream about wolves killing moose. That will be insufficient. Nor is the aim to dream that you are a wolf killing a moose. That will risk projecting too much of one's own bravery or fear onto the wolf. The best we might do is to imagine what it is like to be a wolf having a dream about killing a moose. To dream of a dreaming wolf will generate essential intimacy, yet leave just enough personal space between you and the wolf's experience to limit undue projection of your own self onto the experience, but only so long as our dream is adequately anchored to sturdy facts.

In daring attempts of empathy, imagination pulls against knowledge and knowledge against imagination, balanced by the tension of a suspension cable. That's what empathy is—an under-engineered suspension bridge to another's understanding. To prevent your imagination from pulling you into a solipsistic fantasyland, you need a firm grip on knowledge that is no less powerful than your imagination.

First, know that wolves dream.

How do we know?

Dog owners know because they've seen it—sharp tremors and tiny seizures accompanied by muffled moans and muted grunts issued by that distant relative of wolves as she sleeps on your living room floor while you eat popcorn on the couch.

And why do we believe those are signs of a dreaming dog?

Because dreaming is widespread among our mammalian brethren.

How do we know that?

The better question is, *What do they dream about?*

Most mammals cannot tell you about their dreams, but they can be made to act them out. To prevent dreams from being acted out on the stage of the real world, unconsciousness orchestrates a suite of bio-chemical reactions that leave muscles paralyzed during sleep, not merely relaxed but paralyzed every time we surrender our minds. Were it not so, dreams would result in considerable embarrassment, danger, and on some occasions, prison time. This paralysis, aka REM sleep atonia, is prevented if your locus coeruleus (located deep in the brainstem) is damaged. Without a properly functioning locus coeruleus, your muscles are engaged during dreams.

Researchers of the 1960s surgically destroyed the locus coeruleus of some cats and watched their dreams. They observed unconscious cats, sound asleep, prowling and pouncing on imaginary mice. We mammals dream about the important stuff in life. For cats that means dreaming about hunting mice. The dreams of a blue whale must be astounding. Wolf dreams? They certainly include hunting.

Lie down, enjoy the content sensation of a full belly, and the satisfaction of muscles that ache from having accomplished a physical feat. Melatonin rushes your body, your lungs give up a deep sigh, and consciousness falls away like the tide.

You know those experiences as well as every mammal. In this case, the full belly and aching muscles, you can imagine, are from having killed a moose two days ago.

I cannot imagine the dreaming wolf for you. No one can. I will provide some natural history, empathic triggers, and notes from the field. I'll provide the knowledge to brace against the imagination that serves only those who conjure for themselves. Interspersed and in italics, I'll also offer toeholds for imagining this exotic truth.

Evening shadows bleed together in a pool of darkness that covers the forest. Your parents sleep, her muzzle resting on his back. Your littermate tries to rouse your two younger siblings, each asleep in their own place. Your oldest sibling plants his haunches firmly to the ground, tenses his front legs, stretches his mouth to the sky and howls. You and the others wake to join the vespers song.

Warriors crush their fears with battle cries. Demonstrators drive their own civility off with riotous hooting and hollering. Football players prepare to smash and be smashed with primal vociferations. Boisterous vocalizations enable base intensions. Inhibitions retreat, and physical capacities are enhanced. You can practice the experience in diminutive form with an experiment in the exercise room. Work out today while Barry Manilow (sorry) eases on through those ear buds. Tomorrow, workout to the throbbing beat of AC/DC. Grunt, groan, and grumble like Lou Ferrigno (sorry). Compare the results.

Wolves howl for the same reason.

From their beds, your parents allow your brother to incite commotion. After the commotion rouses their consciousness, they lead the chorus themselves. Wagging tails. Rising blood pressures. Chasing, tumbling, tackling. Racing hearts. Paws, teeth, fur. Sweaty paws. Minds dissolve and meld after just a few minutes of ecstatic chaos. Your parents trot—not the gait of a happy dog on a brisk walk, more like gladiators entering the arena—down and off the open ridge top and into the forest. You and your siblings follow.

Somewhere in the forest a moose has just finished ruminating and begins to refill her rumen.[12] She heard the howls echo off ridges and soak into the cedar swamp where she feeds. She is aware, but not especially concerned. She continues feeding in the dark. For most of their lives, moose have little to fear over wolves. Wolves do not lord over the forest taking whatever they want, whenever they want, because they aren't able. The essential task in killing a moose is not finding one or catching up with one. The burden is to find a moose you can kill without being maimed or killed yourself.

Will you contribute much on this hunt? Last time you only ran along side, got in the way, and didn't even try to get close until it was injured, until it was safer. Even then, you almost got clocked in the head with a hoof. Your sister lost her breath after she was thrown against the trunk of a tree. After that you stayed back.

Do not confuse words for experience. Wolves do not experience the words. They experience memories, intentions, and images. Moreover, your mental experiences are not a teleprompter. They are, to a large degree, rekindled or anticipated feelings and images of what was, what is, and what might be. This much you have in common with wolves. For the dreaming wolf that you imagine, the scenes are physical and violent, and the emotions are fueled by hunger and risk of injury.

The hunt begins with no indication of when or how it will end. Five hours later we pass the little round lake—the one with the gigantic beaver lodge that we always climb atop. Then over the ridge to the shore of the big lake, beyond the corner of the big square harbor where we always cut inland to the lake surrounded by a little ridge, through a break in that ridge where we killed a moose last month. We walk single file because the snow is deep. Except in places where the snow is windswept and hard. In those places, my young brother leapfrogs to the head of the line as if he's the decider. He lies down in the snow as soon as he's unsure of what direction to go next and resumes his position toward the end of the line. My sister walks with me, brushing shoulders. She's bored and would rather play.

Their evening walk is as familiar to them as walking through your own neighborhood with your own family is to you.

Somewhere in the forest is a mother too inexperienced to protect her calf. Somewhere else is a 16-year-old bull whose arthritis has progressed to the point that his femur is now completely dislocated from his hip socket. For several weeks, he has eaten every miserable twig his mouth can reach without having to take another step. These are the moose that wolves search for.

Wolves also fare better with moose who suffer jaw necrosis. As moose age, their teeth loosen up, allowing twigs and needles to get lodged between. Moose don't floss, so the debris leads to infection. Moose don't take antibiotics, so the infections worsen. Feeding is impaired and vitality fades.

In the worst cases of necrosis in the lower teeth, the bacteria rot a jaw bone right through into two pieces. The worst cases of jaw necrosis in upper teeth rot a hole through the palate admitting masticated vegetation into the nasal cavity, where it ferments and obstructs breathing. If not for predation, most cases of jaw necrosis would become worst cases.

Allow me to take us away from our dreaming for a bit to review salient details of wolf predation, colored with some insight about how we learn of such things.

When observing wolves from the *Flagship*, I regularly find sites where wolves have killed moose. I mark the spot with a dot on a map and wait until the wolves finish feeding and leave the site, which typically takes three to six days. Then the pilot lands the plane on the nearest frozen lake, and I strap on snowshoes and hike in to perform a necropsy on the moose.

These kill sites are routinely deep within a tangled mess of forest heaped onto a rough and wild ground. The alders are alive. They grab and pull at my pack and snowshoes. Using both hands to pull through the tangle, I feel like I'm crawling even though I'm upright—most of the time. After the alders are exhausted, the cedars step in. Every meter forward includes one up, another over, and then one down, crawling over one fallen cedar after the next. As I get close, and to guide me closer, wolf tracks appear. Like the braided streams of an estuary flowing atop the snow-covered ground, these tracks are a week's worth of comings and goings. The next guidance is provided by an unkindness of raucous ravens. Invariably, we hear each other at the same moment. A tornado of black wings burst upward and around in every direction from the ground and trees that surround the carcass. In a trice they are gone and the forest is silent. Ravens never lie, but they never tell their secret twice. To be of directional assistance, I have to anticipate the black tornado so I can see where it came from. Otherwise, I know I'm close but have no idea in what direction. I might have covered 90% of the distance but only half the time required to find the carcass.

Final guidance to the carcass, just 20 meters before the anticipated arrival, is a Pavlovian moment, a blend of two smells—several hundred pounds of flesh and blood on the snow that smells a little like a frozen meat locker mingling with the sour smell of fermenting vegetation from the moose's rumen. During the necropsy, at a point when I'm close enough to have a good look at the teeth, a third scent is added to the medley—the funky smell of rotting flesh, the waste of bacteria who began eating this moose weeks ago. Jaw necrosis.

A wolf's nose is packed with 280 million olfactory receptors. You and I have only 2% that number of olfactory receptors.[13] The portion of a wolf's brain dealing in smell is 40 times greater than ours. Their nostrils and inner nose are the envy of any NASA engineer; we have the Ford Pinto model. A wolf's sense of smell is 10,000 to 100,000 times more accurate than our own. Wolves don't smell with their noses—they see the world with their noses. Every adult dog learned some time ago to overcome their befuddlement at our inability to smell past our noses.

If I can smell jaw necrosis at a kill site, then a wolf can smell it many times over. Very likely, wolves can smell jaw necrosis before seeing the moose with the rot in its mouth. Wolves hunt into the wind, not only to conceal their intent and to locate moose, but also to know what kind of moose lies ahead.

A sniff test is indicative but not determinative. The definitive test by which wolves know they have found a moose they can likely kill involves charging the moose. If the moose lowers her shoulders, plants her hooves through to the frozen ground, and releases a roar from lungs twice the size of a lion's, if she holds her ground, ready to meet any attack from any direction with a swift kick, if she is ready to explode, then panicked paws skid and slip through the snow. The wolves pivot from offense to defense in less than a heartbeat, retreating to the edge of safety, just a few meters from the moose. Because the response could be a thin concealer, the wolves charge a second time. If the moose maintains more fury than fear, the wolves retreat farther, mill some, and move on before long. The intense portion of the encounter often lasts only a few minutes, sometime less.

Nineteen out of every 20 encounters end with passing excitement for a moose or the heart-pounding terror of a near escape. Occasionally, wolves startle a moose who has momentarily forgotten that he has nothing to fear, as illustrated by these notes from the field.

March 4, 2007. Flying along the shoreline of Siskiwit Lake, we found and began to follow two wolves who had been traveling into a southwest wind. They may have been the same two who killed a moose a few days earlier, and without the assistance of their parents or siblings. If so, they are making a bit of a show in being on their own this far from the security of their natal pack.

Abruptly, their pace rose to a determined trot as they traveled through a stand of large, open-grown spruces in the savannah-like country not far from Malone Bay. Within a few moments we came to see what the wolves could smell. Up ahead, a large moose—probably a bull—had bedded himself on the windward side of a particularly large spruce, whose impenetrable canopy grew straight to the ground and spread 20 feet across.

He may have been enjoying the cool breeze. Moose are easily overheated, even in winter. Quite aside from any possible climate comforts, the windward side of a thick spruce is a bold place to rest given the strategic liabilities of doing so.

In quick silence, the wolves followed their noses to the leeward edge of that spruce canopy. The moose unaware. The wolves crouched low and crept around opposite sides of the spruce.

The wolves had to have been as surprised as the moose when they got to within six feet. The moose jumped out of his skin and ran like a banshee, seized by convulsive kicking and thrashing. Seconds later and about 30 yards from the spruce, the moose halted as suddenly as he had first erupted. He spun around and faced his assailants. The wolves froze. The bull stared. The wolves reconsidered the prudence of pressing the issue any further. After about a minute, the wolves walked away. The moose walked off in the other direction, and we flew back to basecamp for refueling.

For wolves, all those attempts end, I suppose, in some bittersweet tincture with equal parts of anxiety and relief. Wolves have much to teach human hunters about the doctrine of fair chase. On a 20th try, often after three nights and 30 miles through the forest, an encounter will unfold differently.

Now we can re-submerge into our imagined dream of a young wolf's hunt.

The moose panics and bolts. You never have run so fast or breathed so hard. You dig in to keep up with your parents. They are just a foot behind a pair of hooves that snap inches from their faces with each pace. Your siblings follow at a safer distance.

The objective of committed members of the pack is to bite into the massive muscle of the upper rear leg and hang on. That won't stop the moose, only slow it.

Your nose and left eye are pressed tight against a thousand pounds of rage. Somehow, you got there first. The moose dekes left, your mother falls for it, and your father stumbles. But you are right there. Your mouth is full of moose fur and your canines only precariously past the skin. All your weight is held by your teeth—a strain you've never felt before. With each forward lunge of the moose, you are kicked in the chest by the lower portion of the rear leg to which you attached. Your own back legs are dragged over logs and rocks larger than you.

Wolves have crazy powerful teeth, and jaw muscles to match, but they are still just teeth.

The aim is to stop the moose from moving forward. The moose will try for thick cover where it can protect its rear by backing against a tangle of fallen cedars. The wolves will try to force the moose into a forest opening or onto a frozen lake.

The moose slows to a stop. Your siblings catch up. The moose is surrounded. Even one of your youngest siblings, who has seen fewer than a dozen kills, is blocking an escape—precarious but effective, at least for the moment.

Look up to the upper corner of the room. Your neck muscles tighten as you tilt your head. That is the wolves' perspective on a surrounded moose. Neither injury nor exhaustion obviates the plain, physical fact that this moose is 10 times larger than a wolf. The hooves are still deadly. The maximum penalty for misreading a cue is death. Common penalties are broken ribs and concussions. Wolves confront this beast with resolve to kill it with their teeth.

Your father alternates between hanging from and pulling on the moose's rear. The moose swings. Your father's legs lift from the ground with centrifugal force. His canines haven't held as well since two of them broke last year. He lets loose and is tossed against the trunk of a large spruce. You reconsider.

Here, we can conclude our meditation and break to easier and more familiar modes of understanding.

While that dream covered the basic themes, each encounter between wolves and moose is an improvisation, performed once and never again. To better understand some of the variations, here are accounts of five tragic victories.

February 20, 2009. Don Glaser, the longtime pilot for this research project, dropped me off at Mud Lake. I snowshoed a mile to necropsy the moose that Middle Pack had killed a week earlier. The hike was easy, across open terrain, with only an occasional spruce or birch. Tracks in the snow confessed how Middle Pack had forced the female moose up against a bouquet of birch trunks. At one point, it seems, she reared and flailed her hooves. Off balance. Her front left foot landed and caught in the crotch of two trunks, several feet off the ground. It ended quickly. The favors played by Fortuna have nothing to do with fairness or deserving.

This next account reminds us that heartrending grief often shadows triumph for a moose who survives a wolf encounter.

February 19, 2008. For the entire flight Don wrestled the control stick, matching every push and every shove as turbulent wind bullied the *Flagship*. We made the rounds, looking for each of the four packs that comprise this year's wolf population. We never found Paduka Pack. East Pack remained at the site northeast of Lake Eva, where they'd killed a moose on the 13th. Six miles south, Chippewa Harbor Pack fed from a moose that they had killed on the 17th, a little northeast of Lake Mason.

Middle Pack had traveled from Houghton Point to the north side of the Big Siskiwit Swamp and were now traveling through thick stands of spindly spruces and twisted cedars. Because of the wind, we could only fly large, fast circles.

Then a moose appeared, kicking and stomping. With each pass we caught only a glimpse through the thick canopy. On one pass we saw a wolf being tossed as the moose spun. The wolf's teeth let loose of the moose's hind end. On some passes we saw nothing at all. Then four wolves surrounded the moose, lunging teeth first. Four more wolves raced across an opening in the swamp vegetation to catch up with the action. Within five minutes of first observing the attack, the moose was dead.

Three days later, I snowshoed to that site. Only then did I realize that the wolves had killed a nine-month-old calf. The mother was undoubtedly present—panicked and then anguished. We just never saw her through the thick forest cover.

When moose lose, it is never clean. This next episode, however, is about as clean as any moose vanquishing.

February 12, 2006. Don picked me up from Lake LeSage, where he had left me to perform a necropsy on a moose that Chippewa Harbor Pack had killed not far from the lake and just a few days before. With the sun sagging low and the fuel gauges leaning toward empty, we did not have a great deal of time to make a final check on each of the packs.

We quickly found East Pack as they crossed to the west side of Mc-Cargo Cove. After noting the number of wolves and direction of travel, I suggested looking for Chippewa Harbor Pack. As the words came out, I noticed a bull moose feeding in a thinly forested area just upwind of East Pack. We could wait to see what would happen, but then we wouldn't have time to find the other packs. And, I'd seen it plenty of times: a pack confronts a moose, the moose escapes, the wolves carry on.

Don, having observed wolves for longer than I've been alive, suggested that we stick with this group. The moose was about 200 meters ahead of East Pack. Within a few minutes, East Pack worked through a very thick stand of cedars. They saw each other when perhaps 50 meters apart.

The bull reeled and fled. The wolves were there. The moose spun again to make a stand. The wolves skidded to a stop and then lurched back a few steps. In a moment the bull was surrounded. Wolves grabbed any hold available at the moose's rear. The moose tried to face each lunge. In the tumult, every turn exposed one flank or the other to some wolf. In another brief moment, there was a break in the circle of wolves. The moose bolted through that opening, heading for the thick cedar stand from which the wolves had come. As the moose passed one of the wolves, it lunged and bit deeply into the moose's right hamstring.

The moose dragged the 80-pound wolf through two feet of snow. After 20 meters, the moose tore free. He didn't quite make the edge of the thick cedar stand when the wolves caught up, and one managed to bite and hang onto the moose's hindquarter. The moose slowed considerably. A second wolf leaped and latched onto the moose's rear.

The moose slowed enough for the alpha male to run to the moose's front end. Front hooves are more dexterous and better aimed than rear hooves. But biting and hanging from the bull's nose gains considerable leverage and tends to hasten the end. The alpha male waited and maneuvered for opportune angle and timing. The moose yielded no such convenience.

Four sets of teeth were pulling on the bull's rear end. His hind legs collapsed with more than 300 pounds of gnashing wolf. Amazingly, the moose's front remained upright, and still no wolf could bite his nose. Beyond amazing, the moose shook himself free from all four wolves and stood up. He was, however, surrounded by the nine wolves of East Pack. Some were focused and waiting for the right moment to attack; others milled around just waiting to feed.

After several minutes, one wolf attacked, then a second, third, and fourth. The moose's rear was brought to the ground once again where it remained for several minutes. The pounded-out snow turned pink and then red. Again, the moose managed to stand and shed its wolves.

This cycle of being brought half-down and then recovering repeated itself twice more. Forty minutes after the wolves first chased the bull, his rear legs collapse for the last time and then his front legs. All nine wolves began tearing flesh from all sides. From the ground, he could no longer kick. For a few moments, the wolves fed while the moose was not quite dead. At some unknown moment, a life passed.

A wolf's life is mostly occupied with walking, socializing, and sleeping. The idyllic-sounding routine is punctuated by the violence and terror of killing a moose. It happens about once a week or so, as frequently and consistently as you might get groceries. A fortunate wolf will be present for the death of about 300 moose.

January 24, 2008. While oatmeal burbled on the stove in the light of Rolf's[14] headlamp, I organized flight gear. After breakfast, we prepared the *Flagship* for flight at dawn's twilight in the sharp crispness that accompanies −8°F. Takeoff at 0830.

We found Middle Pack just south of Lake Desor. Six, curled into tight little balls, all asleep. They looked peaceful.

After a few more circles, we noticed a moose standing just 50 meters to the southwest. She stared like a zombie in the direction of the wolves, her rear end firmly planted against a spruce tree. Almost certainly, Middle

Pack wounded this moose last evening. Likely, it's too dangerous to kill at this moment, but Middle Pack thinks the wounds are mortal. With every passing hour, the wounds will stiffen and the moose will tire. Eventually the wolves will kill it. In the meantime, and not far off, they slept.

A light northwest wind blew across Lake Superior, stirring it, like the three witches of *Macbeth*, into a steaming cauldron.[15] The lake gave portions of herself up to the sky. Snow-laden clouds began to pour over the ridge tops. Seductive, but not conducive for flying in a small aircraft. We flew back to basecamp. Cocoa, some chores, and by 1400 the sun shoved the clouds aside. With the opportunity for another flight, we learned that Middle Pack had not killed that moose by day's end.

January 25. Shortly after takeoff, while the sun was still hovering above the horizon, we found Middle Pack. Sometime during the night, they killed the moose.

They had already partially dismembered the carcass. Ravens had a turn feeding while all eight of the wolves slept with bellies full of organ meat, muscle, and fat.

She was strong and dangerous, and able to breathe and stand and fight well past sunset. But not into the light of the next day. The beginnings of jaw necrosis had tipped the balance against her.

She felt minutes that lasted an hour and experienced entire days that had been poured into an hour.

With this next, and final, account, we'll see that sometimes eternity comes only after the passage of too many days, each of which had become their own eternity.

February 2, 2010. Sometime in the night, Middle Pack abandoned the scattered remains of their last meal. Now, at the edge of a swamp, deep near the heart of Washington Creek, they stood watch over an adult moose. They had wounded it earlier this morning or late last night. The nine wolves engaged the moose in rounds, just often enough to keep it from feeding or lying down. This went on all day. It's not vicious or gratuitous. It's unavoidable. The wolves are unable to kill, and the moose unable to escape. With the passage of each hour, the moose's injuries stiffen and the impasse softens.

February 3. Middle Pack spent another day with the moose they'd wounded. The wolves allowed the moose to lie down, and it did. If the moose was unable to stand, the wolves would have killed it. Apparently, the moose was tired enough, yet confident enough, to lie before the wolves.

The unease is palpable from our cold seats in the cockpit of the *Flagship*.

By nightfall, the wolves leave.

After our dinner, we studied photographs we'd taken from the *Flagship*. They suggest that one of the wolves may have been injured by the wounded moose.

February 4. Throughout the night and into the morning, most of Middle Pack traveled south across the Grace Creek drainage, over Red Oak Ridge, and on through the swamps south of Feldtmann Lake. Unable to kill the moose they'd wounded two and a half days ago, they searched for an easier prospect.

Thinking little of that prospect, one of the Middle Pack wolves remained with the moose in which they had already invested much. The wolf ate blood-soaked snow. The moose is apparently too injured to chase that wolf off or escape to some other part of the forest, but it was strong enough to keep one wolf at bay while it fed and ruminated. It is plausible that the moose was also lucid enough to understand that this wolf's attendance portends the pack's return.

February 5. Low clouds were developing with the rising east wind. We'd watch that trend carefully and would likely have enough time for a short flight.

Middle Pack returned to the wounded moose from the southwest. But they stayed only briefly and continued northeast toward Lake Desor. They were checking on a casserole in the oven. The comparison, not the circumstance, is cruel. The circumstance is without comparison to any culinary experience in my life.

The clouds lost their loft and sagged toward the ground. We flew home as the flyable airspace was compressed to a film between the cloud bottoms above and the ground below. At base camp, I helped Don with a few mechanical items on the *Flagship* and put our tools away long after the feeling had left our fingertips.

February 6. The wolves of Middle Pack did what all wolves do best. They traveled. Their tracks stretched through 18 miles of boreal forest in a large circuit around Lake Desor, past a site where they'd killed a moose a few weeks ago. We find satisfaction in observing that they revisited a kill site to stir the scattered bones of a moose that had been snow-covered and previously unknown to us.[16]

The rescattering of bones was hardly so satisfying for the wolves. The remains were thin and only reminded the wolves that it had been a few weeks since last eating. By day's end, 10 minutes before we landed for the day, we watched Middle Pack return to the wounded moose.

February 7. The dawn sky was a hundred shades of gray and black, like a Rembrandt. The clouds held all the snowflakes they could. One would occasionally fall to the ground. We slept in.

Snow built and continued through the day. We were grateful for the first snow in a month. The island had been covered in a tangle of tracks—wolf, moose, fox, and otter tracks—confused and difficult to sort out. Snow buries the disorientation that often accompanies too much information. On some occasions, and perhaps ironically, fewer clues are easier to decipher.

In the warmth of the woodstove, we studied more photographs from the past few days of flying. They show the alpha female walking with an awkward posture. We wonder if she had been injured in last week's attack. Injury would explain Middle Pack's reluctance to kill that moose. She has been leading Middle Pack for almost a decade. Big changes are in store if her best moose-killing days have passed.

February 8. Snowflakes tumbled from the sky and time passed lightly. The weather eased with the last hour of daylight. Don and Rolf took off for a flight with enough time to ascertain only the basic facts. After six days, the wounded moose was still alive. Middle Pack was three-quarters of a mile to the north—just 15 minutes away, 10 if there was occasion to hurry. At this repositioned location, they had wounded another moose.

The investment strategy is uncommon and reflects, I suspect, considerable frustration.

February 9. Ravens roosted. Foxes slept. Moose, unable to listen for what lurks in the forest, were anxious. Snow didn't fall. Instead, it rose from the harbor, 30 and 40 feet into the air. The day belonged to a deafening 30-knot wind that blew from the northwest.

February 10. The trees screamed. *Flagship* tugged at the ropes that held her to the ice. Another day taken by the wind. I worked on a scientific manuscript, Rolf reviewed field notes, and Don counted his toes (practicing for the moose census).

February 11. The wind subsided and we were hungry for time in the air. For the first time in a week sunlight reverberated between a brilliant sky above and blinding snow below.

We found Middle Pack feeding from the remains of two moose. Judging by the extent of disarticulation, both moose had died within the past day.

Predation compressed the experience of those nine days more compactly than matter is crammed into a spacetime singularity.

In a typical year, most moose on Isle Royale are tested by wolves about once a month. No moose fails more than once, and most moose eventually fail the test. They die from circulatory shock, hypovolemia, insufficient blood pressure in the brain, or injury to some other vital organ—all preceded by an ocean of exhaustion, a lifetime of anxiety compressed into a moment held by a fast and shallow heartbeat, a light and vertiginous feeling in the head brought by low blood pressure, and a dissociating numbness that is a true gift, offered by a cocktail of biochemicals that alter the response of nerve cells and suggestive that evolution may be more compassionate—if only by accident— than her reputation. Teeth, swinging hooves, bloody snow, spinning sky, faintness, a once proud and still massive shoulder hitting the ground hard, tearing of flesh, fading, and then nothing.

Now the reasoning and rationalizing begin. We try to squirm away to easier thoughts: he was about to die anyway. He might have lived a bit longer, but his quality of life would be so poor. We might even tell the story by replacing the pronouns, he and his, with it and its.

We reason that the only other option is death by starvation or the infection that spreads from jaw necrosis. We praise death by predation for outmaneuvering those alternative vehicles of death. Mercy killing. But there is no reasoning here and little rationale for picking one horrible end over the other.

We pivot for other escape routes: Wolves keep the moose population healthy by killing weakened moose. To some limited extent, sure. In any case, these are all excuses for absolution from failing to give tragedy due attention.

These excuses diminish the majesty of a moose's life and the depth of his or her suffering. Excuses obscure the horror of the death and the celebration of a dangerously brave life. They hide persistence to the point of absolute exhaustion, the extension of one life by the taking of another, and transformations of tissue and energy from one species to another—processes that have occurred each moment over the past billion years. Excuses conceal the flash ignited by vulgarity colliding with sublimity, the mundane smashing into the divine.

Our Paleolithic ancestors were the first in the history of all life to imagine the circumstance of their providers. That vital juxtaposition between living and dying demanded reconciliation. We developed ceremonies and rituals. Religion was born along with our capacity to imagine the circumstance of our victims.

We no longer care for those primitive observances. Warm, pumping, bloody muscle is reduced to "meat" that is stored in "lockers," packaged into casings, and formed into patties. We eat, live, and kill without a thought.

Why wolves? Because they remind us to think.

2 |

Thoughts of a Moose

January 24, 2014. A new wind rose during the night, tearing through the forests and clawing at the bunkhouse. We knew we'd be earthbound for the day.

It would be a good day to perform a necropsy on the remains of a moose that is reachable by skis—about six miles from the bunkhouse, past the end of Washington Harbor, across Grace Harbor, almost to Middle Point, and then inland from the shore by just 50 meters or so. West Pack killed the moose last week, fed from it for several days, and then left to patrol territory and hunt moose along the south shore.

After oatmeal and coffee, we gathered some gear—garbage bags, ziplock bags, ax, notebook, camera, and some candy bars—strapped on our skis, and glided from the bunkhouse, down the hill, and onto the frozen harbor where our faces met the roar of that southwest wind. We leaned into a near gale all morning. The lasting memory was incessant noise, something like pointing a hair dryer to each ear for a couple of hours. Thankfully it wasn't cold, about 10°F.

Within an hour, we reached the end of Washington Harbor. The ice extended another mile beyond the harbor and onto the open waters of Lake Superior. From that distance we saw waves, the height of a single-story building, crashing over the ice's edge. Waves traveling beneath the ice made it heave and moan.

We expected it would be unsafe to ski on the ice at Card Point, and we wouldn't be able to ski the shoreline because there isn't a shoreline on which to ski—it's a bit cliffy.

So, we crossed overland on a spit of land that separates Washington Harbor from Grace Harbor by just 150 meters. Here, the dormant winter vegetation is thick in a way that only a snowshoe hare would consider the environs spacious. A moose would just lift her legs up high over the small stuff and barrel through what would not give way to a human. If we had been on snowshoes and with packs, then the crossing would have been laborious, though straightforward. But we didn't have packs or snowshoes. We were on skis pulling sleds. Deep, uneven snow and subnivean branches twisted our skis, knees, and ankles into orientations that turned our ski into a game of Twister. Our sleds were sideways and upside down as often as not. If a Picasso could ever to come to life, this was it.

For 45 minutes we traversed the spit at an average rate of nine feet per minute.

We would have skied across Grace Harbor were it not for sloshing, open water between tightly packed sheaves of ice. Some stretches of the shoreline were easy going. We removed our skis a few times to walk gingerly across slippery boulders of ice the wind had piled up weeks ago.

The map suggested that we leave the shoreline and enter the spruce forest at the point closest to the kill site. We recognized that entry point when we were abeam the southwestern tip of Grace Island, which lies across the harbor. Fresh snow covered the wolf tracks that would otherwise have led us to the kill. The spruces were packed tight, and we could see just 20 feet in any direction. It'd be easy to walk right past the kill. Soon enough, however, we found some snow-covered tracks of a fox. We couldn't be too far off.

Five or 10 minutes later, I hollered to my colleague in these affairs, "Rolf, it's over here. I found it." The snow was padded down over an area the size of a large living room. And this had been their living room for about a week. The snow was mottled pink with blood and brown with hair and rumen. They had killed a large, old cow moose.

Rolf swung his ax and cracked open a femur. The marrow was flaky and firm, like shortening. She hadn't starved. Blood on the lower legs indicated that wolves had killed her, not found her dead and then scavenged her carcass.

We were unable to find any hide, ribs, vertebrae, or scapula. Two of the four legs were missing. The carcass had been consumed, disarticulated, and scattered to an extent that suggested this meal had not been especially easy to come by for the wolves.

We collected the skull, mandibles, a metatarsal bone, and the pelvis. Much of the necropsy is performed on these samples later in the lab. The purpose is to understand the condition of this moose at the time of her death. Would she have soon died, even if wolves had not killed her? If so, her death would be of little consequence to the moose population. If moose about to die for other reasons were the only kind of moose that wolves ever killed, then wolves would be best thought of as scavengers too impatient for death, preferring warm juicy meat to something frozen or slimy—according to the season. However, the necropsy may indicate that this moose would have gone on living and reproducing, had she not been killed by wolves. Too many deaths of that kind would be of considerable consequence to the moose population. In any case, the answers lie in the bones.

We skied home with the wind to our backs and made the bunkhouse as the day's last light filtered through low, thick clouds.

The details of that particular necropsy are documented under record #4935. We never pass the opportunity for a necropsy. Of the nearly 10,000 moose to have lived and died on Isle Royale over the past half century, we have necropsied about half.

These examinations of death are never macabre and always a celebration of life. Each necropsy, like its subject in life, is its own experience. But uniqueness does not deter the human proclivity for pigeonholing. Some necropsies are like scavenger hunts and begin by shuffling through the summer vegetation to find bones scattered by wolves and foxes. Some necropsies are more like archeological digs, with bones that have not seen life in more than a decade and are now slowly consumed by the earth. Some necropsies are feats of athleticism, long treks by snowshoe, hauling an 80-pound pack loaded with a selection of frozen remains, racing to arrive before sunset to the lakeside extraction point from where I'd be flown back to base-camp by ski-plane.

Occasionally, we conduct a necropsy on a warm spring day just a few weeks after death. These inquiries are obscenely aromatic. Checking for arthritis requires reaching in and separating a femur from the pelvis. Most of the necropsy requires reaching in, which places your ear close enough to hear a million maggots tumbling and whirling in a Dionysian festival of rotting flesh. Their celebration of life sounds like soft radio static, the kind of white noise that would invite sleep in any other circumstance.

Not only maggots but also ticks. Yes, maggots and ticks. But the ticks aren't happy, and they definitely aren't celebrating anything. Just a month ago, all had been going according to plan. They had just molted into their adult outfits and had begun a last blood meal before consummating their life's purpose. Some who had been sucking next to each other for the past week would have soon become irresistibly attracted to each other. Others would have set out in search of a stranger, somewhere on the other side of the moose. The entire continent would soon fall into debauchery. Soon it will be mating time on the moose.

Then life turned for the worse. The entire world rolled sidewise and heaved her last breaths. For those on the lee side, the daylight that had filtered down through the forest of fur was extinguished. The ground grew cold and blood that had flowed, like oil from the sands of Arabia, stopped. The ticks cannot know the terrible contribution they made to this death, which quickly became their own fiasco.

I have no idea why, but some ticks hang on, mouths firmly connected to her ground, long after she dies. Others, many thousands, slowly abandon ship. They crawl away and then onto the tops of nearby twigs. They hang on like survivors surrounded by an ocean of nothingness, inhospitably vast, no less than the space between galaxies.

They cling. Each day the sun sets later. The last piles of snow melt. And they wait. The bluebead lilies shove their way through the soil and the thick mat of last year's brown leaves. They wait. Marsh marigolds explode like all the newborn suns in a far-off nebula.

They can wait for weeks. Unless tick faculties include a sense of time (stranger truths exist), this waiting is, in every sense of the word, an eternal experience.

On rare and lucky occasions, at some point, eternity snaps. A vibration, the sensation of warmth, a waft of carbon dioxide, and a breath of life. With her third and fourth sets of legs anchored to the twig, the tick stretches her front pairs up as high as she can, an arachnidian sun salutation. A moving sack of blood is close—it may be a scavenging fox, a lone wolf, or a necropsy-performing biologist. Whatever its form, if it passes close enough, then in a choreographed instant her sticky front legs grab on and her back legs let loose.

This creature whose acquaintance we've been making, *Dermacentor albipictus*, appears to the undiscerning eye like the familiar dog tick in size and overall appearance. The similarities are genuine but distract from remarkable differences. Most invertebrates are active in the warmth of summer and dormant during the wintertide. Not so for *D. albipictus*, whose nickname is the winter tick and who enjoys the vivified portion of its life throughout the short, cold days of winter.

From the first spring peeper in May to the last withering hawkweed in September, the winter tick exists only in anticipation of its animated self, as an egg in the soil. The eggs are laid in clusters of many hundreds. In late September, the eggs hatch. Infantile ticks, each about the size of a period, crawl as a collective jumble to the tops of grasses and wait. Ticks at the bottom of the heap latch their legs to a blade of grass. Ticks in the middle of the heap are latched onto each other. And ticks at the top wait with legs extended upward. Many of these sibling-groups die with the first hard frost. Some win the lottery because their mother laid eggs in the spring right where moose will pass by in the fall and the entire monkey pile is swept aboard as one passes.

They crawl to the good real estate: the moose's arm pits, rump, and shoulders. After those neighborhoods are full, newer immigrants take residence in districts that are on par with Baltic Avenue, such as the legs, belly, and neck. Shortly after settling in, each larval tick takes half a drop of blood. It's a tiny meal for a tiny creature. Each larval tick sheds its exoskeleton to emerge as a nymph—not a nubile, forest-dwelling spirit, but the same tick in the next stage of life.

In early winter, the nymph ticks enter diapause, a quiescence of living sustained on just one breath per minute. They remain for several months in this physiologically enforced abeyance. Ticks cannot leave the moose until spring—so it's prudent to eat lightly, and it's easier to control one's appetite when the thermostat of life is turned down. The essence of a tick's life is quiet waiting.

In January, the nymph ticks reinvigorate themselves and feed, this time sucking more than 10 times the blood they took as larvae. They shed another exoskeleton and emerge as adults in late January. They feed, find mates, and do their thing. After mating, the females feed one last time in April. Filled with hopes of laying their eggs in the soil, the females launch leaps-of-faith that leave all other leaps-of-faith standing beside acts of perfectly calculated reason: the females drop from the moose into a cold universe their male mates will never see, as the males stay to die at home on their moose.

You do not think of ticks often, but on those occasions when you do, you think of the ticks you know, such as the dog tick or maybe a deer tick. These ticks happily feed on the blood of deer, raccoons, dogs, humans, or pretty much anyone. However, of the 900 or so different tick species that crawl on this planet, most have finicky hypostomes. Isle Royale is home to only a few tick species, one of which specializes in the blood of snowshoe hares and another that is content only when sucking from the skin of a red squirrel.

For *D. albipictus*, at least those living on Isle Royale, moose blood is the only acceptable meal. That behavior earns this tick another nickname, the moose tick. When a moose dies and the ticks evacuate their home, they wait for another moose, who is almost certain not to come.

Some ticks will be issued what only seems a second chance, perhaps in the form of a fox scavenging from the carcass that the ticks are fleeing. Later in the day, the fox beelines it down the trail, her little vulpine brain preoccupied with some secret destination or some urgent fox affair. Then her mind is hijacked by an itch. She freezes, turns her narrow muzzle back on herself to nibble and grind with her incisors. The life of another tick is crushed, laying on the snow in teleological ruin. A biologist, who had conducted a necropsy on that same moose that same morning, reaches around to the back of his waistline and shudders with disgust on realizing what he's pinching. That tick just began the great diapause that never ends.

When winter ticks find themselves aboard someone other than a moose, they wander far and wide, unable to find a suitable place to bore down. I don't know how ticks experience it. I suppose they can "smell" the blood beneath their legs, and it must smell foul. In all that crawling and searching, the tick gives itself away to the unsuitable host.

The Latin name *Dermacentor albipictus* translates into English with a bit of literary license as *skin pricker who paints in a palette of white.* "White" because the broken hairs of a moose are white in cross section. Each winter an individual moose routinely hosts tens of thousands of ticks. A heavily infested moose houses 100,000 ticks, something like seven ticks for every square inch of skin.

Moose respond by rubbing vigorously against trees. They scrape with their lower incisors and scratch where they can with their hooves. The result is large white patches of broken hairs. Where large patches of hair have simply let loose from follicular stress and irritation, bare, black skin is exposed. (Black is the normal color of moose skin.) By spring, moose are a mosaic of chocolate brown patches of undamaged fur, bare black skin, and large patches of white, broken hairs. Heavily infested moose lose or break hairs over 85% or more of their skin. These are the ghost moose—an epithet reflecting their appearance and, often, proximity to death. The patchwork of brown, black, and white is the community art of 100,000 skin prickers painting in a palette of white.

Parasite is an exquisite name, taken from a Greek word, *parasitos*, meaning "person eating at another's table."[1] *Host*, however, leaves much to be desired in its service as a name—as if the relationship were some kind of cordial affair. There is nothing in the soul of cordial, nothing that cordial could stand for, nor any liberal or ironic application of that word that offers a reasonable description of the relationship between a moose and his winter ticks.

But the deer mice of Isle Royale participate in a host-parasite relationship that just might be considered congenial. I need to tell you about an ectoparasite of these mice, not to satisfy any peculiar preoccupation with parasites, but because the deer mice of Isle Royale are important to this story.

In the 1990s, Rolf Peterson was leading the research on Isle Royale's wolves, the wolf population was foundering, and Rolf wondered whether inbreeding had been playing a role. He knew that many of the islets that surround Isle Royale would be inhabited by deer mice, whose name, by the way, is apt because of their agility and quickness, as compared to other species of mouse.

Abecedarians will say that *individual* wolves and mice are a study in contrasts. But the wolf *population* of Isle Royale and the mouse *populations* of the surrounding islets are an opportunity for comparison. Some of these islets are just a few hectares in size.[2] Each islet population of deer mouse is isolated and composed of just a few dozen individuals, and for those reasons each is likely inbred. In this regard, islet populations of mice may well resemble an island population of wolves. The shorter life of a mouse also means that inbreeding progresses faster, possibly allowing us to anticipate the future of inbreeding in the wolf population.[3]

Rolf found a new graduate student to lead the search. Leah Cayo (later Vucetich) would focus on assessing degrees of isolation, rates of inbreeding, and any adverse consequences of said inbreeding for a handful of small populations of mice living beside a small population of wolves.

At the time, I had just begun a PhD project focused on the mathematical facets of inbreeding and its impact on animal populations. With the recent advent of laptop computers, I could do that work from anywhere, and I wanted to be on Isle Royale. I had worked on the Isle Royale wolf project for each of the past four summers as a student field technician. I believed that continuing would require me to keep my value fresh. So I offered myself as a reliable and sturdy assistant, able and willing to lug gear through the forest and find connections among our related research interests.

A first step in Leah's research was to determine which islets had populations of deer mice, including an evaluation of Passage Island, which lies about three and a half miles off the northeast tip of Isle Royale. The next closest landfall is 20 miles to the north.

Passage Island is about a mile long and 500 feet wide, making it larger than the other islets we studied. The islet is nearly cut in half by a small but deep harbor, whose icy, clear water reflects enough light of the high summer sun to create a double exposure of cumulus clouds floating above and jagged boulders submerged below. The harbor is further decorated with basalt cliffs on the east and a dilapidated boathouse on the southwest. From the boathouse is a footpath that disappears into a forest of spruces and firs, tall and straight, dripping with moss and quenched by frequent fog.

Before setting camp or getting to work, we were drawn down the path. The forest was perforated by bright openings covered with sprawling junipers that draped bare basalt. Just before reaching the far end of the island, the forest gave itself up entirely to craggy boulders, sprawled like beached leviathans that gave way themselves to indomitable waves. Off to the right, a 19th-century lighthouse, quiet and unattended, surveilled the horizon. The water seethed and foamed.

We stood at the edge of the largest, coldest, roughest body of freshwater on the planet.

Passage is too remote to have ever been colonized by moose. Without moose, its forest understory is a marvel. Canada yew, spared from herbivory, realizes its potential. A dozen needle-soaked branches radiate from a cluster, up and over in a gentle arch, rising a meter or two off the forest floor and back down again. On Isle Royale, Canada yew is uncommon and rarely grows beyond a stubble. On Passage, the forest floor is packed dense with yew clusters that are too tall to step over and too low to crawl beneath. One doesn't really walk through this forest so much as spring from one arch to the next, often without touching the ground. Off-trail hiking on Passage could pass for Abbot and Costello's rendition of bounding through the bamboo forest in *Crouching Tiger, Hidden Dragon*.

We spent 10 days on Passage Island. One of us (we've agreed not to recall which one) forgot to pack a spoon or any eating utensil. I know I remembered to pack a bottle of wine and two glasses. We shared wine and a spoon.

In our time on Passage, we never spotted a fleeing hare, we never heard the alarm chatter of a red squirrel, we never raised the curiosity of a fox, and we never caught a mouse. A simple, yet far-reaching, fact of nature is that isolated islands tend to have fewer species and larger islands tend to have more species. Passage Island, it turned out, is more isolated than it is large. The expedition to Passage was, however, auspicious beyond expectation. By the next summer Leah and I were married.

Fortunately, we found deer mice living on other islets that were much closer to Isle Royale, but also much smaller. Leah selected five of these islet populations to study in detail. Each evening, just before sunset, we'd cast the skiff's lines and boat to each of those islets. The smallest was 400 feet across at its widest. The largest islet was 20 times that size. Each evening we'd open about 100 live traps that we'd arranged in a grid across a portion of each islet. Each trap was 15 meters from its four nearest neighbors, with all the orthogonal precision that can be imposed by a tape measure and compass on a rambunctious forest. Foisting those grids was as unnatural as sighting a Pollock painting with a T-square.

Just before sunrise the next morning, we set out in the skiff, boated to each islet, and revisited each trap. By opening and checking all 500 traps, we aimed to estimate the number of mice living in the area by the pattern of new and repeat visitors. The opportunity for repeat visitors arose from opening the traps on each of 28 nights throughout the summer. We identified caught individuals from ear tags and then let them on their way. Each trap was provisioned with cotton batting for warmth and seeds to compensate for our occupying their evening. We entertained enough repeat visitors to believe they enjoyed the stay. The entire operation had the feel of a bed-and-breakfast.

Leah and I worked parallel lines on the grid. Whenever one of us found a trap with a closed door, we would call the other over to see who had spent the night. Leah is especially adept here, holding the trap in her palm and pressing the door open with her index finger. A little nose pokes through the batting to meet the innkeeper. The skill lies in removing the mouse from her comfortable accommodations and reading the number on the ear tag without harm or premature escape.

No cuter creature lives in the night forest of Isle Royale. Deer mice are tan on top and white below. They are built for life after dark, with ears that are soft but perky and as big as their face. Their eyes are black saucers, proportioned like the subject of a Margaret Keane painting. Stiff whiskers convey supplemental nocturnal stimuli. If you had whiskers like theirs, they'd reach as far as your outstretched arms. They can wiggle their nose with an exuberance that is positively bewitching.

But when you see one for the first time, you may be taken by disgust. That is, when you see up close a deer mouse hosting the larva of a bot fly.[4] When this maggot turns into an adult, he is a little larger than a house fly, but with noticeably small wings. His thorax and lower abdomen are dull brownish yellow. His eyes and upper abdomen are black. You shouldn't mistake him for a bumblebee, but that is certainly the gestalt.

These adult flies cannot bite or eat. The sole purpose of an adult male fly is to consummate its life before starving to death. An adult female fly has the added responsibility of laying eggs. She knows how to find the secret paths of deer mice. She will lay a few eggs on a favored resting spot, or at a twist in the trail where a mouse frequently pauses to sense the danger of owls, foxes, and weasels, or maybe at the doorstep of a den itself. The eggs hatch with the warm presence of a mouse. The comma-sized maggots do all they can to wiggle onto the mouse. A successful maggot wiggles from where ever it makes landfall to the nose or mouth, where the skin is easy to penetrate. From there it migrates beneath the skin to somewhere on the torso, often near the abdomen or an arm pit.

The maggot feeds on body fluids and grows without burrowing deep. The skin stretches outward. In a few weeks, the larva weighs a gram. Just a gram—three and a half one hundredths of an ounce. Sometimes a number conveys nothing. If you hosted a maggot beneath your skin, proportional in size to that carried by the sweet, little deer mouse, you'd have a maggot the size of a six-pound medicine ball.[5]

After a few weeks, the maggot is ready for the next stage of its life. By way of a breathing hole it had been maintaining through the mouse's skin, the maggot pulls itself out of this womb and back into the world. It burrows into the soil, suspends all sign of life until spring, and then metamorphoses into an adult fly.

The maggot leaves behind a flaccid sac of skin that would be, in human terms, the size of that medicine ball. The gaping hole would be vulnerable to infection were it not for the maggot having coated the wound with secreted antibiotic. Even parasitic maggots manage to leave their children a world of healthy mice as rich as the one they inherited.

Scurrying across the forest floor has been more exhausting during those weeks, and these host mice have been eating for two. (Sometimes for three or four, as a heavily parasitized mouse will raise two or three maggots.) To host the developing body of another's offspring is like a surrogate pregnancy, notwithstanding the anatomical misplacement.[6]

Every parent knows raising a brood is exhausting and demands time and energy that cannot be spent elsewhere. Child rearing is an opportunity cost that permeates every facet of life on earth. Female elephants give birth once just every four to six years. That pace that is amply slow to leave enough life energy to sustain an elephant on the savannah for more than half a century. Rabbits reproduce, you know, like rabbits, but each one is lucky to survive their first winter.

Reproduction is inviolably traded against longevity because longevity requires costly investments in the long-term maintenance of one's body to fend off cancers, arthritis, and tooth decay, and to maintain muscle mass, eye sight, hearing, memory, and immune systems. Spend your limited energy on reproduction and it's unavailable for maintaining your soma.

Each of the millions of species that inhabit the earth is its own unique solution to manage that trade-off between reproduction and longevity. That diversity in ways of living, driven by opportunity cost, is manifest not only among species (from mayflies to giant tortoises) but also among individuals within a single species. The most telling evidence is that humans who live longest—we future centenarians and Methuselahs—tend to have fewer children than the rest of you.[7]

That trade-off can be bent by conscious choices, such as when a human decides not to reproduce, or by imposition. Bot flies impose. When all goes well for a deer mouse, there is enough summer to raise two litters of pups. She gets started with the first litter early, before the evenings have been set free of past winter's chill. The second litter is sent on their way while next winter's chill reaches back to infiltrate the nights of late summer. The energetic cost of raising a bot fly or two severely diminishes the likelihood that a mouse can successfully rear a second litter during late summer. Ironically, the mouse ends up with more energy than if it avoided the bot fly and raised its own second litter. Consequently, the mouse is more likely to survive the upcoming winter and more likely to reproduce in the following spring before the next generation of bot flies emerge. Mice infested with bot flies tend to live longer.[8] Of all of the uninvited guests, none seems more giving in return than the maggot of a mother bot fly. This relationship between mouse and maggot was molded over millions of years of evolutionary history and provides relief for understanding why ticks are so seemingly cruel to their moose.

On Isle Royale, winter ticks feed on moose because moose are the only suitable host. On the mainland, however, winter ticks also feed on the blood of mule deer or white-tailed deer (neither of which live on Isle Royale). Deer are far less gracious as hosts. They begin grooming themselves as ticks climb aboard in October. The vast majority of ticks are evicted shortly after arrival. A typical deer ends up supporting a few hundred, sometimes only a few dozen, ticks throughout the winter. By devastating contrast, moose don't groom with any diligence until January, by which time they've accumulated tens of thousands of ticks that have grown from period-sized hellions to couscous-sized demons. January is too late.

The elite cadre of scientists who study this sort of thing—few in ranks, distinguished in ambition—hypothesize that mammalian grooming happens by one of two strikingly different mechanisms. First up, and plainly enough, is stimulus-driven grooming. You have an itch, so you scratch. The second possible mechanism is programmed grooming. Cats are the archetype, with 10% of their waking lives devoted to grooming.[9] Cats tend to groom just after waking and at other habitual times of the day. They even groom in a more or less prescribed order—first the face, then hind legs, flanks, neck and chest, nether regions, and finally the tail. Cats are not that itchy. Their obsessive grooming is mostly preventative. We say it's programmed.

An extremely rare stripe of scientist focuses a cultivated interest in mammalian grooming on moose and deer infested with winter ticks. They hypothesize that deer exhibit programmed grooming in the fall, but moose are limited to stimulus grooming in the winter. One reason for thinking so is that larval ticks may be too tiny to be felt by moose.[10]

Science had long given the impression that itches are registered as some gentle stimulation of nerve cells otherwise devoted to sensing pain. By that belief, moose would seem insensitive brutes for being unable to detect the presence of tens of thousands of ticks.

But those beliefs about itch perception were born from limited imagination. Scientists, with imaginations big enough to see what others had overlooked, discovered in 2013 that mammals perceive itches through a special kind of neuron exclusively devoted to sensing an itch. It is at least plausible that deer have far more, or far more sensitive, itchy neurons than moose.[11]

So don't scoff at a moose. Don't think being a sophisticate requires the capacity to experience an itch. Bats don't laugh at our inability to echolocate. Elephants do not gossip over our incapacity to speak through the earth in infrasonic voices and listen to those seismic sounds with the soles of our feet.[12] Dogs don't jeer our inability to smell the cancer growing deep within our own bodies. Bees don't mock our blindness to the ultraviolet colors of a flower. And we shouldn't think less of a moose for not scratching an itch it has no way of perceiving.

This physiological account is a proximate explanation for why moose suffer so much at the barbed hypostomes of winter ticks. The physiological explanation is important, but inadequate. Inadequate in the same way that observing a bullet hole in a victim's lung is an inadequate explanation for the murder.

In ecology, the ultimate explanations must be approved by Darwin. On that docket, what we know is that deer coevolved together with winter ticks in North America. Deer negotiated for few enough ticks to guarantee their own success. Ticks negotiated enough blood for enough hypostomes to assure their success. The terms of agreement took a million years to hammer out, and each party's commitment has been tested every fall for the past 8 or 9 million years.

Moose, however, negotiated the terms of their lives in Europe and Asia, where there are no winter ticks or anyone that even remotely approaches the habits of a winter tick. A tick-free life would waste its energy by investing too much in sensory instrumentation so specialized as to only perform the peculiar task of announcing an itch.

Moose arrived to North America only 10,000 years ago, after crossing a bridge of land that forms from time to time between Siberia and Alaska when the oceans drop.[13] Moose have been living with ticks, in terms scaled to that of a human lifespan, for just the past few months.

Every tick and every moose is an individual organism. Some have magnificent births, tragic deaths, and triumphant lives in between. Others have tragic births, magnificent deaths, and dreadful lives. You also experience life as an individual organism. We spot the experience of organisms with such a bright light as to darken what lies just beyond the beam—the simple and essential reality that life is layered.

We organisms inhabit just one of life's layers. Every organism is also a collection of genes, each playing their own part in composing the life of an organism, like members of a symphony orchestra. Most genes have clones—identical replicates of themselves—that play in other orchestras, performing the same part, but arranged differently to produce different symphonies. The genes who recite the notes to color your eyes also perform in the orchestras of your close relatives. The genes for basic functions like protein synthesis perform in the orchestras of your distant cousins, including your cousin the winter tick.

Genes inhabit a layer of life nestled within *organisms*. We *organisms* do our own composing for *populations*—the layer of life into which we organisms are nestled. Organisms of the same species (let's say, moose) are born and then die in an immortal sequence that gives meter to the life of a population. When births play fortissimo and deaths are pianissimo, the population waxes. In the next movement, births and deaths exchange parts and the population wanes. The volume by which births and deaths pound out their notes is cued by the next layer of life, *food webs*.

Genes inhabit organisms, organisms inhabit populations, and populations inhabit food webs. Food webs are inspirited when populations of one species play off populations of another, as when moose populations are cued to wolf populations, and vice versa. These reciprocal cues emanate from various kinds of relationships, such as predation, herbivory, parasitism, symbiosis, commensalism, and amensalism. When predation—the relationship between predator and prey—cues a philharmonic food web with a downbeat, prey populations diminuendo and predator populations crescendo, moose are killed and wolves are fed.

The reciprocal cues for moose and ticks emanate from the baton of parasitism. The repertoire of reciprocal cues in a food web is a panoply of styles of eating and being eaten—granivory, frugivory, insectivory, fungivory, piscivory, and planktivory. That's a short list in broad categories. The full list of dining relationships is nuanced enough to include, for example, sanguivores, creatures who live by consuming blood. They are a diverse lot themselves—leeches, mosquitoes, vampire bats, vampire moths (!), vampire snails (!!), and vampire finches (!!!). Cringe if you like. Then reconcile your disgust with black pudding, red tofu, sanguinaccio, or any of the myriad ways humans have for making a meal out of blood.[14]

The reciprocal cues that compose a food web—parasitism, predation, and so on—are thrice expressed. First, to narrate at one level, the gnawing, biting, and sucking connections between the lives of *organisms*. Second, to convey, one level up, connections between the lives of *populations*.

The third expression? We won't find this harmony line on another level, but in an entirely different dimension, superimposed across the set of nestled layers. In this dimension, a food web directs the *flow of energy and matter* into itself, through itself, and out its other side. In this dimension, organisms are not composed by genes, nor are populations composed of organisms—rather, life is energy and matter poured into permeable little packets that we perceive as organisms, subsequently grouped into stores that we recognize as populations.

This ethereal dimension is bizarrely unfamiliar to our own experience of life, but the weirdness renders it no less real or binding. We can explore this alternate reality through questions that make no sense from our native experience of life. For example, of the moose blood that circulates through an ecosystem, how much is ushered through the mouthparts of ticks?

What?

Yes, I know, in this dimension it's easy to get vertigo. We'd better find the horizon: the blood, the red fluid itself, is certainly limited to moose vessels and tick guts. The question is about the energy and matter of which the blood is composed. How widely across a food web does that circulate? Enough to matter only to some individual moose and ticks? Enough to be of consequence for whole populations? The entire food web? Could the movement of that energy and material knock on and reverberate throughout the moose population, and the wolf population supported by moose, and the forest that supports the moose?

Toeholds for these questions poke through into the layer of life that we more readily recognize. Each female tick, in preparing for her great leap of faith, consumes nearly a cubic centimeter of blood. The large community of ticks will deprive its moose of about 12 gallons of blood over a four-week period. To keep up, a moose must replace most of her blood supply.[15] Ticks' greatest impact hits at a particularly inopportune time. By winter's end, after the snow has melted, the world is browned with the corpses of last year's vegetation, the foraging has been meager for months, and starvation creeps behind each moose. That peril is compounded by the cost of innumerous micro-withdrawals from the blood bank. Many moose are also trying to nourish fetuses throughout the winter. Life and death depend on whether spring green-up arrives this week or next. Starvation and ticks are joined by a third motif, played by wolves on the hunt for moose, especially moose weakened by tick-enhanced malnourishment.

Given the time of year and the ensemble of players, ticks cannot bear sole responsibility for causing the death of any moose. Causes flow simultaneously through life's many motifs. Each cause is also the result of some prior cause. *By predisposing moose to predation, ticks might indirectly contribute to a reduction in moose abundance. Reductions in moose abundance are followed by reductions in herbivory, which can subsequently result in the forest growing more food for moose.*

Eventually the sequence of causes and results folds back onto itself. *More moose food can contribute to the subsequent growth of the moose population. More moose mean that more tick larvae are likely to find a moose next autumn.*

The loop is closed, beginning and ending with ticks. No result ever has just one cause, there is no original cause, and every result has many causes, even when the cause is a tiny little tick. Life's layers fold onto themselves with more convulsions than any fugue that Bach ever wrote. The hierarchically nestled layers of life are the "order of holy beings," as implied by the etymology of the word hierarchy. If ecologists have a gift for the world, it is seeing and sharing this view of life that is otherwise obfuscated by our personal experience as organisms.[16]

Some ecologists look down from the top layer, and others look up from the bottom layer. Our work on Isle Royale has tended to look out from the middle layer. We focus on the lives of two populations (wolves and moose). To understand their biographies, we regularly look to the organisms inhabiting the next layer down, to the food web inhabiting the next layer up, and occasionally to that other dimension of fluxing energy and percolating matter.

Populations fluctuate over time, without exception. They rise and fall as assuredly as organisms respire. Sometimes slow and deep, other times shallow and erratic. Every population explosion, every decades-long increase, and every erratic, upward throb is the consummation of so many births and a few less deaths. Every kind of population decline is the denouement of the dreadful eclipsing the glorious. Populations are living ledgers of the miraculous weighed against the heartbreaking.

When wolves eat enough moose, wolf births tend to outpace deaths and the wolf population grows. Otherwise, the population declines. The tipping point occurs when a pack kills and eats about three-quarters of a moose each month for each wolf living in the pack.[17] When wolves kill moose too frequently, moose births cannot outweigh the deaths and the moose population declines. Otherwise, the population grows. The breakpoint for moose occurs when wolves kill about 10% or 12% of the moose population in a year.

Those are not strict thresholds so much as tendencies that are mitigated by disease, weather, genetic health of the wolves, and nutritional condition of moose.[18] We have quantified the strength of those tendencies. Of all the year-to-year fluctuations observed in the abundance of Isle Royale moose over the past six decades, about two-thirds are attributable to year-to-year fluctuations in the proportion of the moose population killed by wolves that year.

That last idea, where fluctuations in some force (predation rate) account for a quantified portion of fluctuations in some state of affairs (moose abundance), is one of our most basic scientific ambitions. This idea deserves another example. Consider a fluctuation in wolf abundance from one year to the next. With your mind's eye, hold at just one fluctuation, say, when the population grew from 25 to 29 wolves between 1999 to 2000 (see the graph on page 146). Now, gather up all of the year-to-year fluctuations that have occurred over the past six decades. Toss those fluctuations into a pie pan. All of them. Mix 'em up. Stir vigorously until the fluctuations break down into a homogenized goop of dynamism. I imagine it to look like pecan pie filling.

Stay with me. Don't skip ahead. This little daydream is the reason we each lug our bulging neocortex everywhere we go.

When it's time to serve this fluctuation pie, cut one largish piece equal in size to one-fifth of the entire pie.[19] That is the portion of fluctuations in wolf abundance that is explained by how frequently wolves kill moose. It's the single largest piece in this pie. The other pieces will be much smaller, and we'll have to serve about a dozen—including one for disease, one for inbreeding, one for years when the wolves killed more calves than adults. Many of the pieces belong to unidentified influences.

Those breakpoints, tendencies, and strengths of tendencies are all crafted from simple measurements that began with a pencil and notebook, binoculars, and an airplane just large enough to carry a pilot and one observer. Count the wolves, count the moose, and count how often wolves kill moose. This counting will require about two months of winter when the trees have shed their leaves and snow covers the ground, making it easier to see wolves and moose from the airplane. You'll need to reserve these two months each winter for counting for at least a few decades. It's laborious and time consuming but simple.

Transmute those counts with some simple arithmetic thaumaturgy. Divide 19 moose killed by the 24 wolves who did the killing, then by the 42 days during which we observed those wolves. The quotient is 0.56 moose per wolf per month. With that arithmetic, our crafted sense of tendency tells us to expect wolf abundance to decline. If decline is what we end up observing, then we'll believe lack of food is an important part of the explanation. We made those counts (19 dead moose, etc.) during January and February of 2000. A year later, wolf abundance had declined by a third.[20] The wolves were suffering, but not in a way that was readily detectable with a look through binoculars.

Arithmetic is only the beginning. We have maximum likelihood estimators, Akaike's information criteria, decomposed variances, and a shop full of number-working tools for carving knowledge and insight. But the foundation is simple. It's just counting animals from an airplane. All in the service of our main scientific purpose, which is to understand wolves and moose as populations by describing and explaining the fluctuations in abundance from one year to the next.

When Durward Allen began observing the wolves and moose of Isle Royale in the late 1950s, the ambition focused on counting wolves, counting moose, and doing so every year. Six decades later, we still count wolves and moose. Allen retired in 1974 and left the project to Rolf's leadership. Rolf immediately expanded the scope of the project with two obsessions. The first was never pass an opportunity to perform a necropsy. (Allen and his crew performed moose necropsies, but not obsessively, just enough to get a general sense for the kind of moose that wolves kill.) The second idée fixe that Rolf added was to estimate how frequently wolves kill moose each year. (Allen's crew noted sites where they opportunistically found that wolves had killed moose, but estimating kill rate requires the systematic diligence of aiming to find every kill site.) Today, we continue to estimate kill rate and perform necropsies with the same fixation.

The growth of knowledge is greatly favored by gradually increasing the scope of observations. Sticking to only the same small set of observations, year after year (say, just wolf and moose abundance), risks a diminishing return on investment. Leah and I wondered what new category of observation we would add to the research as we assumed increasing responsibility for the wolf-moose project toward the end of the last century.

We also needed a place on Isle Royale from which to base our summer field seasons. Rolf and his wife, Candy, had been working their summers out of Bangsund, a little cabin named for the Norwegian fisherman who built it in the 1930s. The cabin sits on the protected shoreline of Moskey Basin, on the east end of Isle Royale and in the heart of a forest so lush that the untrained eye would struggle to find evidence of its being continuously chewed at by hundreds of moose. Across the basin from Bangsund is land whose ownership is routinely disputed by two packs—literally a pissing match.

Bangsund is an inviting little cabin but too cozy to lodge two more. And it made geographic sense for Leah and me to station at Windigo on Isle Royale's west end, in the heart of another pack's territory, and in the middle of a forest scarred by decades of incessant gnawing by moose.

Windigo is also inhabited seasonally by about a dozen park employees who occupy the existing housing. That first summer we worked out of a storage shed, 8 feet wide by 20 feet long. The place was packed from floor to ceiling with equipment for the winter field season—ice augers, axes, ropes, shovels, tarps, cooking supplies, crates of cold-weather clothing, mechanical sundry for the aircraft, a drying oven to process bone marrow of necropsied moose, and more. We slept on a shelving unit. My office was my computer balanced on the handlebar of a snowmobile. The next summer, park staff allowed us to live in a cabin tent, but the allowance was limited to setting the tent in a depression that filled with water after every rain.

We eventually appeared needy enough and of sufficient value to the mission of a national park. The district ranger negotiated with park leadership and secured for us permission to set up a yurt, 20 feet in diameter and set on a wooden deck. It's been our base of summer operations every year since. The yurt is an eclectic mixture of modern and premodern living. Internet but no running water or heat. Spring mornings regularly bottom out in the upper 30s.

The yurt's best feature is its fabric walls, which offer no barrier to the night sounds. The day's last conscious aural experience before falling to sleep is often the Jurassic-styled call of sandhill cranes retiring to their evening roost. When the evening temperature is consistently above 40, spring peepers begin chorusing. A week or two later, the toads take over. An insomniac white-throated sparrow calls at 0300. A barred owl occasionally issues a baritone warning to other owls that all these hares and mice belong to her. Often enough, we are pulled from sleep by a distant wolf howl soaking through the walls of the yurt. The summer nights here are as vibrant as in a happening neighborhood.

In those first summers, we explored the forest for moose to necropsy, we hiked the trails for wolf scats, we broke the night to monitor radio-collared wolves, we listened for howling that would signify pups (high-pitched yipping howls intermixed with long low howls), and we tried to invent better protocols for assessing herbivory. We played researchers in the way that young children play house. We had all the trappings and habits but felt we had not yet quite arrived.

Into the woods and through our minds, we wandered and wondered, looking for a decidedly rare species of question—one that is unanswered but answerable, and interesting but heretofore unrecognized.

Leah took to spending time at a spring that flows into Washington Creek just a mile from the yurt. Moose go there to suck, slurp, and lick the mud that borders the spring, grinding the muddy broth of the earth's grains between their teeth. The behavior is *geophagy*, another code word for the kinds of stuff we stuff in our mouths. Butterflies, parrots, chimpanzees, elephants, mice, bears, bats, and about 300 other animal species have been observed exhibiting geophagy.

More than 400 independent accounts have been gathered of human communities, in various places and times, whose diet included earth.[21] These are not pathologic-leaning addictions to salt. Rather, the most common case is pregnant women in tropical environments consuming specific kinds of clay-based soils. A misplaced sense of pity leads some to believe the practice is a nonadaptive epiphenomenon rising from hunger or nutrient deficiency. The craving is more likely, especially given the particular kinds of soil involved, an aid to digestive health. Clay-based soils can soak up toxins and poison parasites, which is useful because pregnancy lowers the immune system.[22]

Geophagy may also aid the digestive health of moose. Mud cravings take hold in early June when diet and digestion undergo upheaval, shifting from brown winter twigs to fresh spring greens. Throughout the winter, moose harbor in their rumens a community of microbes whose specialty is fermenting twigs, but less so fresh greens. With the seasonal change in diet, that microbial community is evicted and replaced by another whose expertise is fermenting leafy greens. The mud cravings also coincide with incisive nutritional demands brought by lactation, antler growth, and seasonal changes in plant toxins intended to dissuade a moose from browsing.

Those circumstances are pertinent, but the decisive reason for sucking mud is sodium. Without sodium, muscles cannot contract and nerve cells cannot fire. When sodium drops too low, hyponatremia brings lethargy, confusion, and then coma. Mammals cannot produce sodium, so we ingest it regularly.

Carnivores ingest enough sodium by consuming the muscles of their prey. Herbivores get some sodium from plants, but you may notice that lettuce greens are not particularly salty. Plants don't need sodium, so their tissues contain little.[23]

A moose needs to consume about three teaspoons of salt a week throughout the summer. Taking account of plants' sodium content and the rate at which a moose stuffs plants into her mouth, she would have to eat for 64 hours of every day to meet her sodium needs.

Except, there is another source of sodium. Isle Royale's landscape is lightly sprinkled with patches of mud where sodium-rich spring waters percolate to the surface. We call these places, not for what they are, but for how they're used. "Mud licks." Small ones are bedroom-sized, big ones mansion-sized. Notwithstanding sweaty t-shirts on a hot day, mud licks are the saltiest substance on Isle Royale. At a mud lick, a moose can slurp 10 gallons of gritty water with a three-day supply of salt in eight minutes. A moose can ingest much of a year's supply with occasional visits to the mud lick throughout the summer.

The mud cravings are greatest when moose begin eating spring greens. Those fast-growing shoots and leaves contain high concentrations of potassium that cause a mammal's body to lose grip of their sodium.[24] Every spring, many moose pilgrimage from their regular home ranges to one of the island's mud licks.

Some mud licks are goofy with moose in the springtime. Moose are drawn to these meccas from miles around. Moose learned the locations from their mothers, who brought them as calves. For creatures so content with solitude, the annual pilgrimage must be a social shock.

When Leah began spending time at the mud lick just upstream from the yurt, I suggested that she write down what she saw.

"Like what?" she said.

"Everything that you see."

Moose gathered at a mud lick. J. A. Vucetich

She repeated, "Like what kinds of things?"

"I don't know, everything."

Leah may have understood, or she may have just been agreeable. Either way, she is not merely a capable observer but also excellent at seeing. She is a good listener but excellent at hearing. I'm not referring to the sensitivity of Leah's retinal or aural neurons.

The difference between hearing and listening is that the latter requires considerable filtering. The rumble of a car, the hum of a refrigerator, the tick of a clock, the creak of a floor joist—they are all filtered out as one *listens* to a conversation. You have to listen *for something*. Listening requires knowing in advance what you're listening for.

Observing, as opposed to seeing, is selecting one of the myriad sights in your field of view, aligning that sight with one of the myriad beliefs crammed in your head about the way the world is, and inspecting the conjunction between that sight and that belief for the purpose of confirming, rejecting, or adjusting the belief. Observation offers its rewards but so too does awareness of what can be seen when we notice and peer around prejudicial filters that separate our conscious minds from our retinas.

Moose are brown. But it is possible to look at a moose and see that its *essence* is little daubs of brown light. Claude Monet showed us how. I showed you how to see the essence of a moose as a sodium-concentrating vessel and as a community shelter for ticks. Moose are also, in essence, just furry people with big noses and gangly legs. To see a common thing in uncommon light requires dissolving filters that reshape stimuli into perceptions that comfortably match a restricted catalog of Platonic Forms.[25]

Every evening back at the yurt, Leah shared her notes from the field. They included behaviors, times of occurrences, detailed maps of movement through the mud lick, antler shapes and sizes, and so on.

One night, I asked Leah, "Why do you sketch pictures of all the moose you see?"

Smugly, "Just trying to write down everything that I see."

"No, for real, why do you do it?"

Leah's response was pragmatic, "To help me keep track of one moose from the next."

Each sketch was distinguished by antler shape (for bulls), a tattered ear, or the pattern of hair that had been lost or damaged by that winter's herd of ticks. Then it occurred to us—if we could sketch the pattern of hair loss on a bunch of moose and then estimate the amount of bare skin on each, and if we did that each year for many years, we'd have data to assess year-to-year fluctuations in the abundance of ticks. That's a big step in quantifying their impact on moose. We'd be on our way to understanding how big a piece of fluctuation pie belongs to ticks. Within a few years, small digital cameras obviated the need (and benefits) of drawing moose by hand. Before digital cameras, we carried pocket-sized data sheets with us in the field that depicted the profile of a moose. Anytime we came upon a moose we drew the pattern of hair loss on the datasheet.

Recording images of moose is not the demanding element of this endeavor. The challenge is to find enough moose to make a representative

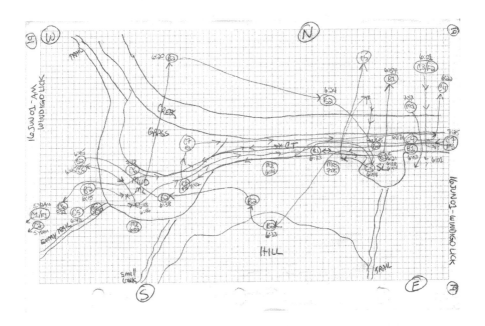

A page from Leah's field notes showing the map of a mud lick with times and locations of different moose. L. M. Vucetich

sample in the short time that is available. We arrive to Isle Royale each spring in early May, typically as the last ice floats out of the harbors. By June 15, bare skin and damaged hairs are covered by a freshly grown coat of hair. To see enough moose by that date, spring field work includes time at a mud lick or two to see moose fulfilling their urges to slurp the salted, earthen slurry.

We roll out of our tent by 0500, raise eyebrows at the skim of night ice that formed in our cooking pot, heat the water, eat some mush, and then stumble over the hill to the edge of the mud lick. Except for the time it takes to eat a bowl of oatmeal in the morning and a dehydrated dinner in the evening, we keep watch from the edge of the mud lick until about 2200. To maximize the effort, Leah and I find different vantage points from which to monitor different portions of the mud lick. After two or three days, we have usually seen most of the moose that frequent any particular mud lick.

Field Notes- Moose

16JUN01-AM- WINDIGO LICK	2hrs 15min	13 Moose
5:00 Alarm	5:42A-8:00A	2 Bulls
5:20A leaving camp		7 Cows
≥ ¼ moon, cool, windy, clear		3 Calves
		1 Unkwn

5:42A C1 on entry trail w/me; browsing low.
    I approach across opening.
5:44A she sees me & I see F1. C1 quietly grunts.
    C1 hairloss only on shoulders.
5:46A I wait for them to move off trail→W.
5:48A C2 W of SW entry trail near edge of woods.
    C2 aware of me. Hair on shoulder blades erect.
    Hairloss: neck & sides of shldrs.
5:50A looked up from notes & C2 gone - poof! (??)
5:56A I reach staging area.
6:01A I am @ cedar clump @ SW entry by ML.
    C3/F2 cross creek S→N just E of SL; gone.
6:03A B1 facing me on CT E of SL. I am @
    cedar clump near SW entry trail.
6:07A B1 stays on CT as I get net, gloves, watching me.
    B1 hairloss= chest, shldrs, back, rump
    largest so far   B1 sucks mud @ SL; watches me.
6:12A ears forward/up
6:14A head up; look @ me; then look →N across creek.
    B2 appears from SW trail→ML. Sees me.
    Hairloss= chest, neck, shldrs, back
6:17A B2→ML & B1 @ SL and me in between.
    C4 yearling appeared @ SL.

An excerpt from Leah's field notes from the mud lick. L. M. Vucetich

Time at a mud lick is mostly quiet, waiting and listening for the grunt of an old bull, the shuffling of hooves through the leaves, or the splash of a young cow crossing the creek up around the bend.

We share the mornings with a shivering boredom that requires some management. Readjust to chase off pins and needles that tingle

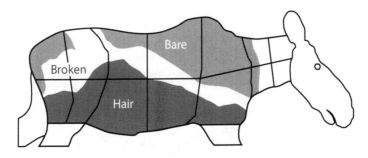

Before digital cameras, we carried pocket-sized data sheets with us in the field that depicted the profile of a moose. Anytime we came upon a moose, we drew the pattern of hair loss on the datasheet.

a calf muscle. Think of reasons to find a new position along the perimeter of the mud lick. Become mesmerized by the sight of pirouetting water striders. If I move over there, the sun will warm my face and chest. Yes, that's a good reason. Make the move. A hoot and a yodel, and a loon rockets across the sky. More pins and needles. Nibble trail mix at a controlled rate to meter the time—one peanut or raisin at a time. A red squirrel vaults across the branches overhead. If I move over there, I'll get better pictures because the sun will be at my back. At an unscripted moment boredom is shattered. A moose. Be very quiet.

Wolves know these mud licks and their gravitational pull on moose. And moose know that wolves know. So moose approach with vigilance. It is okay to let a moose see you as it approaches the mud lick. They readily see the difference between humans and wolves. But a moose should see you before they hear you because they are not quick to discern whether a crackled leaf or snapped twig is announcing a wolf or human. Absolute quiet until the moose sees you. Then some talking in a deep, gentle voice is useful. Most moose seem to understand that humans are harmless and that they speak English, whereas they've never heard a wolf speak one word of English.

With four firmly planted hooves, a moose can reach its mouth across a four-foot arc. Not impressed? Try it yourself. That large sweep is also useful, especially during winter when another step means pushing aside knee-deep snow and each bite offers just a gram of nutrition. A moose's mouth does not, however, reach the mud quite so naturally. Some moose reach the mud by spreading their front legs like a stilt-legged clown. Others reach the ground turning their toes and feet backward to rest on their wrists.

Five or 10 minutes of more or less nonstop slurping is typical. We sit behind the camera shutter waiting for a moose to offer a complete profile of the hair loss on their left side and then their right side. Some moose slurp on and off for an hour or more. Sometimes three or four moose are in attendance at once. But much of the time is just waiting for moose.

By early afternoon, the morning chill has dissolved into light delirium induced by a tincture of swamp quietude and solar brilliance. Tired of sitting, I reposition and take a turn at standing, less standing and more draping over the tripod and camera.

Moose on bent ankles to reach mud. J. A. Vucetich

Then a peculiar sensation, a little push from beneath the sole of my foam shoe, worn thin from the wilderness. I straighten up, feigning to myself that I have been alert. Was I already that bored with standing? Whatever. I shift my weight and return to a hazier state of mind.

Last summer's swamp sedges, dead and desiccated, stand tall and rustle in a light breeze. Green shoots are still only an accent to the beige hues of early spring. I stand right at the swamp's edge, where the soil is soft with moisture but not saturated. Nothing particularly advantageous about this position, except being just dry enough to keep water from percolating through the pinhole in the sole of my foam footwear.

Again, that same funny sensation on the bottom of my foot. I shift weight again. With a third push, I compare the observation against ideas, suggesting the various ways by which I could be losing my faculties.

I am hungry, not really from hunger, just bored-hungry. The six-year-old in me is tired of standing. The grown-up in me isn't going to reposition, especially not so soon after the last repositioning. Then comes another push. Exasperated, I abandon my post in a huff, take one step aside, get down on my knees, and push the grass aside to meet the pointed nose of a baseball-sized toad. She had been pushing her head against my thin-soled shoe with all the force she could muster with toad push-ups. She was trying to stretch out after a long winter's nap. Her black eyes were wide open. Eye contact was unavoidable. I do not know the limits of toad cognition, and I fully expect that I am projecting the thought, but she seems disgusted, wondering whether I am more ignorant than insensitive or the other way around.

A year before, at the same mud lick, in the same disordered state of mind, a thought had occurred to me. I hadn't been looking for thoughts. Nevertheless, the autonomous narrator in my head proffered one. I was sitting in the wilderness counting what was left of the day's allotment of M&M's. And then, *How many other moose does a moose know?* Then the words, *Is it boring to chew cud for eight hours of every day?* I wondered, *What is it like to be a moose?*

The questions were not warmed with sentimentality. No, they struck me, cold and horrifying.

Let's say you select 100 people, all at random, from all parts of the world and all walks of life. Make that 100 million people. Line them up in order, from most to least knowledgeable about moose. I would be pretty close to the front of that line. But I had no idea how to answer those questions.

Soon afterward, I realized that moose are not merely alive but that each moose has a life. A moose has memories of yesterday, hopes for tomorrow, joys and fears, and a story to be told.

If a moose has a life—so I reasoned with the awesome powers of deduction I acquired from all those years in school—then I bet that a wolf has a life too. This gift of awareness was first presented to me by an anonymous mud-sucking moose, and I regifted it to you when we imagined the dreams of a wolf in Chapter 1.

If moose and wolves have lives, then the chickadee and squirrel that live in town just outside our house have lives too. Being less familiar with the details of their lives in no way diminishes the fact. Those miserable ticks! They have lives, too.

Our minds are obstinately anchored to our own experiences and perspectives more than we appreciate.

It feels awkward to say, but I owe those ticks gratitude. They led me to this mud hole on a warm spring day, where in a mild state of delirium I wondered for the first time, *Are the thoughts of a moose beyond my imagination?*

We convert the photographic image of each tick-embattled moose into pixel counts, representing the proportion of a moose's skin that is bare or covered with damaged hairs. Each year we calculate average hair loss. During that first year (2001), the average moose had lost or damaged hairs over 30% of its body. That loss was more than prior years, according to our unpixellated, undocumented memories. The tick population, unaware of our memories, increased over the next six years. By 2007, the average moose was nearly naked, having lost 80% of its hair to ticks.

We also had the sense that those winters were shorter than typical. The relevance of a short winter is that lingering autumns may favor moose ticks by allowing more time for moose to pass by and pick up more clusters of larval ticks before the ticks would otherwise be killed by the first hard frost. Early springs may also favor ticks because a gravid tick and her eggs are more likely to survive if she leaps from the moose onto dry ground, rather than a pile of melting snow.

While tick abundance was high and rising (2001–2008), predation rates were also exceptionally high, and moose abundance declined to the lowest levels we'd ever observed. We recognized each melodic theme and wondered what polyphony they'd form. We'd have to keep listening.

Perhaps climate change made for warm years, warm years made for more ticks, ticks weakened moose, wolves took advantage of weakened moose, moose abundance fell, and the forest thrived. What a remarkable prospect. The skin-pricks of such a small creature just might resound throughout the forest.

We still haven't listened long enough to know. But I think we'll find out, because we have front-row tickets to the only performance ever of the Isle Royale predatory fugue in tick minor.

Whatever we discover, I am confident the ticks have already offered their greatest lesson.

3 |

Beginnings

It crawls with pedestrians and automobiles, and it foments lobbyists, bureaucrats, and elected officials. The landscape's orienting feature is a white marble obelisk that erupts beside a long reflecting pool, and its topography is marked with neoclassical monuments to American democracy. A block away from the reflecting pool, in an austere monolith of offices devoted to the Department of Interior, two men beheld a lump of plaster.

One of the men, Durward Allen, gaped through horn-rimmed glasses, sandwiched between his large forehead and soft cheeks. He, who was not prone to exclamation, exclaimed, "There are wolves on Isle Royale for sure!"

Earlier reports were not quite credible. But now, Allen held definitive evidence. Wolves had recently crossed an ice bridge that connected Isle Royale to the mainland, likely during the winter of 1948 or 1949.[1]

These were not low-level bureaucrats breaking the monotony of office life with entertaining triviality. Allen was acting assistant chief of the US Fish and Wildlife Service, and his colleague was chief biologist of the US National Park Service. The evidence indicated a critical development and merited direct delivery from one high official to another. The plaster cast, made earlier that year, formed the positive impression of a bold chevron led by a splay of four meaty lobes, each following the trace of a claw. Its heft conveyed as much as its linear width. At four inches across, there was no question the track had been left by a wolf.

The lump of plaster spawned an irrepressible scheme in Allen's gut. He immediately recognized the opportunity. Allen—career government biologist, former Boy Scout, raised by Indianan ways of life—was indisposed to hyperbole or unseemly obsessions. But he could not curb this emerging plan. He brooded it, kept over it as Sméagol would have, and soon named it "the greatest of all experiments in predator-prey relations."

While Allen was marked by conventions that accompany high posts of government, he was not bound by orthodoxy. He also effectively played the role of interloper and was at ease chiding established dogma. In particular, he was offended by government policies that apathetically promoted killing predators and abusing forests and grasslands for the quixotic purpose of growing as many deer and elk as possible, as if they were cattle. That criticism was an important theme in an otherwise docile textbook, *Our Wildlife Legacy*, that Allen wrote while assistant chief.

Allen believed the newly established wolf population on Isle Royale was an opportunity to observe "the pattern of primitive times, now to be replayed in a world where such patterns are confused and obscured by the almost universal hunting of moose and the wiping out or heavy control of wolves." He supposed, "Potentially at least, the wolves could build up, stabilize the moose herd, and bring some protection to the vegetation."[2] But Allen was not a supposition kind of man. He wanted to learn from nature first hand.

While Allen was making plans to study the new wolf population, Lee Smits, a news reporter from Detroit, proposed releasing zoo wolves on Isle Royale, even though it was known that wolves had just released themselves onto the island. Smits' idea was proposed, approved, and implemented all within a year. By contrast, Allen was unable—even as a high-ranking official of the US Department of Interior—to secure funding for his plans. Doing outranked observing, even when the doing had already been done.

In 1954, at the age of 44, Allen left Washington believing he could more effectively serve the world from a different post. He repositioned himself as a university professor at Purdue. For five years, Allen nursed his excitement for studying the wolves of Isle Royale and eventually cultivated an agreement with the Park Service to study the wolves. The Park Service would allow the work, but they wouldn't pay for it. Allen secured funding from the National Science Foundation (NSF). The plan was audacious—a series of three-year studies that would span most of a decade. No one had ever proposed from the outset to investigate the interactions between predator and prey for such an extended period of time. Allen convinced himself first, and then NSF, that a marathon investigation would answer all the significant questions.

Allen raised research funds and cleared administrative blocks, but he still needed a PhD student with the intellect and physicality to perform the work. Much of the field work would occur during the winter when Allen would be tied to teaching university classes. Allen met a young undergraduate, in the fall of 1957, who was presented to Allen as a "highly recommended scholar and self-trained woodsman." He had tracked fisher* for recreation and gained early experiences with a black bear researcher in upstate New York. Allen saw that Dave Mech was "hale and eager" to learn about wolves. Allen may have foreseen that this young man aspired—or would soon aspire—to acquire more knowledge about wolves than any other human ever had.

Allen and Mech began the first set of studies in the summer of 1958 with a seven-week field season. They, just like the wolves when they first arrived, knew next to nothing about Isle Royale. They hiked and camped. They interviewed park staff and visitors. They observed whatever presented itself for observation and recorded every recordable

---

*A member of the weasel family, but larger than most weasels, weighing upwards of 12 pounds.

happening. They became as familiar as they could as quickly as they could.

That summer, Allen and Mech did not see a single wolf. They did not report any tracks or howling. They did gather a few second-hand reports. A hiker had seen wolf tracks last week on that trail. Some campers heard wolf howling a few nights ago from across that lake. That's all—a handful of reports. The professor and his student could say little more than "yup there are wolves." Except they also collected 70 wolf scats found from hiking over 100 miles of trail.

Nearly six decades after that first summer, we still collect hundreds of scats each year. The methods of collection and reasons for collecting are different, but scat has this enduring power to provide fresh insight.

As a scat is formed and given to the world, it scrapes off microscopic bits of intestinal wall. The scraped cells are pressed onto the surface of the scat, and they contain DNA that is unique to the wolf who left the scat. If we collect enough scat from enough wolves, we can document the individual identity of each wolf and know which two wolves are brother and sister, which two are mother and daughter, and so on. If we end up collecting a scat from the wolf with genotype A during each winter field season from 2001 to 2005, then we reasonably infer that wolf was born in April (they're always born in April) of 2000 and died shortly after the winter of 2005.[3]

The only other way to get such detailed information would be by live-trapping each wolf and making them wear a radio collar for the rest of their lives. It is gratifying to learn so much by such unimposing means.

Making such use of fecal DNA would have been science fiction in 1958. Only a few years prior, Watson and Crick discovered the chemical structure of DNA, and decades would pass before geneticists developed the facility to use DNA for discerning siblings and other close relatives. Extracting DNA from a scat requires technology that was even farther into the future.

Today, we do most of our scat collecting in the winter when the scats are frozen rock-solid. We invert a gallon-sized ziplock bag, don it like a mitten, pluck the scat from the snow, reinvert the bag, zip it, and drop it into a sack with all the rest.

During the summer, while hiking down the trail, we sometimes come upon a particularly fresh scat. Something about it piques our interest. The attraction is intuitive, as opposed to being based on some formal criteria. Nevertheless, when a scat speaks to us in the summer, we kneel beside it, pull a sterile cotton swab and a small vial of ethanol from our knapsack, swipe the surface of the scat with the swab, drop the swab in the vial, and continue down the trail. Valderi-valdera.

That technique is just a few years old, before which ziplock bags had been the best method for transporting the tarry, fetid remains of a wolf meal in the summer heat, until we could get it to a freezer. In the late 1950s, before the advent of ziplock bags, my predecessors collected scat in brown paper bags. At day's end, they'd open the bags and line them up in the late afternoon sun to dry, at least a little bit. It worked as nicely as you'd imagine.

Allen's last graduate student, Rolf Peterson, began working on the project in the early 1970s and continues five decades later. He considers ziplock bags to be the most important of all innovations applied to wolf-moose research. Allen, whose time in the field ended before ziplocks, recognizes winter jackets that zip rather than snap and tents with mosquito netting.

Mech and Allen collected scat to confirm the wolves' diet. They presumed that moose would be the primary fare but were uncertain enough to merit a look. Some scats were composed of the dark, glossy hairs of a beaver. On rare occasion, a scat was stuffed with soft fluffy hairs that once belonged to a snowshoe hare. Sure enough, most of the scats were filled with the stiff, wavy hairs of adult moose and the softer blond hairs of moose calves. The professor and his student could say with scientific certainty that there were wolves and they ate moose.

In August 1958, Mech and Allen left Isle Royale and returned to Purdue's campus. Allen had classes to teach, and Mech had classes to take. Notwithstanding the 70 scats, the summer's accomplishments were inauspicious. The start was as lackluster as the overall plan had been bold. They hoped for a fulfilling winter.

A few winters earlier, a government biologist, James Cole, had taken several flights in a small aircraft over Isle Royale during the winter. From those flights, Cole figured that between 15 and 25 wolves inhabited the island. He also found eight sites where wolves had killed moose. Allen hoped that more flights, more effort, and more focus would yield a more precise scene than Cole had obtained.

Cole also reported that the plane sometimes frightened the wolves. Allen's plot required the wolves to remain indifferent toward a plane circling overhead. The plan—well not so much a plan, but more of a hoped-for idea—came with no assurance.

In the fall of 1958, Allen met Arthur C. Tomes, who had been a pilot during World War II. Tomes had captained 35 missions in a B-29 over Japan. After the war, Tomes returned to the Iron Range of northern Minnesota and started a flight school, Northeast Airways. Tomes was used to being a part of a big plan and was enthralled by Allen's big plan. Tomes promised to supply Allen with all of his aviation needs, planes and pilots, for the winter studies of wolves on Isle Royale.

The pilots of Northeast Airways had a reputation for shooting wolves from their aircraft. From a small two-seater plane, the pilot and passenger would find a pack of wolves and chase them out of the forest and onto a frozen lake. Then the pilot would fly low, less than 50 feet above the ice, line up a shot, and the passenger would fire away.

Not long ago, hundreds of thousands of wolves inhabited our nation—"*from sea to shining sea.*" As a nation, we devoted ourselves to the complete extermination of wolves from the contiguous United States—"*crown thy good with brotherhood.*" Even Teddy Roosevelt, the environmental president, despised wolves as "beasts of waste and desolation." We almost achieved total destruction—"*till all success be nobleness, and every gain divine.*"

By 1958, we had shoved wolves into a corner of northeast Minnesota, pressed against the rugged shoreline of Lake Superior. With only a few hundred surviving wolves, the land of 10,000 lakes was the last place in the contiguous United States where a man (it was mostly men), with his gun and his plane, could act out hatred against wolves. For some pilots, the motivation was not so much hatred but violent indifference. Some would claim they did it for the money. Each wolf was worth a bounty plus what a fur trader would pay for the pelt.

The office walls of Northeast Airways, like all aviation offices, was cluttered with the photos of proud pilots standing beside their planes. A conspicuous number of the photos at Northeast displayed doubly-proud pilots standing beside the empty skins of wolves draped over the struts of their planes.

Allen believed these pilots would have the skills to find and observe wolves from the air. He only hoped they would be interested enough to trade guns for binoculars, skinning knives for notebooks, and methodical murder for ordered observation.

Allen's research crew arrived at Isle Royale for that first winter study on February 3, 1959. Their ranks were spartan—a willful young scientist, a wolf-killing pilot, and an assistant provided by the National Park Service. (Allen's professorial obligations kept him in Purdue during that first winter study.) The crew would live in a cabin just up the hill from Washington Harbor, located on the southwest end of Isle Royale.

The cabin was minimally outfitted with a few oil lamps for light, a propane stove for cooking, a woodstove for warmth, a table and chairs for eating, a two-way radio for communicating with the mainland, bunks in a common room for sleeping, and a human-sized bucket of galvanized steel for occasional "bathing." But the root cellar had been well stocked with canned goods. And the dock had been amply lined with 55-gallon drums of aviation fuel. They'd also stashed food and fuel at additional sites across the island in the event of an emergency landing far from Washington Harbor.

Upon landing on the frozen harbor, the first task of winter study—now, as it had been 60 years ago—is to freeze ropes into the harbor ice to secure the plane and prevent extraordinary gusts from flipping the light plane onto its back. Next, we knock a small hole in the ice for an endless supply of fresh water.

From my own experience, there is no breaking to rest while scurrying from one top priority to the next as we unpack gear and prepare equipment to be ready for a first observation flight by tomorrow morning. The sorting and arranging pushes well into night. Before the woodstove has fully warmed the cabin, we cut the 3 kW generator and extinguish the lights on the first day of winter study. I lie on my bunk and stare into darkness. Actions yield to thought and emotion.

It's been 10 months since I last saw the wolves and immersed myself in their lives. Catching up will be wholesome. Will I fulfill my obligation to learn some lesson worth sharing, any lesson at all?

No work satisfies quite like counting moose—it is a simple answer to a simple question (how many moose?) that swirls out of a thousand circlings of the plane and all attention devoted to a search image honed by years of experience. Will we be thwarted by weather or mechanical failure or paralyzing bureaucracy from the mainland?

Will the same alphas be at the helm or will we see changes? Will moose abundance keep rising? Will we see past the particulars to recognize broader insights?

Bright anticipation colored with the slightest tint of disquiet, followed by the robust sleep that follows a physical day. For the next 42 days, propelled by privilege and duty, I'll trade the warmth of civilian life for devotion to science on a winter, wilderness island.

Mech had no protocol to follow or example to emulate. His first task was not to learn about the wolves but to learn how to learn about the wolves. Allen crafted the opportunity and counted on Mech to convert opportunity into discovery. The expectation would have been grossly unfair to most 22-year-olds.

A few days into winter study, while Mech was still learning how to tell a fox track from a wolf track at highway speeds and from a few hundred feet away, while still figuring which of his observations to record in his notebook (I assure you, you cannot write it all down), while developing a reliable sense for when to stop looking *at* the wolves beneath the plane and start looking *for* other wolves, while making observations in the distracting presence of nagging doubts, at that inopportune moment, Mech recalls, "The plane almost conked out and we almost had to go down."[4]

Under certain atmospheric conditions, teensy amounts of ice can form on the inner workings of a plane's carburetor. A good pilot is sensitive to those conditions and the sound of the engine as the first crystals form. Too much ice can result in engine failure. The problem is easily remedied, if detected in time, by flipping a switch that shunts warm air from the engine casing to the carburetor. The only critical element is remembering to do so in a timely manner. Mech's pilot neglected to flip that switch. This is not an arcane bush pilot trick. It's Aviation 101. Knowing when to flip that switch is not a finely honed skill. It's simply a matter of paying attention. That level of inattention was more than disconcerting.

The more demanding skills of wilderness aviation are landings and takeoffs. These little planes only need several hundred feet to take off. On an airport runway, with a few or more thousand feet of runway, no pilot ever wonders whether there is enough runway. There is always enough. On a frozen lake, however, you have to ask the question. And you have to be damn sure of the answer. And you cannot ask the question just before takeoff; you have to ask the question before landing on that lake. Moreover, the length of runway needed varies considerably with cargo. Will we be loading up with 70 pounds of necropsied moose parts? How much fuel is in each of the two 20-gallon tanks? At seven pounds a gallon, it's important to ask. The required length of runway depends on snow conditions (wet, sticky snow calls for a longer runway than hard, icy snow) and wind (taking off into a 15-knot wind will shorten the length of required runway). It's not enough to merely have gotten off the ground before the lake ends; the pilot must also have enough space to clear the hill and trees at the end of the lake. This is not a game of darts with cheers and an attaboy for landing a bull's eye once in a while. The pilot has to get it right every time—not by accident or dumb luck—but on purpose.

A couple of flights after the carburetor icing incident, Mech recalls being shook up when "the take-off was a little too short [and] we almost hit some trees." Mech knew that he was no expert in aviation, but he felt good about his instinct for self-preservation. Back in the bunkhouse, Mech radioed the mainland to request a different pilot.

Before winter study began, Allen and Mech knew that an airplane could be useful only so long as the lakes and bays were capped by a sturdy sheet of ice. Without good ice, the plane cannot safely take off or land. The plan had been to follow the wolves by airplane until ice break-up. Afterward, Mech would track the wolves by snowshoe.

The shortcomings of that two-part plan appeared as Mech was also realizing the pilot's shortcomings. Wolves routinely walk 25 miles or more a day. They continue to use lakes and bays for travel routes as the ice is breaking up. Any snowshoed human who could command the endurance to keep up with wolves would surely drown. Every expectation for studying the wolves was followed by an apparition of complete failure.

Mech desperately needed a pilot who could deliver. Otherwise, he'd be considering an alternative topic for his PhD dissertation. Maybe the fox squirrels of central Indiana's farm country. Allen had a sincere and cultivated interest in the topic and could easily have led Mech toward such a goal.

The second pilot that Tomes sent to Isle Royale was Jack Burgess. In Mech's memory, Burgess was "a real aircraft jockey." Burgess hunted wolves from his plane and liked to tell a story about how he had once "come in on a wolf and hit it with a part of the aircraft . . . the wolf jumped up and pulled the plane out of the air and Jack crashed." Mech turned his focus from the crash part of that story and appreciated

Jack's competence in wilderness flying. As with any exceptional pilot, the plane was an extension of Burgess's body and mind. He shined with the mere thought of wind washing over the wings, driven by a prop spinning in time with the strokes of four deafening cylinders. Heavenly. As Mech recalls, "It was just beautiful to watch him fly."

No less beautiful than Burgess at the control stick was the fact that Mech was beginning to make outstanding observations, and doing so safely. Disappointment came when Mech realized that Burgess wasn't much interested in the work. He did not relish spending all that time away from family and friends just to follow some wolves around. Jack left after little more than a week.

Tome's bench of relief pilots was at least deep enough to send a third pilot to Isle Royale. This latest pilot, Don Murray, had been working as a truck driver out of Mountain Iron, Minnesota. He subsidized his flying obsession with some crop dusting on the side. Murray confessed to occasional barnstorming, but there are no witnesses. Murray earned three bronze stars during the Korean War, and eighth grade was the last time he ever bothered himself with a classroom. He was forged from the same stock of ideals and iron from which Hemingway built men.

Murray was stout with a round face covered by a dark beard. His hands were thick and toughened, always ready for work. Murray was jocular as often as dead serious. He was as rigidly bound to his presumptions as he was open-minded. And his steely blue eyes said, "Yeah, I'm your man."

Murray knew wolves as well as anyone. As Mech recalls, many years after the fact, Murray's knowledge came from having hunted wolves. And Murray recalls, also many years later, "The first year we watched the wolves kill moose and I thought, geez, this isn't very nice, the wolves tear moose apart." Over time, Murray came to know the edifying pleasure that comes with just "look[ing] at them as they are making their own way." Murray cherished living in a winter wilderness with wolves, moose, a scientist, and an airplane. He became *the* winter study pilot for the next 23 years. Murray made it his calling to know exactly how many wolves there were each year and which packs they lived in. For the next decade, Murray was the backbone that aligned the intentions of a succession of four youthful and inexperienced graduate students. Winter studies would not have lasted more than a few years without Murray.

Years later Murray also recalled, "I had never done any *real* wolf hunting. I did it once with Ty Murphy . . . but I don't know if we ever got any wolves. He said we never got any, but the bounty on a wolf at that time was 35 dollars." Maybe Murray never hunted wolves. If he did, he left that violent indifference behind with his younger years.

Either way, Mech and Murray perfected techniques for observing wolves from the air. The first principle is simple: fly whenever possible. If the wind is not howling, if snow is not falling, if there's more than a few hundred feet of ceiling above the ridge tops and five miles visibility, if the sun has not yet set, then it is time to fly.

From the leeward end of Washington Harbor, with the plane pointed into the wind, Murray would pull back on the control stick with his right hand and press the throttle open with his left hand. Both motions executed with the deliberateness and confidence called for in separating one's self from the solid earth. The machine screams and careens across the uneven, frozen surface of Washington Harbor. The first bumps are the biggest, backed by the full force of gravity. Another 75 feet and the bumps soften as aerodynamic lift begins its magic. A moment later the skis separate from the earth.

Murray's plane, an Aeronca Champion, had two seats. One for the pilot and the second, just behind, for an observer. The fuselage is about two and a half feet wide. The narrowness allows an observer to easily see out either side of the plane. The view of the forest below is not obstructed by wings because they are mounted above, rather than beneath, the fuselage. A Champ is especially maneuverable and capable of flying slower than most planes. Slow and maneuverable is precisely what is called for.

On taking off, if you do not know where the wolves are, the first objective is to find their tracks in the snow—the footsteps of a dozen or so animals in an area the size of Chicago. The challenge is amplified by needing to see past the distracting clutter of moose tracks and to discern wolf tracks from fox. While a fox track is only about two inches wide compared to a wolf's four-inch span, that difference can be difficult to judge at 60 miles per hour and 500 feet up. With something like 100 or 200 of the little tricksters flitting about the island, their tracks are more common than wolves.

One clue for distinguishing tracks is to know that foxes do not travel far and wide, as wolves do. If you think you are looking at wolf tracks, but then you see that they deke over to that fallen spruce, and turn back to inspect whatever might be on the far side of that boulder, and then walk down to the shoreline, only to about face and walk up over the hill, these tracks are more likely to have been left by a fox. If the tracks that have your attention, however, cut across the landscape, mile after mile, if the tracks go somewhere, or if the tracks extend far enough ahead that you have to keep flying on to see where they end, then you're probably looking at wolf tracks.

From an airplane, the tracks of a lone wolf are easily confused with an otter's tracks. If you follow wolf tracks that suddenly end at a small hole in the ice, then you've been following otter tracks. If you follow wolf tracks to the top of a gentle rise, at which point paw prints are abruptly replaced by what looks to be the run of a small toboggan, then you've been following an otter that just slid down the hill on her belly.

An especially clean set of tracks from a single wolf—tracks laid not too long ago in snow of just the right consistency—has the appearance of a series of comets, each about a foot and a half long, each one strung immediately after the next. They stretch on for miles. The comet's nucleus is a paw print, punched into the snow with the full weight of a wolf's body. The dust tail is the slight drag in the snow as the paw comes out of the print. The comet's tail points in the wolves' direction of travel.

Often enough the comet is distorted, and it is not possible to discern the direction of travel. But there are other indicators. If you follow the tracks to where you'd seen them the day before, then you've been backtracking. If the tracks become fainter with each mile and eventually fade into oblivion, then you've probably been backtracking. If crisp, clean tracks enter beneath the thick cover of cedar and do not come out the other side, then circle back, because you've been fore-tracking and you just left the wolves behind in that swamp. We want to know all the places where the wolves have been, so we take the time to follow the tracks in both directions.

While the revolving door of pilots was deeply disturbing, Mech still managed to make valuable observations during those first weeks. Two days after arriving to Isle Royale, Mech and the pilot (the first pilot whose name is not remembered) followed tracks of a large pack. They led to a south-facing hillside along a deep inlet on the island's north shore. That hillside was smeared with evidence that 15 wolves had killed a moose, feasted on it, and moved on. Mech and pilot found and followed the tracks past the kill site. Three days later, and about four miles to the west-northwest, Mech and pilot found another site with all the tell tales—blood-stained snow, scattered bones, and the paw prints of 10 or more wolves in a swamp nestled between two ridges. The same wolves had killed a second moose. After another three days, and 35 miles to the southwest, Mech, now flying with Burgess, witnessed a third site at the base of a ridge, about 50 feet from shore, where the wolves had killed a moose. The wolves seemed to be killing with the frequency of a metronome.

Winter offered more treasure than Mech could ever have hoped. Permissive weather allowed for frequent flights. Mech and his retinue of pilots regularly found tracks before the tracks were blown in by the wind or obscured by new snow. They followed tracks forward and back, keeping up with the wolves on an almost-daily basis, reconstructing nearly complete travel routes. That winter Mech likely registered every kill made by the big pack over a 28-day period.

Burgess and Murray could fly slow circles over the forest, keeping any portion of the ground in constant view, allowing Mech to observe wolf behavior without causing any concern to the wolves. In addition to observing how often wolves killed moose, Mech also documented other key features of wolves' lives—hunting behavior, howling behavior, courting behavior, territoriality, and social relationships within and between the packs. Much that Mech witnessed on a regular basis had been documented by others on just a handful of occasions. Concern for modest returns of the summer field season melted like freshly fallen snow on the warm cowling of an airplane.

Mech observed wolves and moose on Isle Royale for three summers and three winters and then transformed his notebooks into a PhD thesis entitled, straightforwardly enough, *The Wolves of Isle Royale*. When published in 1966, fewer than 20 scientific papers had been published on the ecology or behavior of wild wolves. Today, about that many are published each month.[5]

The Wolves is uncharacteristic of today's scientific writing, which may be why it is still read today by professional and lay aficionados. Part of the attraction is its readability. Today, scientists are encouraged to write by surrounding plain ideas with impenetrable prose and drowning them with imposing jargon. Another attraction to Mech's dissertation, which many readers may only intuit, is its tone. In many ways, it is an ode to the innocence of scientific ignorance. Take the following example.[6]

According to Young and Goldman (1944:120): ["]The pack is generally a pair of wolves and their . . . offspring.["] . . . Olson (1938) also asserted that packs are family groups. . . . Murie (1944) . . . ascertained that the "family theory" of the pack held for at least two groups of wolves in Alaska. . . . Theories conflict regarding the status of the pack in summer.

The phrases "according to," "asserted," and "family theory" betray the tentativeness assigned to even the most basic elements, like an answer to the question, what is a pack? Mech's treatment of each topic was similar. He described what he saw, compared it to the precious few previous observations made by others, and offered some tentative reasoning for what it might mean. Always tentative.

Wolf packs are territorial, meaning there is a patch of earth they defend from all other wolves. Wolves sometimes lose their lives defending territory and violating their neighbor's territory. Playing it safe is not an option. Slack territorial boundaries are a slippery slope to starvation. A pack aims to hold a territory just large enough to contain all the moose necessary to keep the pack going. Violating a territory is like raiding your neighbor's refrigerator. Territoriality is on the mind of the alpha pair much of the time and influences where they hunt, where they rest, and where they travel.

In the early 1960s, the territorial nature of wolves was suspected but far from certain. After spending 16 months in the field, spread out over a three-year period, Mech summarized his sense of the matter: "Isle Royale's packs also seem to be territorial, at least in winter."[7] It seemed so, but he was apparently not entirely certain. To document an observation is only part of the job. A good scientist also accurately indicates how much weight their observations can support.

"It is well established that wolves travel where the going is easiest."[8] That sentence contains the only occurrence of a phrase anything like "well established" in The Wolves.

Wolves tend to kill moose at night. When Murray and Mech found that wolves had killed a moose, they generally did not see the actual killing. What they saw, more often, was the aftermath—an unkindness of ravens, lazing wolves, bloodied snow, scattered moose limbs, and the large, emptied shell of a rib cage.

Occasionally, just before sunset, a pilot and observer see much more. On February 24, 1959, about three-quarters of a mile northeast of the outlet from Siskiwit Lake and about 600 meters in from shore, Mech observed the following:[9]

> At 6 p.m., 10 of the 15 wolves were traveling along the shore of Siskiwit Lake about 1 mile ahead of the others. Suddenly they stopped, and several pointed more or less crosswind for a few seconds toward three adult moose three-eighths of a mile away. Heading inland single file to an old beaver meadow, they traveled downwind a few hundred yards, veered, and continued for 250 yards until directly downwind of the moose. Then they ran straight toward the animals, which were still browsing when the wolves

were within 150 yards. Two of the moose sensed the wolves 25 yards away and began running. The wolves gave chase a few yards until they spotted the third moose, which was closer and had not left. They immediately ran the 50 feet to this animal and surrounded it.

A few seconds later the moose bolted and the wolves followed in its trail. Soon five or six animals were biting at its hind legs, back, and flanks. The moose continued on, dragging the wolves until it fell. In a few seconds the animal was up, but it fell a second time. Arising again, the moose ran through the open second-growth cover to a small stand of spruce and aspen, while the wolves continued their attacks; one wolf grabbed the quarry by the nose. Reaching the stand of trees, the moose stood, bleeding from the throat, but the wolves would not attack.

Within a few minutes most of the wolves were lying down, including the last five, which had caught up. Two or three continued to harass the moose without actually biting it, and the moose retaliated by kicking with its hind feet. Whenever the animal faced the wolves, they scattered. Although the moose was bleeding from the throat, it appeared strong and "confident." At 6:30 p.m. we left because of darkness.

The next morning at 11:15 a.m. the wolves were gone. The moose lay within 25 feet of where it had made the stand. After we made several low passes, it finally arose and moved on. Although walking stiffly and favoring its left front leg, the moose was not bleeding and seemed in good shape. The wolves were 16 miles away feeding on a new kill.

Mech documented a similar scene just two days before and would see similar events four more times that winter. Mech felt all the excitement of a proud young man who'd worked hard, took a risk, and now knew that he had something special. Over the three winters that Mech observed Isle Royale wolves, he would see close encounters between wolves and moose on 71 occasions.

Those close encounters also suggested that only five or six of the pack's 15 wolves expended most of the labor and endured most of the risk. So why, Mech wondered, do wolves live together in such large packs? If a pack split into two smaller packs, the work might be divided more evenly and the return (moose meals) on investment (hunting effort) would be greater for each wolf. In the decades to follow, other researchers documented similar inefficiencies among lions, African wild dogs, and other carnivores with social tendencies. An explanation did eventually present itself, but not until after Isle Royale wolves had been observed for more than four decades. To honor the four decades that Mech waited for an explanation, I will wait a couple of chapters before sharing that explanation.

Mech and Murray did not limit themselves to tallying events and marking locations on a map. They did more than follow life and death from the privileged perspective of the plane. Each time wolves killed a moose, maybe on the second or third day after a kill, Murray would land the ski-plane on the nearest frozen lake. He and Mech would strap on snowshoes and hike a few dozen meters or several kilometers. Whatever it took. Mech wanted to inspect these kill sites up close. He wanted to know what kind of moose the wolves had killed.

During his three winters on Isle Royale, Mech documented 36 instances where wolves had killed moose. Of those, he snowshoed in to perform necropsies on 33.[10] For every 10 moose that he necropsied, four had suffered in life from jaw necrosis, starvation, a wheezing number of hydatid cysts in the lungs, or some combination of those ailments. Many ailments go undetected because wolves eat much of the evidence before a necropsy can be performed. Four of every 10 moose is almost certainly an underestimate, for wolves' tendency to kill weakened moose.[11]

Of 71 close encounters that Mech and Murray observed, only six ended with wolves killing a moose. Wolves killed moose often enough, but only with great effort, many misses, and the frequent-enough advantage offered by weakened moose. Mech was providing the science to know wolves are neither rapacious nor gluttonous. Every meal is a Herculean labor. Failure is death. Not at all uncommon.

This sense of the life of a predator has been confirmed and reconfirmed on many occasions since, for many species of predator, feeding on many kinds of prey, across many kinds of environments. Today, these insights are widely appreciated, taken for granted, and considered common knowledge. Every bit of common knowledge had a time before it was so. Isle Royale is the birthplace for this idea that wolves depend on weakened prey.

The wolf behaviors that Mech had been observing were fascinating, but Allen's mission entailed a broader set of questions. How did the number of moose, as a source of food for wolves, influence the number of wolves? How did the number of wolves, as a source of mortality for moose, influence the number of moose? Questions of that nature require knowing the number of wolves and moose and knowing how those numbers change from year to year. No one had ever before attempted to count all the wolves and moose living in an area as large as Isle Royale. Knowing total abundance requires finding wolves who have no last known location and wolves whose existence is in doubt.

The method that Murray and Mech developed for counting wolves on Isle Royale has been used ever since. The difficulty does not lie with enumeration—we have all been good counters. The challenge is finding wolves whose existence heretofore has evaded our awareness.

After observing a pack of wolves for a spell, a decision is necessary. Continue observing or move on? Continuing to observe is routinely productive and interesting. Moving on will initiate a search that has a chance of producing some unforeseen and worthwhile observation, such as the discovery of a new kill site, evidence left behind in the snow of a skirmish between two packs, and occasionally the discovery of an undiscovered wolf or two or three.

Moving on might begin with flying along one of the half dozen major ridges that run for miles with the long axis of Isle Royale, scanning the ridge top at 60 mph for wolf tracks. The lowlands on either side of the ridges are heavily forested, except for frequent and irregularly spaced openings that extend down from the bare ridge tops. Each opening is less than a one-in-a-thousand chance to see a wolf, asleep and curled beneath a spruce tree.

This scanning is not a casual, laissez-faire, if-I-happen-to-see-something-great kind of scanning. It's an absorbing, focused, my-purpose-for-being-in-this-plane, my-purpose-for-being-on-this-island, my-purpose-depends-on-seeing-something, I-don't-know-what-but-I'd-better-notice-it-when-it-passes kind of scan. But the scanning cannot be frenetic either. With a typical flight lasting four hours, this scanning is a marathon not a sprint.

Often there is not enough time to consciously scan an entire opening. On these occasions, one depends on some inarticulable gestalt suggesting that we should turn the plane back for a closer look. But there are thousands of perforations in the forest canopy, and many more rocks that very much resemble a sleeping wolf. It takes about a minute or a minute and a half to circle back for a closer look, and we do not have enough minutes to confirm the location of every rock posing as a wolf.

After following 5 or 10 miles of ridge tops, we might slide over and begin following a shoreline using the same scanning technique. Next up, the beaver drainages. Same routine, scanning their length for tracks, eyes regularly darting up the adjacent hillsides, hoping to find something before flying over the Big Siskiwit Swamp, a featureless, tangled, desolate expanse of black spruces and cedars that sprawls across a fifth of Isle Royale. Even the wolves avoid it most of the time.

The sun is always about to set or snow is just about to move in. Time is the provision in shortest supply. Every decision about how to spend time in the air is accompanied by nagging doubt about the unknown cost of lost opportunity. Often enough a search ends only with inappropriate regret for moving on from the wolves we had been observing.

Undiscovered wolves do not always reside in far-flung corners of the island. Sometimes a lone wolf is everywhere that the pack had been yesterday, skulking in the pack's tracks, and scrounging from moose carcasses with too little meat to hold a pack's attention.

Weather also contributes to decisions to search for undiscovered wolves. On a bright sunny day, the deep furrow of pack tracks and the postholes of a lone wolf are like a canyon or crater beneath a desert sun. The rim shines in the light, and the walls are dark in the shadow. The contrast is arresting, and tracks are as conspicuous as words printed on a page. When the blue skies turn gray, the contrast dissolves and tracks evaporate. When clouds fill the sky, do not waste one minute looking for wolf tracks.

Over a fortnight without significant snowfall, tracks accumulate. Wolves begin to retrace their tracks. They walk in and across the ever-growing maze of moose tracks. The ground looks like the work of an Etch A Sketch artist gone mad.

Four to six inches of light snow is deep enough to make a brilliant track and cover old tracks, yet not so deep as to discourage wolves from traveling and thus laying down new tracks. Over the few days that follow such a snow, tracks accumulate to the perfect amount. Then we hope for a bright sunny day—an eye-squinting day, showcasing the sparkle of individual snow crystals. That will be an excellent day for counting wolves.

At some unexpected moment during a flight, independent of light and snow condition, and by happy accident, we will be in the right place at the right time. We will see, in plain sight, a lone wolf or maybe a pair of wolves trotting down some remote drainage. Then it's time to figure out whether those wolves really are newly discovered or just taking a break from life with one of the other packs of which we already know and, consequently, have already counted.

On February 9, 1959, just six days after arriving to Isle Royale, Mech saw the pack of 15 wolves at McCargo Cove. That same day, Mech and the pilot also saw a wolf five miles to the southwest near Todd Harbor and another wolf about four miles east of McCargo Cove. Two weeks later, they found the big pack near Davidson Island with a 16th wolf traveling behind the pack. On that day, they also saw a pack of three wolves near Five Finger Point. From those observations, Mech concluded that the wolf population was composed of 19 or 20 wolves. That winter, with more than 100 hours of observation from the airplane, they never saw anything to suggest any additional wolves.

None of those big questions about whether, how, and why wolf and moose populations ebb and flow could ever be answered by knowing the number of wolves at a single point in time. No pulse can be known by hearing just one beat.

The following winter, Mech and Murray did it all over again. They observed the same large pack. It still had 15 wolves. The pack was still trailed by a 16th wolf living on the fringe. On several occasions they also saw a group of three wolves, another pair, and another lone wolf. Believing that some of those observations were the same wolves seen on different occasions, Mech concluded that the population was, for a second year in a row, composed of 19 or 20 wolves. That winter, Mech and Murray also followed the big pack of 15 wolves consistently enough to enumerate their kills. The tally was 15 moose in 45 days, or one moose every third day, as had been the case in the previous year.

With considerably more wind and snow than previous years, flying conditions were far less accommodating during February and March of 1961. Estimating wolf abundance was further challenged by the big pack regularly splitting up into smaller groups that were more difficult to track. Ultimately, Mech felt comfortable estimating that the population included a total of 21 or 22 wolves. Over a 37-day stretch, the big pack killed 12 moose. Again, one moose every third day.

During the first winter, Mech had observed three wolves that were light in color, lanky in stature, and more playful in disposition than the other wolves. No such wolves were observed in either of the next two winters. Mech supposed those wolves had been born in the April before the first winter field season and had since grown out of their pup-like appearances and behaviors.

Mech observed wolves copulating each winter, but he never saw evidence of denning or the successful rearing of pups during the summer. The wolf population's abundance appeared unchanged for three years running. Without evidence of reproduction, Mech surmised the population did not experience any births or deaths. Even the number of moose killed was a consistent one moose every three days for each of the three winters. The pervading impression was stability and constancy.

The work for estimating the abundance of moose is fundamentally different than it is for wolves. There are hundreds of moose. You can see another moose for every 5 or 10 minutes of flying without even trying. To estimate abundance, Murray flew straight lines orthogonal to the long axis of the island. Each flight line was three to six miles long and, depending on the visibility afforded by the terrain, spaced about an eighth of a mile from its neighboring flight lines. With that spacing, no patch of land would be overlooked. Mindful of the risk of counting some moose twice, Mech attempted to recall the locations of moose seen on the previous transect. Moose are not social, but when you find one moose there is a tendency for another to be browsing not far off. Upon finding a moose, Murray would circle back and search the area for others. Mech had hoped to cover the entire island with transects that would take at least several days of plane work, enumerating every observed moose. They knew they had little idea how many moose would slip by undetected. They also knew that no one else would have been able to do better, given how little anyone knew in the late 1950s about how to count moose in a boreal forest.

While executing their plan, a miscalculation presented itself. They had stocked about 600 gallons of fuel at Washington Harbor during the previous fall, and another 400 gallons at Mott on the island's northeast end. Observing wolves had been more productive than expected and the weather had been accommodating. They didn't save enough fuel to count moose across the entire island. Murray and Mech were stopped short at 194 moose counted on transects covering about two-thirds of Isle Royale.

Murray and Mech were better prepared the following winter. In February 1960, they devoted 45 hours of flying, spread over a 10-day period, to counting moose on transects that covered every bit of the island. They counted 529 and concluded with a dose of caution that the island was inhabited by about 600 moose.

During the winter of 1961, and with two years of experience, they were more proficient at counting wolves and moose. But proficiency was overtaken by poor flying conditions. Frequent snow and windstorms kept Mech and Murray from making as many observations as they would have liked. Nevertheless, the evidence Mech gathered did not hint at any big changes, so Mech and Allen believed that moose abundance had remained about the same.

Mech and Murray had pretty much figured out how to watch and count wolves during the first few weeks of the first winter. But the ability to detect moose was varied. If it was too windy, the turbulence stole the focus required to stare at the ground looking for moose. Under some cloud conditions, the lighting disguises moose in the shadows of large spruce trees. When the snow becomes crusty, moose tend to spend more time in thicker forests where they are less likely to be seen. It would take most of the next decade to develop an effective method for counting moose.

For three years running, each verse was the same as the first—about 20 wolves, about 600 moose, and about one wolf-killed moose every third day. Mech believed the wolf population had simply been static, experiencing neither births nor deaths during the three years of observation. The moose population, however, he took to be constant in a dynamic way. To demonstrate, Mech concluded *The Wolves of Isle Royale* with an exercise in actuarial accounting. Suppose the Big Pack's tendency to kill a moose every third day applies not only during the winter, but throughout the year. Further suppose that kill rate applies to other wolves in the population. If those rates apply, then of the approximately 600 moose in the population, about 163 would be killed each year. The population would quickly be driven to extinction unless the losses were offset by the birth and survival of new calves.

The next line of the account sheet is drawn from summer observations. Mech recorded the sex and age of every moose that he saw throughout the summers—adult bull, adult cow, or cow with a calf. Of those sightings, every fourth moose was a calf. Exposing that observation to a little arithmetic suggested that the moose population produced about 188 moose each year, which is a pretty close match to the 163 moose that might have been killed by wolves each year. Mech conceded the imprecision of the calculations. The reliance on supposition was conspicuous. But it is simply irresistible to acknowledge the calculations' suggestion that losses to predation are just about evenly matched against the gains rising from moose reproduction. That balance is exactly what you'd expect if the population was essentially constant over those three years.

To further interrogate that expectation, Mech developed a semi-independent line of calculation. It begins with the cold acknowledgment that calves are fortune's fools. Some are conceived just ahead of mild winters, born from endowed wombs, and well prepared to survive their first year on the ground. Others are conceived on the eve of terrible winters. Their mothers have little to spare on a developing fetus. These calves are born small and disadvantaged. Many die before the first anniversary of their birth.

Distilling off the stochasticity introduced by the unpredictable lives of calves might produce a clearer pattern. To do so, one could devise an account of lives and deaths that skips past springtime births and begins with midwinter, by which time the unpredictable fates of the newborns have largely been determined. In other words, compare predation losses among moose older than eight months with the number of calves that survive each year to at least eight months of age.

For this appraisal, Mech employed observations made during winter flights. By late January, moose calves are eight months old, still living in the company of their mothers, and are about three-quarters the size of an adult moose. Of all the moose observed during winter flights, about 17% were calves. This suggested to Mech that about 85 calves survived and graduated onto adulthood. Mech also estimated that wolves killed 83 adult moose each year. This line of figuring indicated, within just a few percentage points and well within the margin of precision afforded by the observations, that predation losses were balanced against reproductive gains.

The most provocative explanations unleash two hounds. One demands belief until the explanation is proven wrong, and the other disallows belief until the explanation is proven beyond doubt. The skeptical hound, upon seeing Mech's explanation, will be impressed by wide margins of error and firm reliance on supposition. But the hound looking for a plausible explanation to the plain perception of constancy will be impressed.

After the balance sheets, in the closing pages of *The Wolves of Isle Royale*, Mech declares the "obvious result" that wolves had been maintaining a "healthy herd" of moose. That grand and overarching result also rested on the simple observation that at least 39% of the moose killed by wolves had something seriously wrong with them. Thanks to wolves, the moose of Isle Royale were, as Mech put it, "one of the best 'managed' big-game herds in North America."[12]

Mech also believed the wolves had been limiting the abundance of moose with enough force to provide relief for the forest. Moose had devastated Isle Royale's forest in the decades that preceded colonization by wolves. But Mech had observed, just a decade after the arrival of predation, "new stands of balsam fir and aspen a few feet high" growing "for the first time in decades."

As for the wolf population itself, Mech concluded: "Apparently the Isle Royale wolf population has increased to its maximum (under present conditions), for if it were going to increase further, it would have done so years ago." Putting it all together, the only sensible conclusion was that "apparently the Isle Royale wolf and moose populations have reached a state of dynamic equilibrium," a state of health for the moose population and forest, provided by the wolves.

The editor of *National Geographic* came to know of the observations from Isle Royale during those first few years. And he believed everyone should know. In a 1963 issue of the magazine, millions of readers were introduced to the wolves and moose of Isle Royale, where Allen and Mech concluded that "the wolves and moose of Isle Royale have struck a reasonably good balance."[13]

4 |

Balance of Nature

When Allen and Mech explained how wolves and moose had struck a reasonably good balance, they invoked one of the most venerable, consequential, agitated, and tangled ideas ever conceived about nature: balance of nature. An adequate understanding will require some careful unknotting.

It's best to ease into this idea. Think for a moment of a lake that we polluted, a population of fish that we overharvested, or a forest that we overcut. Now imagine that we realized the error of our ways and stopped polluting, fishing, or cutting. Do you think the lake, fish, and forest will recover from our errant ways, or will they never be quite the same again?

A cautious person might resist answering because the question does not provide sufficient context, and the answer almost certainly depends on the severity of the polluting, overfishing, and overcutting. Extreme cases are easy to evaluate: pull weeds from a garden, they return quickly, as though never plucked. Just as assuredly, the Aral Sea—now an anthropogenic wasteland—is not coming back.

But in many cases, we do not know whether our impact is severe or not. Sure, the impact could be put to numbers. So many tons of pollution dumped. These many million fish netted. That many board feet milled. But, in too many cases, the numbers deliver no insight. So, without any more information, do you think those systems would recover? The answer is intuitive and to be found in your soul.

This balance of nature question is as easy (or difficult) to answer as, "Are you a glass-half-full or glass-half-empty kind of person?" Does nature have a delicate balance, easily upset by humans? Or is nature resilient, with an unambiguous tendency to restore its prior balance when the disturbance passes? Or are you among those who believe that nature has no balance—delicate, resilient, or otherwise?

Your belief on these matters implies much. For example, believing in the delicacy of nature's balance is often accompanied with a conviction that nature should be better protected from the impact of human enterprises.[1]

The ideas we harbor are not entirely our own. They are inherited, as much as discovered. Of course, you can claim your ideas for your own, and you should. But those claims are greatly strengthened by understanding the heritage of an idea—how it's been handed down through history and across cultures.

"Balance of nature" is as cross-cultural as any notion has ever been. The Cherokee people, whose home had been southern Appalachia, believed the following:

> In the old days, the animals and plants could talk, and they lived together in harmony with humans. But the humans spread over the earth, crowding the animals and the plants out of their homelands and hunting and killing too much. The animal tribes called a council to declare war on the humans. They each selected a disease to send to the humans that could cripple them, make them sick, or kill them. When the plants heard what had been done to the humans, they agreed this action was too severe and called a council of their own. They agreed to be cures for some of the diseases the animals had sent.[2]

For the Cherokee, humans can upset the balance, and nature can restore the balance.

Greek mythology shows how humans can be part of restoring nature's balance. Long ago, Hades, god of the underworld, abducted Persephone, goddess of the Spring. Her mother happened to be Demeter, goddess of harvest. In vengeful sorrow, Demeter "made that year the most terrible one for mortals, all over the Earth, the nurturer of many. It was so terrible, it makes you think of the Hound of Hadês. The Earth did not send up any seed. Demeter, she with the beautiful garlands in her hair, kept them [the seeds] covered underground." Keleos, first among Demeter's mortal priests, "assembled the masses of the people . . . and he gave out the order to build . . . a splendid temple, and an altar too, on top of the prominent hill. . . . At this moment, she [Demeter] could have destroyed the entire race of . . . humans with harsh hunger . . . if Zeus had not noticed" the human prayers for mercy.[3] Upon noticing, Zeus intervened for Persephone's return. Hades agreed to her spending a portion of each year on Earth, during which time flowers blossom in the warm summer sun. For the remainder of the year, she lives with Hades in the underworld. That time each year is marked by withering plants and frozen ground. The most basic of all nature's cycles—the seasons—depended on human supplication.

The cultural breadth of balance of nature speaks to the idea's antiquity. About 75 years before the birth of Christ, Marcus Tullius Cicero was rising quickly through the political ranks of Rome's Republic. In midcareer, Cicero found himself on the losing side of a power struggle that ended with his exile and banishment from politics. He turned his energy to what he considered the second-best preoccupation. He wrote philosophy. Most of that writing applied and extended Stoic principles of virtue to politics.

The exiled Cicero also forayed into theology and authored a book, entitled *De Natura Deorum*, or as we would say, *On the Nature of the Gods*. Here he found occasion to write about the balance of nature. By this time, balance of nature was a venerable idea with Plato and Herodotus having had written on the topic 300 years earlier. The science of Cicero's day largely presupposed that nature was harmonious, constant, and balanced—especially the heavens and the human body. Cicero's contribution to this corpus of thought was to propose underlying mechanisms by which balance of nature was maintained, not just in astronomy and medicine but in the earthly world of nonhuman life. According to Cicero, the ecological processes from which balance arose included mutual relationships between species, differential rates of reproduction between predator and prey, each species being endowed with different physical attributes, and each species having its own diet and habitat preferences.

Cicero circumscribed much that is contained within the scientific field of ecology—2,000 years before its advent. His thoughts make the details of Allen and Mech's explanation for wolves and moose on Isle Royale seem late in coming. The comparison is interesting, but the richer relevance of raising Cicero stems from these words of his: "In order to secure the everlasting duration of the world-order, divine providence has made most careful provision to ensure the perpetuation of the families of animals and of trees and all the vegetable species."[4] Cicero wrote about balance of nature in a book about theology because balance of nature was taken as empirical "evidence for the wisdom and benevolence of the Creator." Theology was served by plain observations from nature.

One year after publication of *De Natura Deorum*, in 44 BCE, Brutus murdered Caesar. Political confusion ensued, and Cicero returned to Roman politics.

Fast forward 250 years. We find Plotinus reinventing and reinvigorating the philosophy of Plato, by this time 750 years old and largely forgotten. In resuscitating Plato, Plotinus became distracted by an ancient theological conundrum. How could God be benevolent and all powerful but still allow so much evil in the world? Plotinus saw an answer in predation. That predation had a dark side was incontrovertible, but Plotinus also saw it as a manifestation of the balance of nature—essential for creating the greatest diversity and quantity of life. The goodness of that abundant life more than justified the death needed for sustaining other life.[5]

The comparison with Allen is irresistible. Allen's motivation was to understand how nature worked before its patterns become "confused and obscured by the almost universal hunting of moose and the wiping out or heavy control of wolves," and Mech's conclusions agreed with Allen's original supposition that "potentially at least, the wolves could build up, stabilize the moose herd, and bring some protection to the vegetation."[6]

Within that comparison is a subtle contrast. Kudos to Plotinus for anticipating Allen and Mech 1,700 years beforehand. Right on to Allen and Mech and their readers for being properly bedazzled by an idea conceived during late antiquity.

Fifteen hundred years after Plotinus—during the 18th century—balance of nature continued to be a (if not *the*) straightforward and conventional way to conceptualize nature. The idea also continued to inspire theological insight. In 1714, William Derham wrote what would become a popular series of lectures entitled *Physico-Theology*. In it he states, "The Balance of the Animal World is, throughout all Ages, kept even, and by a curious Harmony and just Proportion between the increase of all Animals, and the length of their Lives, the World is through all Ages well, but not over-stored."[7]

It goes without saying that "curious Harmony and just Proportion" is a euphemism for God. Derham certainly knew that nature was not always "well" and was sometimes "over-stored." Rather than neglect inconvenient facts, Derham elaborated on the theology. He explained how pestilence, plague, and other imbalances of nature were simply God punishing us for our offending ways. According to Derham, imbalances of nature "serve as [God's] Rods and Sco[u]rges to chastise us, as means to excite our Wisdom, Care, and Industry."[8]

Like Cicero and Plotinus before him, what Derham did was adjust theology to accommodate an empirical understanding of the natural world. But there was a limit to how much theology could be stretched and molded. At some point, explanations of nature would need to be bent to accommodate theology. For example, because God is a perfect deity and circles are a perfect shape, it was also taken for granted that planetary orbits would be circular. Consequently, we were slow to discover that orbits are elliptical. Because we humans are special in God's eyes, one can readily deduce that the sun and all the other heavenly bodies orbit the earth. Consequently, we were slow to understand the heliocentric nature of our solar system.

Theology also influenced our understanding of life on earth. In ancient times, every aspect of nature was believed to be enlivened with spirit, including the inorganic bits of nature. Consequently, fossils did not require any difficult explanation. They *grew* in place from the enlivened spirit of their parent rock. No more or less improbable than the formation of a fetus within a woman's womb or a crystal within a rock.[9]

By the end of Europe's medieval period, fossils were generally accepted as mineralized bones of once living creatures. However, they were still not acknowledged as evidence that a species could go extinct, because extinction was prohibited by God's balance of nature. God's nature was perfect—perfectly constant like the North Star. To acknowledge an extinction would be admitting to a mistake by God. What's more, God's benevolence logically precludes him from destroying his creations—never mind all that Old Testament smiting.

When presidential affairs were not commanding his attention, Thomas Jefferson sometimes entertained himself in the East Room of the White House, arranging his collection of fossilized mastodon bones. Jefferson is co-credited with discovering a fossilized species of giant ground sloth from a toe bone that he originally misidentified as belonging to a species of giant lion. Jefferson's most remarkable (mis)understanding of fossils is reflected, however, in his asking Lewis and Clark to keep a weather eye for giant ground sloths and mammoths, alive and well in the American West.[10] For learned people of Jefferson's day, fossils were not evidence of extinction. Rather, they were previews of what would be found alive on some future expedition to an unexplored region.

By 1830, Mary Anning was leading the way to the emerging field of paleozoology. From the British countryside, she had unearthed some of the first specimens of ichthyosaur, plesiosaurus, and pterosaur ever to be discovered. It was beyond credulity that such fantastical monsters might someday be discovered alive in some remote corner of the planet. As such, Anning's discoveries contributed mightily to the hard fact that extinction is real.[11]

The theology of many scientists had also changed by the mid-19th century. They had abandoned venerable ideas professed by the Old Testament and Greco-Roman classics, whereby God(s) regularly meddled in the lives of mortals and continually involved themselves with the unfolding natural world. In casting those beliefs aside, scientists tended toward Deism, a theology whereby God created the universe, set it to work by the laws of nature, and thereafter did not intervene. From this view, the purpose of science is to discover those laws of nature, thereby explaining nature without referring to God, except to say that he created the laws of nature.

The problem is that no one could envision a law of nature that could account for extinction. Charles Lyell, a respected geologist of the 19th century, rejected extinction. He accepted the fossil record as evidence of evolution but not extinction. Lyell believed that lineages of creatures were transformed from one kind into the next without there ever being any extinctions. Such an adroit way to save constancy: there are no extinctions, and new species are not new so much as merely transformed.

Others believed extinctions were genuine terminal events caused, not by God, but by humans. The explanation had certain theological appeal—God wouldn't destroy his creations, but humans, in our aberrant ways, might. Fine enough, I guess, except that explanation could only apply to mammoths, mastodons, and other Pleistocene mammals. Humans could not have, for example, caused the extinction of trilobites or Anning's dinosaurs.

Those earlier extinctions could, however, be explained by God having destroyed the earth.[12] Some believed fossils represent species that had not made it to Noah's ark. This tack of thought navigated past the unthinkable—that extinction was some kind of Almighty error. Instead, extinction was merely collateral damage from God's righteous punishment of humans.

All this groping and stumbling over the nature of extinction came to a final and sudden end in 1859. In that year, a new law of nature was discovered. This law decreed extinction's legitimacy and put it in charge. Charles Darwin's *On the Origin of Species* builds from two essential ideas. First, all life traces back to a common ancestor. Take any two creatures—no matter how similar or how different—and go back in time. Go back far enough, and you will find a third creature that is a common ancestor to both. That idea was controversial and most scientists rejected it, but the idea wasn't even all that new. Variants of that idea have existed for millennia.

The second building block in *Origin* was, however, entirely novel. It proffers an explanation for how life forms change over time in a way that led from that common ancestor to the diversity of life seen today. Darwin's radical thought was a natural process for evolution that did not require continuous or even occasional intervention by God. The mechanism was a "struggle for existence" that followed from animals producing more offspring than could be supported by the available resources, food, and space. In that struggle for existence, creatures with superior traits would survive and the rest would be driven to extinction. Darwin explained extinction as the result of a machine-like process, blind, without any purpose or meaning. Darwin explained the diversity of life without leaning on God, except as the force who set the rules of the game in motion.

The inventor of these wild ideas was, nevertheless, very much a conventional balance-of-nature kind of guy. The following is from the chapter of *Origin*, entitled "Struggle for Existence":[13]

> I have also found that the visits of bees are necessary for the fertilisation of some kinds of clover; for instance . . . 100 heads of red clover ([*Trifolium*] *pratense*) produced 2,700 seeds, but the same number of protected

heads [protected from bees] produced not a single seed. Humble-bees alone visit red clover, as other bees cannot reach the nectar. . . . Hence we may infer as highly probable that, if the whole genus of humble-bees became extinct or very rare . . . red clover would become very rare, or wholly disappear. The number of humble-bees in any district depends in a great measure upon the number of field-mice, which destroy their combs and nests. . . . Now the number of mice is largely dependent, as every one knows, on the number of cats; and Colonel Newman says, "Near villages and small towns I have found the nests of humble-bees more numerous than elsewhere, which I attribute to the number of cats that destroy the mice." Hence it is quite credible that the presence of a feline animal in large numbers in a district might determine, through the intervention first of mice and then of bees, the frequency of certain flowers in that district!

Darwin was not merely predisposed about nature's balance. He was exuberant and marked by exclamation for balance of nature. Earlier in the same chapter, he writes the following:

[In some places] cattle absolutely determine the existence of the Scotch fir; but in several parts of the world insects determine the existence of cattle . . . this is caused by . . . a certain fly, which lays its eggs in the navels of these animals when first born. The increase of these flies, numerous as they are, must be habitually checked by some means, probably by other parasitic insects. Hence, if certain insectivorous birds were to decrease in Paraguay, the parasitic insects would probably increase; and this would lessen the number of the navel-frequenting flies—then cattle and horses would become feral, and this would certainly greatly alter . . . the vegetation: this again would largely affect the insects . . . [and subsequently] the insectivorous birds, and so onwards in ever-increasing circles of complexity . . . in the long-run the forces are so nicely balanced that the face of nature remains for long periods of time uniform, though assuredly the merest trifle would give the victory to one organic being over another. Nevertheless, so profound is our ignorance, and so high our presumption, that we marvel when we hear of the extinction of an organic being; and as we do not see the cause, we invoke cataclysms to desolate the world, or invent laws on the duration of the forms of life!

Extinction did not threaten the balance of nature. Rather, it was essential for maintaining balance. Furthermore, balance of nature had nothing to do with constancy. Rather, balance of nature was the dynamic economy of nature, driven by predation and competition. Darwin saved balance of nature from its own extinction by reconceptualizing it.

We are beginning to trace the genealogy of an idea. Plotinus cogitated and articulated the idea's ancestral form: death by predation is a creative force. Darwin cerebrated and enunciated the first progeniture of Plotinus' idea: life evolves from the play between two equal and opposite forces, extinction and growth beyond bound. We are well on our way to passing this developing lineage of thought to Allen and Mech, who were engrossed with a balance between losses to wolf predation and gains by moose reproduction.

Darwin was deliberate and thorough, and he spent more than 20 years developing the theory of natural selection. His caution was driven largely by fear that his idea would be poorly received. He shared the ideas with only a few close friends and dragged his feet in publishing *Origins*. In June 1858, Darwin received a terrifying letter from Alfred Russel Wallace. Wallace was about to scoop, albeit unwittingly, Darwin's theory of natural selection. Wallace's ideas, to boot, depend not on reframing the balance of nature, but rather disposing of it.

In 1848, Wallace booked passage on a ship for the Amazon in search of answers to questions raised by the fossil record—questions about how creatures evolved. As Wallace prepared for the trip, contemplating how many plant presses, insect mounting pins, and blank journals to pack, Darwin had already been thinking—unbeknownst to Wallace— about natural selection for a decade.

Wallace collected tens of thousands of specimens during that voyage. Beetles, plants, the skins of primates—if it once had a life and if its remains could be preserved, then he "collected" it. Most of the lives had belonged to butterflies, birds, and fish. Many of the specimens would be sold to finance his trip. The remainder would become a storehouse of memories and evidence for his developing scientific ideas about evolution. On the homeward sail, the ship caught fire and sank 700 miles east of Bermuda. Wallace and the crew floated in life rafts for 10 days before being rescued.

Later Wallace would recall how his constitution during the Amazon trip had been "upheld only by the fond hope of bringing home many new and beautiful forms from those wild regions; every one of which would be endeared to me by the recollections they would call up."[14]

Now every sample had been lost, along with all his notebooks. Wallace had traveled farther up the Amazon than any other European had up to that point in history. Now, there was not a shard of evidence that he'd ever left home—save one diary. Wallace had been seized by the idea that new species were formed by some natural process and not by the occasional, direct intervention by God. Species were marvelous but not miraculous. Wallace's faith that God refrained from occasionally intervening in earthly affairs was, apparently, not rattled by his survival at sea. That trauma only affirmed his outlook that nature is indifferent, often cruel, and occasionally ironic.

Eighteen months after the Amazon ordeal, Wallace regained resolve and boarded a ship set for the Malay Archipelago of the southwest Pacific. Four years into the trip, Wallace found himself on a tiny, remote volcanic island, sprawled on a bunk, lost in a malarial delirium. He was suddenly overtaken by a flood of ideas that poured from his mind and through his pen, soaking one page after the next. Wallace's most important observations and thoughts from the past decade rushed from every recess of his frontal lobe and formed a single torrent. In just a few days, Wallace produced an essay outlining the essence of evolution by natural selection. The essay included these lines:[15]

> The life of wild animals is a struggle for existence. The full exertion of all
> their faculties and all their energies is required to preserve their own

existence and provide for that of their infant offspring. . . . Perhaps all the variations . . . must have some definite effect, however slight, in the habits of or capacities of the individuals . . . a variety having slightly increased powers . . . must inevitably acquire a superiority in numbers.

Wallace was unsure about the quality of his ideas. A few years back, Darwin had written Wallace to express his appreciation for a couple of papers that Wallace had published on biogeography. Based on that positive interaction, Wallace decided to send the new draft manuscript to Darwin for feedback in that letter of 1858. Darwin knew that he was about to be scooped and risked losing credit for two decades of careful thought.

In just the previous year, Darwin had written "The Struggle for Existence" chapter of *Origin* and the following chapter, during which time he also wrote, "All Nature . . . is at war. . . . The struggle very often falls on the egg & seed, or on the seedling . . . any variation, however infinitely slight, if it did promote during any part of life even in the slightest degree, the welfare of the being, such variation would tend to be preserved or selected."[16]

Previously, neither had been aware of the other's thoughts or writings. Darwin sent Wallace a copy of *Origin*. Darwin, with help from Charles Lyell, arranged for an abstracted version of *Origin* to be read at a meeting of the Linnaean Society, along with Wallace's essay.

Wallace and Darwin were of one mind regarding evolution by natural selection and the role that extinctions play in that process. But they lived in different worlds with respect to balance of nature. While Darwin preserved balance of nature by reframing it, Wallace's thoughts on the matter were unthinkable: "Some species exclude all others in particular tracts. Where is the balance? When the locust devastates vast regions and causes the death of animals and man, what is the meaning of saying the balance is preserved. . . . To human apprehension there is no balance but a struggle in which one often exterminates another."[17]

While Darwin was slow to publish his ideas on natural selection, Wallace never published his remarks on balance of nature at all.

Those thoughts languished in one of his notebooks to be discovered many years later by historians of science. Darwin and Wallace agreed on what they saw, and they agreed on the underlying mechanism. Their difference is that one saw the world, in a manner of speaking, as a glass half-full and the other as half-empty. The disparity between Darwin and Wallace is the overture to the discordant cacophony of thought about the nature of nature that began with *Origin* and continues to the present day. Shortly, we will hear exactly how those incongruous notes resound in what Allen heard from the wolves and moose of Isle Royale.

Darwin's "struggle for existence" idea took permanent root and spread in the decades that followed publication of *On the Origin of Species*. It crept from an explanation for evolution into basic ideas of ecology during the first decades of the 20th century. "Struggle for existence" and its sibling idea "balance of nature" were escorted to modern ecological thought by Umberto D'Ancona.

D'Ancona, after being wounded and decorated as a valorous soldier during World War I, resumed his education in biology, completed a PhD on the digestive morphology of eels, and found employment with the Royal Italian Oceanographic Committee. In that post he came across catch data from several fishing ports along the northern Adriatic. Officials from these ports tracked the annual proportion of total catch that was selachian, an antiquated term referring to sharks, rays, and skates. These numbers were tracked because selachians were considered unpalatable and therefore of less value to the fishery. D'Ancona's interest in selachians was that they are predators of the sea.

Prior to the Great War, selachians represented about 12% of the catch. With the onset of war, fishing was severely curtailed and the portion of selachian catch increased threefold. With the war's end, intense fishing resumed and the share of selachians fell to prewar values. D'Ancona believed intense fishing had tipped the balance between predator and prey in the years prior to the war, balance was restored during the war, and then tilted again afterward.[18]

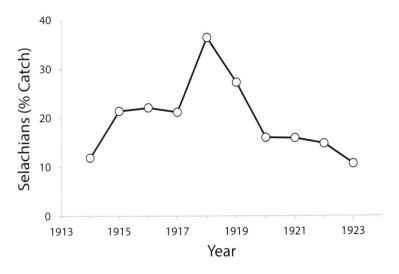

The rise and fall of selachians caught by fishers in the Adriatic Sea before, during, and after World War I. The pattern inspired Vito Volterra's work on predator-prey dynamics. Data from Braun 1983

For all of D'Ancona's excitement about that oscillation, he was short on details explaining how the changes might arise.[19] Conveniently, D'Ancona was courting the daughter of a celebrated mathematician Vito Volterra. D'Ancona piqued Volterra's interest in the problem, not so much for its relevance to commercial fishing but as a vehicle for probing deeper into the "struggle for existence." While Darwin had demonstrated how struggle for existence was a mechanism for evolutionary change, Volterra wanted to know how that rule applied to the dynamics of populations over much shorter periods of time. Volterra wanted to discover the laws of nature that governed interactions between populations of predator and prey. Discovering such laws was, after all, the purpose of science, and the laws of nature are written in the language of math. Need one say more than $E = mc^2$?

Volterra's stock-in-trade, common to every successful applied mathematician, was an adept capacity for translation. First, craft some sentences in English (or Italian) that describe a scene. In this case the scene is of populations of predator and prey. Then, translate the sentences into equations that describe that scene. Finally, translate the equations into a graph, hoping to gain new insights about the scene.

Like poets, mathematicians are not equal in their capacity to paint a scene with symbols. The objective is not to describe every minute detail. Poets do not aim for an exhaustive depiction, and neither do mathematicians. Their shared ambition is to capture an overarching essence with as few words as necessary, but no fewer.

Consider the severe economy of words in a short line of poetry, such as, Marceline Desbordes-Valmore's *Two hearts in love need no words*. That line does not even attempt to be the most complete, detailed description possible. It cannot be, nor does it need to be. Essential truths are best articulated with a sparing application of words. Applied math is similar.

Here's the scene in English as Volterra thought it should be described: Imagine a population of prey (perhaps mice). Suppose the mice live in a place with as much food as they could ever want. They will live full lives, give birth to many offspring, and over time the abundance of mice will increase. Let all those ideas be symbolized by just two strokes, rN. As with all languages, there is a matter of pronunciation. That phrase, rN, when read aloud is "r multiplied by N." N stands for the number of mice at any particular time, and r is the rate by which mouse abundance will increase. N and r could be replaced by numbers if we wanted. If there were 100 mice, and if they grew at a rate of 0.05, then soon we'd have 5 new mice ($100 \times 0.05 = 5$), in addition to the 100 that we had before.

Math can communicate an entire paragraph's worth of ideas, in this case with just two clicks of the keyboard: rN. Every language has its own style. German is loaded with long and harsh-sounding words. French is smooth and sexy. As a language, math is like fruit cake—compact and dense.

According to rN, the mouse population will grow to unrealistically large numbers in no time at all. A real mouse population would, before long, run out of food. Volterra believed he could capture the essence without being bothered by that detail. If you disagree, you can paint your own picture. But Volterra is the master here. We should let him continue painting.

Volterra wanted, instead, to focus our attention on a different part of the scene. These mice, we'll suppose, are living in the presence of predators (imagine foxes). The foxes kill mice at some specific frequency. How frequently depends on how quickly each fox finds its next mouse. More mice will mean more frequent killing—mainly because more mice mean it is easier for a fox to find a mouse. Fewer mice will mean less frequent killing. In the language of math, those ideas may be compressed into aN, read as "a multiplied by N." We already know that N is the number of prey; a represents how quickly predators kill prey, given the number of mice. We can call it the attack rate. If we have 100 prey, and if a is 0.02, then two mice will be lost to each fox in the population ($0.02 \times 100 = 2$). The total number of mice lost is that quantity multiplied by the total number of foxes in the population, aNP, where P stands for the number of predators. If there are three foxes, each killing two mice, then the total loss of mice would be six.

To describe the population dynamics of predators, Volterra painted a similarly simple scene. Predator abundance would increase or decrease at a rate (denoted with the letter e) that is modified (multiplied) up or down according to how much food each predator captured, which we already denoted with the symbols aNP. Volterra also supposed change in predator abundance would be adjusted downward to account for predators dying from causes having nothing to do with food, such as accidents, disease, or old age. The rate of such mortalities can be represented by the symbol m, so the number of predators dying of such causes during a specified period of time is mP.

Volterra then assembled these symbolic letters to form a kind of sentence to communicate a complete thought about how predator abundance changes over some period of time. The idea to communicate is that abundance changes according to the number of prey consumed (*eaNP*) minus the number of deaths due to accidents, disease, and old age (*mP*). The sentence reads *eaNP* − *mP*.

Mathematicians have a peculiar habit of naming their sentences. Volterra named this one *dP/dt*, or "changes in predator abundance over time." So named, Volterra wrote *dP/dt* = *eaNP* − *mP*.

Volterra assembled and named a similar sentence to represent changes in prey abundance over the same time period. We'll skip those details, but the final result is something to behold. All that you've read in the past seven paragraphs—the entire bloody scene—is neatly expressed in two lines: one line for the prey population and one line for the predator population.

$$dN/dt = rN - aNP$$
$$dP/dt\ = eaNP - mP$$

That string of symbols is beautiful in much the same way it is beautiful to read the line *Two hearts in love need no words* in the language in which it was first written: *Entre deux coeurs qui s'aiment, nul besoin de paroles.*[20] Even if you do not speak French, especially if you do not speak French, there is beauty to be felt just in the looking. Some of the beauty lies in rehearsing the foreign line until it becomes just familiar enough to begin making some rudimentary juxtapositions. *Deux coeurs* must mean "two hearts" and *s'aiment* must have something to do with love. That you do not understand every bit of the French doesn't matter because the scene cannot be completely captured with French words or the words of any language, including the language of math.

Volterra is almost finished painting. In our example, we started with 100 mice, then five mice were added, and six were lost to predation, leaving 99 mice. That is not a final answer, merely the change in mouse abundance from one point in time to the next. Suppose the rates associated with the predator (*m* and *e*) are set to 0.1 and 0.39, respectively; then the number of foxes would change by two. (You can do the arithmetic on your own.) Where we had three foxes before, we would now have five foxes. Tracking the number of foxes and mice is like tracking money in two bank accounts: the rates of birth and death are like interest rates earned and interest rates charged.

The changes depend not only on those rates (i.e., the lower-case letters in the equations) but also on the abundance of predators (P) and prey (N), which are also changing from one period of time to the next. With 99 mice and five foxes, the equations tell us that the mice will soon decline from 99 mice to 94, and the foxes will increase from five to nine. The mice decline and the foxes increase. If that trend continues, the mice will go extinct, at which time the foxes would be committed to the same fate. But the rates of predation will slow as the mice decline. At which point the foxes will be eating less. Perhaps mice will persist, and with them the foxes, but only if the rates are balanced just right. It's far from obvious that it would turn out this way.

No human mind, not even a master like Volterra, can just look at those equations and know the outcome. The equations need to be analyzed. One means of analysis is to perform the calculations in recursive sequence—each subsequent calculation representing a subsequent point in time—and then display the results in a graph showing how the abundances in predator and prey change over time.[21] That graph would show two waves traveling across the page and over time, rising and falling like waves on an ocean. At first, foxes decline, while mice rise. Then a shift and both populations rise for a short spell. Another shift and mice decline while foxes continue to rise. Before long foxes reverse course and start to decline. Now both populations are on the decline. But only for a while, after which the foxes begin to rise. The sinusoid is complete and begins afresh.

Volterra's scene involves four rates. One rate is associated with the intrinsic nature of the prey—how quickly it can grow when it has all the food it wants, r. Another rate is tied to the predator—how quickly it dies in the complete absence of food, m. The other two rates describe predatory interactions between the two populations—the attack rate, a, and the efficiency with which predators convert captured prey into their own kind, e. The four rates pull and push on the populations, each in their own direction. The force behind each rate changes continuously as

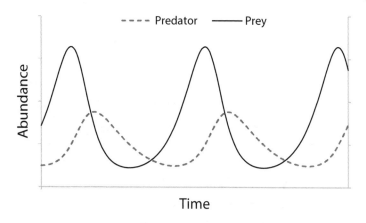

Vito Volterra's mathematical equations predicted that the abundance of predator and prey should fluctuate over time in phase-shifted cycles, as depicted here

the numbers of predator and prey change. Just as one population is about to rocket toward infinity the other population takes hold and the pattern reverses. We have in our minds the most abstract, rarified, and bloodiest game of teeter-totter imaginable.

Change any of those four rates and the dynamics will change. Change any rate *by too much*, one or both populations will go extinct. Professor Volterra assembled an inert string of symbols and found two populations heaving and sighing in synchrony. This is the beat we discerned in Chapter 2 through the immortal sequence of births and deaths. Here, aided by the precision of math, we feel a pulse rise from the undulating balance between dewy newborns and bloodied corpses. Extinction contraposing boundless proliferation. If delicate balance is not an apt description, if balance of nature doesn't capture the essence, we are left only with the inscrutable symbols.

Volterra's equations sired the next generation of the idea that we have been tracing. First, Plotinus with death by predation as a creative force. Second, Darwin with life evolving from the play between two opposing forces, extinction and growth beyond bound. And now, Volterra with populations rising and falling in response to the same play of forces. Allen and Mech are just over the horizon.

Volterra shared his ideas with the world in a 1926 paper that concluded "it is to be *hoped* [italics added] that this theory may receive further verification and may be of some use to biologists."[22] Volterra was smitten, never again worked on mathematical physics, and focused the remainder of his professional life on mathematical biology.

Volterra's "hope" was quickly pursued by the Russian ecologist Georgy Gause. Just a few years after Volterra's 1926 paper, and while Gause was in his early 20s, Gause devised a clever way to evaluate whether the scene Volterra painted bore any resemblance to the real world. Wanting a fair evaluation and believing that most of nature was more complicated than Volterra's spartan equations could convey, Gause needed creatures in an environment whose population dynamics would be adequately described by Volterra's four rates (r, m, a, and e). He needed simple creatures in a simple environment.

To this end, Gause enlisted two species of single-celled creatures that we call ciliated protozoans. Entire populations of these little fellows can live out their lives in a test tube or petri dish of water. Ciliated protozoans are also easy to come by—found in just about any scoop of pond water. Many species of ciliated protozoans make a living by feeding on bacteria. Others get by feeding on other ciliated protozoans. Gause selected *Paramecium caudatum,* which feeds on bacteria, to represent the prey in his laboratory experiment, and *Didinium nasutum,* which feeds on those paramecia, to represent the predator.

To prepare the study, Gause dumped some oatmeal in water and inoculated the mixture with bacteria. Soon the water teemed with bacteria. He strained the oatmeal from the water, leaving a bacteria-rich broth, then he poured less than a teaspoon of the broth into a test tube.

These protozoans live at the edge of our visibility. Hold a glass of water filled with paramecia to the light and you'd detect little flecks scooting about, but that's all. You'd see a bunch of someones and be able to count them, but you wouldn't be able to discern what they are up to. But if you could, then you'd know the potency of their lives. If you could look closely, you'd know how much authentic nature is happening in this Gausian universe.

To see the tooth and claw of Gause's microcosm, we need a better view. We need to shed the biased perspective we inherited with our gargantuan, multicellular bodies.

Let's make like Alice, take a swig of that potion labeled, "DRINK ME," and watch ourselves "shutting up like a telescope." This test tube is smaller than the rabbit hole; we'd better take two swigs. As quick as your imagination can conjure, a teaspoon of broth grows to the size of a large swimming pool. The water is littered with spheres, roughly the size, texture, and color, of grapefruits—these are many thousands of bacteria.

Gause has just dumped exactly 10 paramecia into the pool with us. The paramecia are shaped like fat cigars, about as long and large around as a slightly overweight human. Swim up and touch one—they are harmless—and you'll see that while they have shape, they are also pretty pliable. Their single-celled bodies are covered with the strangest kind of shag carpet with fibers longer than any shag you've ever seen, though, not nearly as dense. A closer comparison might be a balding man in the habit of comb-overs who's just jumped into the water. Think of all those long waving hairs, too many to easily count, but few enough to know that if you really had to, you could. These are the cilia. They pulse in unison, propelling each paramecium through the water in search of dense schools of bacteria.

They are translucent with their insides showing and light passing right through their bodies. The paramecia have an open flap running maybe a third the length of their body. The flap is covered with cilia that guide bacteria down to a sack called a vacuole. I wouldn't put your hand down in there. When the vacuole is filled with bacteria, it detaches itself from the bottom of this oral groove and begins to float around inside the paramecium's body. You can see for yourself right through their skin. Enzymes enter the vacuole and digest the bacteria into stored energy and other compounds that the paramecium uses to maintain and grow its body. When all the life is sucked out of the vacuole, it migrates to the end of the paramecium's long body and slowly erupts from its body. A new vacuole is formed at the bottom of the oral groove as the previous one is detached. The formation, migration, and eruption of these vacuoles is gently reminiscent of bubbles in a lava lamp.

If bacteria are plentiful, one paramecium can consume 100 bacteria in an hour.[23] When a young paramecium eats as much as it likes, then in just five or six hours, about half of the paramecium's inside parts migrate to each end of its unicellular body and the cell membrane begins to contract in the middle and then pinches off. In a process that takes just a few minutes, where there had been one paramecium, we are now staring at two half-sized paramecia. Each scoot off in their own direction in search of more bacteria.

In 24 hours, the population of 10 paramecia has increased to about 30. In 48 hours, it is hard to swim around without occasionally bumping into one. Our swimming pool is now inhabited by more than 100 paramecia.

At this moment, Gause drops three didinia into the tank. They are barrel shaped, slightly smaller than the paramecia, and agile, owing to cilia emanating from two bands located where the quarter hoops on a hogshead of whiskey would be.

You see sacs embedded in the translucent membrane that is their skin. Each of these cysts is packed with needle-like filaments. When a didinium detects a paramecium nearby—I have no idea how; these creatures have no eyes or ears or nerve cells of any kind, because their entire body is just one cell—they eject those filaments, penetrating the paramecium's skin. Oh, right, I forgot to mention the toxins. The filaments are coated with toxins to immobilize the paramecium and begin digesting its insides before the didinium has even swallowed it. The didinium retracts the filaments with the paramecium still attached, and the paramecium may be folded in half as it's pulled into the didinium's "mouth." The didinium had been smaller than the paramecium, until it was stretched on all sides from the inside by the freshly consumed prey.

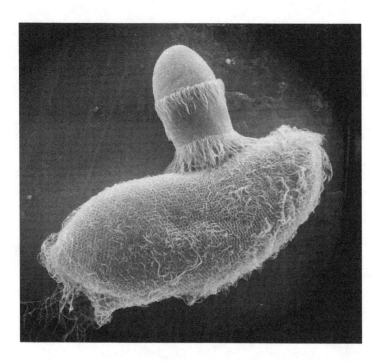

A didinium consuming a paramecium. Gregory Antipa and H. S. Wessenberg, https://doi.org/10.7295/W9CIL39250

Paramecia are not helpless in all of this. They have similar sacs embedded in their skin. If triggered at the opportune moment, they foil a didinium's toxicysts, sidelining the didinium from the hunt for 20 or 30 minutes as it regenerates spent toxicysts. An especially skilled didinium captures and consumes about five paramecia in a day, depending on how easily the next paramecium is found. When didinia have had enough to eat, they undergo, like their prey, the torturous-looking process of binary fission.

These protozoans can locomote, forage, and evade predation. They are more complicated than many of the cells in an animal's body, which could never survive on their own. They are too sophisticated to qualify as bacteria, yet too simple to qualify as animals (-zoans) of any kind. Protozoans have lived, for billions of years, in the taxonomic no-man's-land between bacteria and animal.

Watch out! Right behind you! Don't let that didinium get too close.

While this Gausian universe is a laboratory experiment in a glass tube, don't overlook the raw peril. And do not fail to appreciate the complexity. What happens in here is Daedalian, well beyond knowing how it will turn out unless we keep watching.

Within 24 hours of being dropped into the tank, those three didinia had consumed scores of paramecia and increased their own kind to about 25. The abundance of paramecia plummeted from more than 100 to about 35. A day later, the number of didinia held steady, and the paramecium population struggled against extinction. A day later, the paramecia were extinct. The didinia, now with nothing to eat, went extinct the following day. All that remained were bacteria feeding on didinia carcasses and waste products—recycled parts of bacteria.

In addition to the miniatured space of this tank, time passes differently here. You and I are accustomed to measuring time in hours and days. But life meters time with each passing generation, whose duration varies among species. For humans, a generation is about 25 years, for moose it's about nine years, and for wolves a generation is about four years. Gause's evaluation, lasting only six days, may seem as a flash, but those six days saw the passage of more than 20 paramecium generations.

By comparison, six decades of wolf-moose dynamics on Isle Royale (1958–2020) is just 15 wolf generations. That's right, don't forget Isle Royale. The history of balance of nature is worthy of attention for its own sake. But our particular interest is a rich understanding of Durward Allen's declaration that the wolves and moose of Isle Royale had "struck a reasonably good balance."

Gause's experiment did not echo the constancy that Allen saw in wolves and moose on Isle Royale, nor did it approximate Volterra's equational painting. Maybe the deviations rose from imbalance among Volterra's four fundamental rates, canted by some underappreciated simplicity of the experimental universe. Maybe the attack rate, a, was too high and simply outpaced the growth rate of paramecium, r.

Gause tried again. This time he intended to level the playing field. Rather than strain all the oatmeal from the bacteria-laden broth, he left some oatmeal crumbs to sink to the bottom of the test tube. These crumbs, he imagined, would offer a place for paramecia to hide and thereby lower the attack rate (a).

Into the oatmeal-littered tank, Gause released about a half dozen individuals of each of the two protozoan species. The number of didinia and paramecia remained about the same for a day (which is, as you recall, like several years in the life of a protozoan). After another day, the number of didinia and paramecia both increased, but not greatly. On day three, the paramecia soared to more than 20 and the didinia began to decline. By day six didinia had fallen to extinction and paramecia had grown to more than 40. Gause poured the broth down the drain and terminated the experiment.

Again, the result was not what Volterra's theory had predicted. But Volterra and Gause both knew that nature, even stripped to its simplest form, is more complicated than any pair of equations. Maybe persistence and coexistence is facilitated by some complicating mechanism that Volterra overlooked. Gause remained steadfast, hoping to find some condition under which didinia and paramecia would strike a reasonably good balance.

Gause dispensed with the oatmeal chunks and tried again with a fresh tank into which he set free a few didinia and paramecia. On this third try, Gause changed the protocol so that on every third day he would add three paramecia and three didinia. He'd do so no matter who seemed to have the upper hand. Both populations persisted more than 16 days (the equivalent of 200 years if we were talking wolves and moose). The populations also exhibited cycles not at all unlike those produced by the Volterra equations.

In developing a mathematical theory and then bringing that theory to life, Gause and Volterra represent an archetypal modus operandi for growing scientific knowledge.[24] A century later, dN/dt and dP/dt continue to be the springboard for most research on predation dynamics.

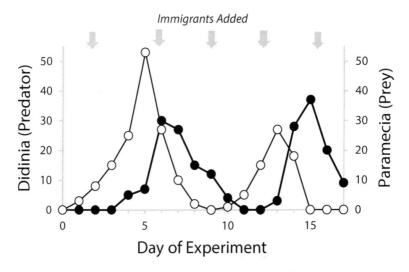

Laboratory experiments conducted by Georgy Gause resulted in a fluctuating balance between predator and prey when immigrants were occasionally added to the populations. Adapted from Gause 1934

While Gause played with test tubes in the lab, North American agriculture was increasingly scourged by insects and plants that had found their way from Eurasia and Africa. Among the worst of these exotic species was St. John's wort. Toxic to livestock, the yellow-flowered plant had taken root across millions of acres of rangeland from Montana to Washington and down to northern California. Some entomologists believed the imbalance exacted by this exotic pest could be rebalanced by introducing the predatory insects that eat these pests on their native turf. To introduce an exotic species as a means of readjusting a balance that had been upset by another exotic species was as controversial then as it is today.

In the 1940s, a team of entomologists acted on the idea. They introduced five species of insect from the Old World that make a living by feeding on St. John's wort, hoping that one or some combination of those insect species would redress the disturbance caused by St. John's wort. Of the five species, four became established. They were two species of leaf-eating beetles, a root-boring beetle, and a gall fly. One of the leaf-eaters had the greatest impact. St. John's wort continues its residency in North America today, but it exists at less than 2% the level of abundance that it had during the 1940s.

Those introductions are one of the first and clearest examples of successful "biological control." The effort was led by Carl Huffaker, who was possessed by the intensity and focus of a champion bird dog, an intensity that even infiltrated his recreational life. As a young boy, Huffaker became interested in pigeon racing and grew up breeding some the fastest racers of his day.

Huffaker was not a one-project-at-a-time kind of fellow. While breeding pigeons and pursuing St. John's wort, he also achieved success in the biological control of olive scales, which had been a serious insect pest for California growers. Being a genius at biological control, he was not content with having altered the balance of nature over large swaths of rangeland or across the orchards of California. Huffaker also wanted to understand *why* he had been so successful— which levers, pressed exactly how far, maintain nature's balance and which ones knock it off balance. Incongruous as it may seem, some of the answers he sought could be found only in the lab.

Huffaker knew that the balance between life and death Gause had observed with paramecia and didinia was influenced by oatmeal debris and occasional immigration from the outside. Huffaker believed more insight would be found in experimentation with other yet-to-be imagined auxiliaries.

Reflecting his expertise in agriculture, Huffaker conducted some experiments involving oranges, six-spotted mites that feed on the oranges, and predatory mites that feed on the six-spotted mites that feed on the oranges.[25] While motivated to understand the influence of nuances that Volterra left to the imagination, Huffaker still knew the value of beginning simply, very simply. He set 20 six-spotted mites loose in a pile of oranges, and then he set loose two gravid female predatory mites. You cannot get too much simpler. Soon after the mites had been released, the six-spotted mite was driven to extinction by the predatory mite, which fell to extinction shortly afterward. The outcome was a rerun of Gause's first observations.

Huffaker wanted to know whether persistence might result from an environment that was more complicated than a pile of oranges. He arranged the oranges on a tray, each orange neatly spaced from its neighbors by a few centimeters, like Christmas ornaments in a box. With a generous application of imagination, Huffaker also dipped the oranges in wax, leaving exposed only a small portion of orange skin from which the six-spotted mites could feed. This was a nice flourish but still not sufficiently elaborate. Huffaker erected wooden dowels, rising high above the oranges, to serve as launching points for six-spotted mites searching for the next orange. Still too simple. Knowing that the predatory mites crawl and cannot hop like the six-spotted mites, Huffaker smeared Vaseline on portions of the trays, creating a maze to impede the rate at which predatory mites could search for their next prey. In this fanciful environment, Huffaker had the mites play their deadly game of hide and seek.

Certain death met any six-spotted mite who fed from a patch of exposed orange skin that happened to be, at that moment, occupied by a predatory mite. However, the six-spotted mites were more mobile and quickly moved from orange to orange. Enough of the six-spotted mites were sufficiently swift to evade the predatory mites, but not so many as to completely deprive the predatory mites of a meal. Where the undifferentiated pile of oranges led to quick extinction, this labyrinth with mites for Minotaurs supported balance. For more than a year, the two populations coexisted, trading the upper hand three times with the passage of as many population cycles. With that result, Huffaker convinced fellow ecologists that persistence does not always require the regular introduction of immigrants. Huffaker's creativity is celebrated to this day as a quintessential example of ecological balance between species that depends on linchpins that appear as mere decorative accoutrements of the environment.

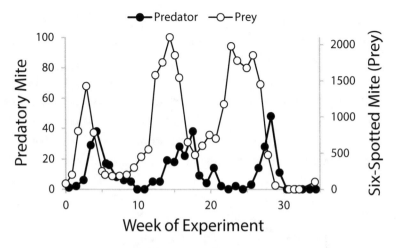

Laboratory experiments conducted by Carl Huffaker resulted in a fluctuating balance between predator and prey, but only when environmental conditions that affect the rate at which predators found prey were manipulated. Adapted from Huffaker 1958

Huffaker's results were published in August 1958, the summer that Allen and Mech launched the wolf-moose project on Isle Royale. A year earlier, the scientific journal *Ecology* published the findings of Syunro Utida, who had been observing predatory relationships involving a beetle that feeds on adzuki beans.[26] This beetle, rather straightforwardly named the adzuki bean beetle, is pursued in life by a parasitic wasp that lays its egg on the beetle. Shortly thereafter, a wasp larva hatches and burrows into the beetle, paralyzing the beetle and eating it alive from the inside out. The beetles have three life stages (larval, pupal, and adult), and the wasps are only effective at attacking beetles in their larval stage.

Utida's beetles and wasps coexisted with each other for more than 100 generations in the lab. Furthermore, they exhibited oscillating fluctuations like those observed by Huffaker, Gause, and Volterra. He supposed their coexistence was favored by differential predation among the life stages of the beetle. Within a few years, Allen and Mech would observe a similar pattern with wolves and moose. Wolves almost never killed middle-aged moose, even though they are more common than calves and older, senescent moose.

Utida's beetles, Huffaker's mites, and Gause's protozoans. In each case, the secret to persistence and coexistence, the secret to nature's balance, was some kind of complexity that was not *explicitly* portrayed in Volterra's equations. Without tending all the vital minutia, Volterra's theory seemed to reflect an essential truth about nature's balance.

What an outstanding platform from which to view the observations that Allen and Mech were about to make with wolves and moose. Isle Royale would be a natural extension, from single-celled protozoans to Acariformes and coleopterans, and now to homeothermic vertebrates. From antiseptic equations and tidy laboratories to a rough and unruly thicket. From test tubes and trays of waxed oranges onto a wilderness island inhabited by wolves and moose. Allen and Mech were about to work in a venue considerably more complicated than any thus far evaluated.

Would the countervailing forces on Isle Royale—mortality and fecundity, predation and evasion—balance as Volterra's theory had shown possible? Would Isle Royale be complicated enough to level the playing field and allow coexistence? Would wolves' tendency to prey more on some age classes of moose and less on others represent a complexity that could stabilize the dynamics on Isle Royale?

Allen and Mech were interested in such questions, but they were not in the least bit motivated by any of the math or the lab work. They do not mention those researchers, discuss their work with contemporaries like Huffaker, or write in venues where scientists working at the forefront of predation research had been sharing their results. Allen's prime interest, we'll see, was focused on another forefront.

While Volterra theorized and Gause experimented, commercial interests were developing a compulsion to understand—manipulate really—the balance of nature out in the real world. The Hudson's Bay

Company of Canada, for example, wanted to better understand erratic changes from year to year in trappers' success in capturing lynx, hares, and foxes, whose skins were used to make clothing. The company had raw data from nature in the form of trapping records going back more than a century. In Europe and the United States, ministers of agriculture wanted to understand occasional and seemingly unpredictable eruptions in the abundance of rodents and insects that ruined crops. They also had extensive records for the timing and severity of past eruptions.

There were hard numbers from forests and prairies bedeviled by all the intricacy that theoreticians and experimenters had been trying to pry away. Never before had such data been available. For the insights divined from those data, Charles Elton was ordained the Father of Animal Ecology. This was his grandest insight:[27]

> "The balance of nature" does not exist, and perhaps never has existed. The numbers of wild animals are constantly varying to a greater or lesser extent, and the variations are usually irregular in period and always irregular in amplitude. Each variation in the numbers of one species causes direct and indirect repercussions on the numbers of the others, and since many of the latter are themselves independently varying in numbers, the resultant confusion is remarkable.

Elton sounds exasperated by his own ideas. Half way around the world, Alexander Nicholson, who spent most of his time teaching entomology classes in Sydney, nearly jumped out of his skin upon reading Elton's announcement. Nicholson requited with a 47-page treatise and asked Elton to publish it in the scientific journal under Elton's editorship, *Journal of Animal Ecology*. Elton agreed. Nicholson's article, entitled "The Balance of Animal Populations," explains "an outstanding feature of animal populations with which all must be familiar," that is the tendency for populations to fluctuate in abundance along with fluctuations in various environmental conditions, especially climate. "These observations clearly indicate that animal populations must exist in a state of balance, for they are otherwise inexplicable."[28]

For Nicholson, the proper evaluation of balance of nature emerged from field data strained through a different mesh of reasoning:[29]

> The claim that animal populations are not in a state of balance is usually based upon the observation that animals do not maintain constant population densities. Clearly this argument is illogical, for, if a population is in a state of balance with the environment, its density must necessarily change in relation to any changes of the environment. A population density that does not change with the environment is evidently not in a state of balance, but is fixed independently of the environment.

Nicholson compares balance of nature in an animal population to a steam engine's governor[30]—that clever little device that limits the engine's speed through mechanical feedback.[31] He quotes Boris Uvarov, a Russian entomologist who explained with quite some literary flair that balance in an animal population resembles

> a cork floating on the surface of a running stream, with its whirlpools, eddies and back currents, while the wind blowing with varying force now ripples the surface gently, now causes it to rise and fall in great waves. The cork rises and falls with them, comes into collision with other floating objects, now rides on a calm surface and is dried by the sun, now is again rolled over and over by a storm and is beaten by rain, but always continues its journey with ever-changing velocity and along a fantastically tortuous course.[32]

Nicholson does not confine himself to deductions and metaphors. He also cites four laboratory experiments similar to those performed by Gause except these experiments involved fruit flies, flour beetles, and springtails, each of which show "clearly a population in a state of balance."[33]

Elton and Nicholson both knew very well that animal populations fluctuated greatly, and they both were aware of the experimental evidence. But they perceived the same world each in their own way, as when two people see the same glass of water differently.

Balance of nature, as an idea, was born long ago when theology constrained most ideas. One might have expected balance of nature to find firmer footing after modern scientists wrestled the idea from parson naturalists. Early indications, however, were not so promising. Darwin and Wallace stonewalled each other. Seventy years later, Elton and Nicholson rehearsed the same script.

Allen's understanding of balance of nature avoided those ruts and did so because his observations of nature rested on different premises. For Darwin and his followers, "living existence" focuses exclusively on organisms—their fitness, their adaptations, and their passing of traits (sometimes with modification) onto offspring. Life is all about being an organism, like a shark, ciliated protozoan, six-spotted mite, or moose.

Might there be other kinds of living things that happen not to be organisms? No, not while in a Darwinian frame of mind. But what if there *were* other ways to be alive? And, if there were, would that alter one's outlook on the balance of nature?

Allen's disposition on what it means to be alive transcends Darwin's and was passed to Allen (with some modification) by an articulate, fast-talking bear of a man named Frederic Clements, who was shaped by the prairies and woodlands of Nebraska. Clements understood the value of cold data a full generation before Elton analyzed trapping records from Hudson's Bay Company. And Clements's obsession with quantifying which plants lived where on the prairies and in the woodlands led him to a remarkable idea about the nature of life.

Clements, working with Roscoe Pound, realized by 1900 how "extremely deceptive" observations of nature could be unless they were quantified: "Actual field experience has shown that species [of plants] which appear most prominent in the constitution of prairies, even to the careful observer, are not necessarily the most abundant. . . . To ensure accurate or even approximate results, it is necessary to resort to some method of actual count."[34]

To redress this shortcoming, Clements revolutionized the science of ecology with a new technique known as the quadrat method. Prepare yourself for something really advanced here. Drive four wooden stakes into the soil (arranged in the shape of a square), then wrap a string around the stakes, count the number of each kind of plant in the quadrat, and, finally, write down the name and number of each species of plant in the quadrat.

Clements's enthusiasm for tallying plants was unstoppable. He practiced this simple method over the next four decades and across the state of Nebraska—the wooded bluffs along the Missouri river, the meadowlands, the sand hill prairies of north-central Nebraska, and the tall grass prairies of eastern Nebraska.

Clements noticed how finding one species—say, gooseberry—in a quadrat meant he shouldn't be surprised to also find Virginia creeper in the same plot, and he'd almost certainly not find any bluestem grass. He could also anticipate, from the climatic and soil conditions of a quadrat, which sets of species would occur where. And quadrats near to one another were more likely to have the same species than quadrats placed farther away.

Clements also got experimental with his quadrats. He removed plants from some quadrats and observed a tendency for those plants to be replaced by other individuals of the same species. In some quadrats he planted species that hadn't previously been present. Those plants tended not to persist. Clements also noticed how the composition of plants in certain quadrats changed over time, and did so in an organized manner. For example, when an agricultural field was abandoned, the first plants to take root were invariably a wide variety of annual herbaceous plants, then perennial plants and grasses, followed by shrubs, then softwood trees (like aspens), and finally hardwoods (like maples). The broad patterns were systematic and repeatable.

Clements accumulated thousands of tallies, lists of plants growing in quadrats across Nebraska and beyond. For his obsessive efforts, Clements is heralded as the Father of Plant Ecology.[35]

To draw a string around that many sets of stakes, to sit down before a small patch of the earth that many times, to get down level with the plants, to take a quick look, catch a gestalt, and then engage in the deliberative task of touching every single plant, recognizing its species name and writing it down, pressing pencil to paper, once for each individual—to do that not for a weekend, not a few dozen times, but to perform that meditation thousands of times over a lifetime—there is no more intimate, more mesmerizing way to connect with nature. Doing so will, without question, generate perspectives that cannot be had any other way.

Clements's notebooks were reductive and apathetic, but they provoked a rhapsodic realization. The plants were not a "mere collection of atomistic individuals." They were not merely coexisting. Rather, they formed a tight community that "arises, grows, matures, and dies." The individual plants living in this community were so intimately interconnected that Clements saw each community as being "like a super-organism." They were not the mere coincidence of things living in close proximity. No, the collection itself was "a live, coherent thing."[36] By "super" Clements did not mean more advanced or more complicated than your standard-issue organism. He just meant big. Clements discerned a new kind of life, here on earth, living and breathing before our eyes—a being that had been living in and throughout his quadrats.

Clements was not the first to advance the idea that ecosystems are living organisms. In the years between Darwin's sanctification of the organism and Clements's first quadrats, other scientists were beginning to divine a broader view of what it might mean to be a living thing. Stephen Forbes studied lakes across the midwestern United States and became one of the first to appreciate the tangled net of relationships that we now call a food web. Forbes described the relationships among organisms as being "like a single organism."[37] At about the same time, German scientist Karl Semper's training in anatomy and physiology led him to see that individual creatures living together in a community are related to one another "like the organs of a healthy living organism."[38]

After Clements, the concept that ecosystems are organisms continued to develop, especially with the vivid vision of ecosystems that began with research during the 1950s on the impact of radioactive fallout on the Eniwetok Atoll in the South Pacific, where the United States government had detonated atomic bombs. The researcher was Eugene Odum, and he certainly appreciated the individual organisms within an ecosystem, but not in their own right. He appreciated organisms, no less, as vessels through which energy and nutrients flow among the different parts of an ecosystem. For Odum, these fluxes of energy and material were no less sacrosanct. For taking in such a broad view, Odum is known to many as the Father of Ecosystem Science. Odum also co-authored the first ecology textbook in the late 1950s, where he connected fluxes within an ecosystem to the balance of nature: "Homeostasis at the organism level is a well known concept in physiology. . . . We find that equilibrium between organisms and environment may also be maintained by factors which resist change in the system as a whole. Much has been written about this 'balance of nature.'"[39]

Ecosystems are alive for the same reason that organisms are alive. Each exhibits homeostasis, which is a process that resists change and maintains the status quo. We humans, for example, maintain our body temperatures. When we get cold, our metabolism revs up to compensate. When overheated, we perspire. When blood pressure drops, the heart pumps faster. When blood sugar falls, the pancreas produces insulin. Being ill is often the temporary loss of homeostasis—an imbalance. Organisms are a bundle of homeostasis stuffed into a skin, or bark, or a cell membrane.

Volterra theorized that populations persist by an intricate set of feedbacks. The empirical demonstration was provided by Gause and his followers. Paramecia increase, then predation increases, and paramecia subside. Didinia increase, then starvation increases, and didinia subside. By Odum's reckoning, an ecosystem's feedback and an organism's homeostasis belong to the same class of phenomena.[40]

Final recap. Plotinus celebrated predation as a creative force. Darwin explained how life evolves from the balance between extinction and unbounded growth. Volterra and his verifiers found populations rising and falling to the same play of forces. Clements tallied plants until inebriated by the thought that balance brings life to ecosystems. Odum put ecosystem feedback on par with organism homeostasis. There it is—highlights in the pedigree of ideas that is the balance of nature.[41] Now we can rendezvous with Allen.

He conspired with Odum. In *Wolves of Minong*, Allen professed that an ecosystem "is an organism. . . . Its metabolism is the flow of energy through diverse forms that are held together by their common need and interlocking functions. The community is hedged against extremes and has seemingly endless feedback mechanisms to steer its fluctuations toward a midpoint . . . often it is the environment (e.g., weather) that produces instability. . . . On this totality of functions, services, feedbacks, and compensations the fate of the whole depends."[42]

If those contemplations seem tenuous, you may find a secure hold by considering your relationship with your kidney. It cannot survive without you. In the same way, a trillium flower growing on the forest floor cannot survive outside its ecosystem. The trillium depends on trees for shade, insects for pollination and seed dispersal, the mycorrhizal fungi growing in its roots for the uptake of nutrients, and soil bacteria to make nutrients available to the fungi. No organism can survive outside of an ecosystem.

Each organism is kept alive by its relationships to other organisms. The *relationships* are as vital as the *relators*. The pattern of emphasis can be understood with a simple sentence: predators eat prey. If the verb (eat) has equal standing with the two nouns (predator and prey), then an ecosystem is more than a collection of organisms, and the whole thing is greater than the sum of all its parts.[43] In this way, and in its own way, an ecosystem is alive.

Then Allen throws a curveball by hailing an inconsistency: "Realistically, there is no 'delicate balance' in the strict sense. It is more likely a seesaw of constantly changing weather conditions, food supply, prey productivity, and predator numbers that somewhere strike a mean that can be determined only over long periods of observation."[44] Later he continues, "We see repeatedly that calamitous extremes of almost any kind are countered by compensations. The high-to-low fluctuations tend to even out in a pattern that spells survival over long time periods for all the kinds of life that fit the local organization. Actually, that 'delicate balance' we hear so much about is not especially delicate, nor is it a precise balance. It is tough and durable, an eternal teetering about a midpoint."[45] Allen embraced the protagonist *and* the antagonist. He accepted the contradiction that led Elton and Nicholson to an impasse.

Allen doesn't explain his ease with the contradiction, but he could have cited two reasons to let the antagonizing ideas coexist. First, while science is an essential way to know the world, it cannot determine the appropriateness of every important belief. Heresy? No. Science can't say who you should love. Science can't distinguish half-full from half-empty, even when the difference between the two is patent. And science may well be unequipped to officiate the disagreement between Darwin and Wallace or between Elton and Nicholson.

Second, the contradiction does not undermine the *relevance* of balance of nature. The contradiction doesn't compromise the obligation that rises from balance of nature: if individual organisms are the only kind of living thing in the universe, then we humans, with our dexterous hands and imposing prefrontal cortexes, are at the top of the heap.[46] For centuries we have been whispering into the ear of God, mistaking our breath for his voice. We see ourselves as the best and most important—supreme among all living organisms.

But if ecosystems are more than the sum of their organismic parts, because relators *and* their relationships share sanctity, then ecosystems are living things in their own right. We organisms are living things and part of another more-encompassing living thing.[47] This understanding threatens the view that humans are at the center of it all. It implies obligations that extend well beyond taking care of our immediate selves. This seems to have been Allen's understanding of balance of nature.

Allen steered clear of others' ruts because he did not limit himself to principles of science in apprehending an idea that transcends science. If all you want is to describe how and why nature is the way it is, then science is generally enough. But Allen also wanted to better understand what counts as a good relationship with nature. For that, one must transcend science.

The best clue we have that Allen thought along these lines is where he wrote "To live with nature on an enduring basis, [we] must sense and believe in the original harmonies that have to be there. Possibly this is more an attitude than a way of thinking. It requires a conceptual base many do not have."[48]

Allen increasingly resigned himself to a world where human culture by and large did not comprehend those obligations. He lamented living in a world where the essential nature of northern ecosystems, those inhabited by moose, had been "confused and obscured by the almost universal hunting of moose and the wiping out or heavy control of wolves."[49] Then in the late 1950s, Allen found Isle Royale—a wilderness island that had once been threatened by human enterprises, but no longer. With wolves having established themselves by venturing across an ice bridge, the island had become an ecosystem where wolves lived free of human persecution, the moose population was not hunted, and the forest was no longer logged. There, on that island, Allen saw a balance of nature.[50]

He found what the Cherokee found in their story for the gift of medicine, and what the ancient Greeks found in Persephone's abduction: knowledge that humans can upset nature's balance, nature can restore the balance, and humans can be part of restoring that balance.

So, is there a balance of nature that can be easily upset by humans? Ask anyone—man or woman, young or old, wealthy or poor, city mouse or country mouse. For every 10 people you ask, eight or nine will agree.[51] The cultural importance of this worldview is also represented through the voice of a different carnivore. Mufasa, in Walt Disney's *The Lion King* (1994), explains to Simba, his heir apparent, "Everything you see exists together in a delicate balance. As King, you need to understand that balance, and respect all the creatures, from the crawling ant to the leaping antelope." Simba wonders, "But, Dad, don't we eat the antelope?" Mufasa answers, "Yes, Simba, but let me explain. When we die, our bodies become the grass, and the antelope eat the grass. And so we are all connected in the great circle of life."

The king is perfectly clear on a point about which Allen had been a little coy. Balance of nature is far more than a description of nature and how it works. The idea also brings broad and demanding guidance on how we should behave and relate to the world around us.

5 |

Exogenous Forces

Mech's three-year stint on Isle Royale ended in June 1961, and Allen replaced him with another PhD student. All according to plan. A few years later, graduation and another replacement. The cycle repeated once more, and again. Each iteration was its own contribution. Phil Shelton provided evidence that beavers might be important to wolves as food and to moose as engineers of wetlands that provision moose. Pete Jordan searched for better ways to count moose. Wendel Johnson took a diversion and pondered foxes, hares, squirrels, and mice. Michael Wolfe resumed course and perfected a valuable technique for determining the age of wolf-killed moose.[1]

Don Murray had been hired for his skill in piloting aircraft, but he was no less valuable as helmsman during that whirligig of students. The quality of each winter's data rested, not with the budding scientists, but with Murray's intense personal curiosity to know how many wolves were alive each year and how they were getting along.

For eight straight years, most Isle Royale wolves lived in a single large pack led by a reproducing alpha pair. There were also wolves living in a few smaller groups and a few wolves living almost entirely on their own. In March 1966, the alpha male of "Big Pack" died. By the next winter the "Big Pack" had disintegrated into several smaller groups and loners. Each wolf's affiliation with a group was fluid, and membership seemed to change on a weekly basis. Approximately half the wolves died that year.

That lurching loss was largely offset, not by the equanimity of nature's balance, but rather by the coincidental arrival of a pack with seven wolves who crossed an ice bridge that connected Isle Royale to Canada during the coldest few weeks of most winters. The pack included four black wolves.

Wow, what Allen must have thought of it all. He described discovering the wolves' arrival in his February 25, 1967, journal entry:

> Out beyond mouth of [Robinson] bay we saw 7 wolves and began to circle. As we got closer we discovered that 4 of them were black! This is obviously a new pack from Canada—the ice is solid and continuous. . . . We circled them and shot some pictures. . . . We backtracked them to the kill on Stockley Bay. Obviously, they made this kill. . . . We buzzed the 7 wolves several times, and they lay down on the ice, curled up to rest—ignoring us. . . . This pack has probably been on [Isle Royale] for several days."

Allen's journal entries from February 1967[2] provide no further detail of what he saw or how he felt about that moment of discovery. On that same February 25, Allen dissected 35 large hydatid cysts from the lungs of a wolf-killed moose. He also noted a paucity of spruce cones that winter and that most of the birch seeds skittering across the snow were inviable. Allen's journal entry of this day offers as much intimacy for cysts and seeds as for discovering a new pack from Canada. He cared deeply for nature as a *whole* in a way that may have made him less comfortable expressing excitement or care for the personal lives of those who compose the whole. A decade later, when Allen wrote his memoir, *Wolves of Minong*, of his time on Isle Royale, the pack of black wolves from Canada goes unrecalled.[3]

In February 1968, Allen and crew saw the wolf population split into two main groups and a few smaller groups. The wolves' social alliances had restabilized, and group memberships were consistent from week to week. Before that dust settled, one of the black wolves left their natal pack to join the other main group. The other two black wolves and their immigrant pack mates were not seen after March 1968.[4] Presumably, they returned to Canada after deciding that island life was too chaotic. With their departure, the population returned to a single main group that included a black wolf who rose to the beta position[5] in 1970 and then become alpha male by 1971. In 1972, another social fissure split the wolves into two packs—this split was permanent.[6]

During the project's first years, Allen and Mech had been too cautious to conclude that packs were essentially family units. Now, Allen and his crew were seeing tumultuous details of pack life, and they bore resemblance to the royal families chronicled by Machiavelli during the Italian Renaissance. In spite of the drama, Allen's attention by and large stayed with actuarial implications of the social pandemonium, such as its effect on the ability to estimate wolf abundance and predation.

Allen and his crew still struggled to count moose in a reliable manner. And not since Mech's time on Isle Royale had they been quite so capable at capturing the influence of predation on all that they observed, though trends began to form during the carousel of personnel of the 1960s. Moose had been on the increase, but no one really noticed. Wolf abundance jerked and stumbled, up and down, and no one raised an eyebrow. Isle Royale's balance exhibited early signs of teetering that drew no attention. The changes were subtle, and Allen may have been indisposed to see such a change.[7]

In time, hindsight would pull the blinkers away.

After guiding five young scientists through their studies of Isle Royale, Allen began to think of retirement. He saw value in continuing the study, but future prospects were hazy at best. Who would have the constitution to endure frozen mukluks in the winter and mosquitoes in the summer that are numerous enough to result in occasional and incidental ingestion? Who would be bright enough to properly collect and interpret the data? Who would be savvy enough to raise the money? Who would endure Park Service staff offering snide reminders that the project, which had been pitched as a decade-long effort, was now in its 12th year?

In November 1969, Allen received a letter from the young and earnest Rolf O. Peterson, who had just watched *The Wolf Men*, a 1969 documentary about Allen and Murray's study of wolves on Isle Royale. Allen easily saw that Rolf[8] had considerable experience living in the wilderness forests of Minnesota, not for work or to exercise his vigor, but for recreation of the deepest kind. He spent time in the wilderness to realize the truest version of himself—someone in love with nature.

Rolf's relationship with nature was shaped early on by the philosophers of wilderness. Rolf knew of Sigurd Olson, who paddled primeval lakes of the remote north country, asseverating wilderness, first and foremost, as an arena for men to prove their manhood. Rolf had also read a little John Muir, who explored states of awe in the misty presence of sequoia trees, professing existential reverence for nature. Rolf had inherited more from Muir than Olson, though Allen would have been satisfied with a dowry of either man's wilderness philosophy.

Rolf also came to Allen with three years of funding from the National Science Foundation, an imprimatur of being more than amply bright. After some follow-up correspondence, Allen offered his mentoring and Rolf accepted without hesitation. Six months after writing the letter that set his life's course, Rolf boarded a Park Service ferry and sailed to Isle Royale for his first summer field season. The summer field notes of 1970 were written by a young man in a constant state of marvel:

"Red squirrel seen just south of Monument Rock on fallen log."

"Aspen . . . along the ridge are completely defoliated by this caterpillar; we and our packs were covered with its silk . . ."

"Found two luna moths . . . just metamorphosed from the pupal stage."

"Discovered 2 hairy woodpeckers and their nest in a dead birch."

Those were the first four days of the field season, and Rolf maintained that intense delight for the entire season, preserving the dates and locations of seemingly everything—blueberries, beavers, butter-flies, goldfinches, snowshoe hares, polliwogs—as if he were getting to know the place by creating an address book of the island's residents.

Perhaps it goes without saying that he religiously documented any-thing related to wolves or moose—every wolf track, scat, and howl, and every moose sighting: 64 bulls, 5 cows with twins, 30 cows with a single calf, and 56 cows without any calf. Most of these records were particularized with some note about each moose's behavior or appearance.

That Rolf would be at home on Isle Royale was clear from the first luna moth. That he could observe nature was apparent from his hav-ing written, on average, one page of field notes for every mile hiked that summer—all 382.2 miles. Time would tell if he could harness his gift of observation to craft broad insight about nature.

Rolf closed his first field season in late August and left the island as the loons began to congregate on the lakes in advance of their own fall migration. Within a few months, the harbors were frozen solid, and Rolf returned to Isle Royale. This time he arrived by ski-plane, sitting behind Murray in the same observer's seat from which Mech had first observed the balance of nature. Murray taught Rolf how to see the world from 500 feet over the tree tops. Murray, who definitely inher-ited more from Sigurd Olson, developed a deep appreciation for Rolf and his indefatigable, mission-focused labor. Rolf reciprocated appre-ciation for Murray's enduring willingness and uncommon ability to indulge that appetite.

Having a year's field experience to his credit, Rolf authored with Al-len the *Ecological Studies of the Wolf on Isle Royale, First Annual Report, 1970–71*. The first sentence of Rolf's first public statement on the wolves of Isle Royale proclaims: "June 1970 marked the beginning of a new phase of these studies."[9] That ever-so-slightly immodest pronounce-ment was at odds with what he and Allen could only have reliably as-sumed would be a three-year stint. Furthermore, enshrining this document as the *first* of its kind is at least portentous, given that each of the seven previous years had been memorialized in annual reports.

In truth, Rolf was skeptical of learning anything about the wolves and moose of Isle Royale that Mech hadn't already figured out. Allen assured Rolf to the contrary, though he was unable to say just what would be discovered or even the topic to which the discovery might be attached. Nevertheless, audacious ink snapped with every keystroke of Rolf's typewriter. From an otherwise thoroughly modest young man, boldness burbled in a way that propels every impassioned faith forward and into the unknown.

Rolf appears to have placed his faith in snow. In the so-called first annual report, in a section that describes the winter field season, the first sentence is not about how many wolves were counted or how many moose had been killed. No. That first sentence is about snow: "Weather during the period 26 January to 11 March was characterized by deep, soft snow." The statement is mundanely credible for an island a little more than half way to the North Pole. While that simple assertion about deep, soft snow would have needed no corroboration, Rolf bestowed the claim with heft by providing evidence in the form of a table showing the maximum snow depth measured on Isle Royale during each of the five previous winters.

The "first" annual report contains an entire section devoted to the topic of snow, "Moose Distribution and Predation Relative to Snow Depth," where Rolf described three roles for snow. First, snow deeper than about 30 inches inhibits the mobility of moose calves, placing them at greater risk of wolf predation. Second, wolves can easily walk on the hard-packed snow of windswept shorelines and tend to use these as travel corridors when the forests are deep in snow. Under those conditions, wolves will tend to kill moose closer to shorelines rather than farther inland. Third, when the snow is deep, moose will be driven to thick conifer stands, where much of the snow is intercepted by the forest canopy. Under these conditions, wolves will tend to kill moose in conifer forests as compared to more open habitats, such as snow-covered meadows.

The reader is instructed that these three conclusions "appear justified" by another table of numbers highlighting the more or less consistent conjunction of those above-stated patterns.[10] This much Rolf had concluded within a year of his arrival.

During his second winter field season, Rolf observed snow deeper than his lanky legs were long and reported that "winter conditions this year were characterized by deep snow . . . similar . . . to 1971." He devoted as many words to snow in his second annual report as he dedicated to moose or even wolves.[11] Snow was neither backdrop nor part of the set; rather, it was a member of the cast and a major character in the island's story. Rolf's third annual report includes this stage instruction, "The importance of snow in the wolf-moose relationship has been impressive in recent years, and the need for a systematic monitoring of snow conditions became evident." Deeper-than-average snow buried Isle Royale again during Rolf's fourth winter.

Snowflakes form high aloft as an infinitesimal kernel surrounded by six crystalline axes, sometimes 12 and occasionally three axes but never eight or four. That variety is a geometric consequence of the 107.5° arc formed by H_2O's two hydrogen atoms. Each exquisite flake flits to the earth, lands in a snowbank, and shortly thereafter becomes a tiny, amorphous lump of ice lightly fused to neighboring lumps. Landed snowflakes are certainly less delicate, but they retain a complexity evinced by all that can be involved in measuring snow.

To measure snow depth, plunge a ruler into the snow until it hits bottom and note the number lying even with the surface. Measuring snowfall requires slightly more foresight. Before the next snowfall, place a board on the surface of the snow, measure snow depth on the board after that next snowfall, wipe the board clean and repeat with every snowfall. Prosaic? Yes, I concede. But we're not done.

Snow deeper than about 18 inches hits a wolf square in the chest. Moose don't experience this burden until storms have piled up more than 40 inches of snow (as much as Rolf had observed in 1971). Depth alone would provide an adequate account if it were not that fallen snow varies so strikingly in its density. Walking through six inches of light, fluffy snow is barely noticeable. Plowing through the same depth of wet, heavy snow swiftly zaps any motivation to carry on.

Rolf's measurements were formal testimony to wide variation in the compactness of snow, which commonly ranges from a light and fluffy 0.03 grams per cubic centimeter to a nearly-thick-as-wet-cement 0.36 g/cm³. The heft of snow varies with conditions of the atmospheric nurseries where snowflakes are born, and it increases over time as it ages in a snowbank.

The concretion of snow is important not only because moose are longer-legged than wolves but also because they have consequentially different foot loadings. Take the mass of an organism (say 150 pounds) and measure the area of its feet (say 70 square inches). Perform some unit conversions, divide the two numbers, and find a foot loading of 2.1 pounds per square inch or, to use the units of science, 150 grams per square centimeter. That foot loading is typical for an adult male human and represents the pressure of an entire body exerted on the ground through one's feet. This kind of measurement is absolutely uninteresting, except in the hands of the few who can see its deadly implications. A snowshoe hare is, for example, about three times bigger than a cottontail rabbit, but the hare's foot loading is about a third lighter than a cottontail's. Hares evade predators by taking flight across the top of the snow. No frantic worry for finding refuge. They escape by being fast on the snow. Cottontails, however, should stay close to their burrows. Even then they won't always cover the short distance to safety before being snatched by lynx. For every 10 rabbits, only two survive two months of snow cover. Hares are three times as likely to survive the same experience.[12]

Moose have much higher foot loadings than wolves. According to Rolf's measurements, the foot loading of a moose is about 400 g/cm² and four times greater than a wolf. When the snow is sufficiently firm, wolves walk right across the top. But snow is never, save the rarest circumstances, dense enough to support a moose. Consequently, snow of intermediate depth and density impedes wolves, thereby offering some relief to moose. However, when the snow is deep and dense, the gaze of the Grim Reaper slinks away from wolves and toward moose.

What could possibly justify this preoccupation with snow, and why would you and I care to relive even this glimpse of what may have been (and continues to be) just one person's obsessive compulsion? You know the short answer is science, but the slightly richer answer is that a scientist's labor is sequenced into two tasks. First, observe nature; second, craft those observations into some broader wisdom about nature. For that second step to qualify, the putative insight must inspire an apperception of nature that is supported by, but also transcends, the particular time and place where the observations were recorded. Science is an alchemical transformation of antiseptic data into breathing insight. We saw this sequence with Mech and Allen. First, they documented constancy in the lives of wolves and moose on a distant island; from that evidence, and in a distinctly separate act, they ideated a balance of nature.

Rolf's enduring enthusiasm for snow is an idiosyncratic obsession, but he couldn't merely continue to chronicle the same kind of observations Mech had. To advance science, Rolf had to catalog some additional phenomenon *and* discern its import. Rolf's irresistible fascination with snow is not unrelated to a hoarder's neurosis, but here it propels the cataloging required for divining insight.

Back to the details of Rolf's commitment to measuring snow. Even depth and density are incomplete characterizations of fallen snow. When the sun warms a south-facing slope, crystals in a snow bank melt imperceptibly, just a bit and only on the microscopic margins of each crystal. The crystals refreeze with the cold touch of twilight and become solidly welded to one another. The result is a crust that is, when broken under foot, strong enough and sharp enough to draw blood from the shins of a moose. The hardness of a snow crust can be quantified (and should be quantified because it's not science without the measurement!). A numerical metric of hardness can be ascertained with a Chatillon "push-pull" gauge fitted with circular plates, the size of which vary according to the hardness. But the preferred tool for the precise measurement of snow hardness is a Swiss-made

Rammsonde penetrometer. The use of this tool involves dropping a weighted stainless-steel cone that has been machined to a 120° point. The cone is dropped along a rod that passes through the cone and a "Ram hardness number" can be calculated from the depth to which the cone sinks upon several drops.

No one can say how much time Rolf has spent measuring the depth, density, and accumulation of snow. He's recorded differences across habitats, from open meadows to closed forests, and differences over time, from early winter to late winter and from one winter to the next. I bear witness from having inspected summarized data representing myriad plunges of the ruler over the past 50 years. I've not even seen all the data in its raw form, composed of many thousands of handwritten numbers and long columns of dates and centimeters. What resonating sagacity could possibly emerge from this litany of measurements?

To find out, Rolf applied the two-step formula that had served Allen and Mech so well. He first observed the dynamics of wolves and moose superimposed onto trends in snow. Moose had steadily increased throughout the 1960s with the momentum of a loaded ore ship, and wolves were unable to arrest the surge. Rolf surmised that snow conditions during those winters afforded moose unassailable vigor. Then over a four-year period during the mid-1970s, Chione, daughter of the north wind and goddess of snow, placed her hand on the balance and unleashed three severe winters with deep snow and temperature regimes that created crusts often enough to advantage wolves. Predation rates increased, calves were killed, wolves were well nourished, and pups were born.[13] Trends of the 1960s doubled back, with moose declining and wolves increasing. That was the *relatively* simple observation harvested from all those measurements.

For the second step, Rolf prayed—as all young scientists do—for inspiration to reform those observations into a recondite realization that shines past the mundane particulars of what had been seen in one tiny place during just a flash of time. That reformation began with acknowledging the obvious truth that wolves and moose are alive but snow is not. Being alive and responsive, wolves and moose influence *each other's* lives. One is food for the other, and the other brings death to the first. That intimate interconnection runs up the hierarchy of life as a sequence of events in time: wolf abundance increases because there are many moose to eat, wolves kill more moose, moose abundance declines, wolves have less food, wolf abundance declines, and the cycle repeats. It's the timeless, self-perpetuating feedback elucidated by Volterra's equations.

But snow's relationship to wolves and moose is canonically asymmetrical. It can affect wolves and moose but cannot be affected in return. There is no feedback. Up to the mid-1970s, most ecologists had neglected the complicating dissymmetry that arises with the influence of abiotic phenomena on biotic affairs.[14]

Imagine, for a moment, the wolves and moose of Isle Royale to be the limit of our interest. Take that island and carve it, in our minds, from the rest of the universe. At that point, snow becomes an external perturbation, meddling in the affairs of mortals with the impunity of gods. Snow is an exogenous force and more than capable of knocking life off balance.[15]

Through compulsive documentation and innovatory inference, Rolf revealed a natural dynamic of interest to ecologists of every stripe. Where Rolf saw snow, others would see an external perturbation that happened to be snow. In other places and times, that exogenous force might be a drought, the arrival of an exotic species, or a hurricane. The substance of Rolf's PhD thesis, published in 1977 and read to this day by young wolf ecologists and aficionados alike, seemed to illustrate a basic process in nature. He suggested how exogenous forces may forever adjust the balance, first in one direction and then at some undetermined moment in a different direction.

Yet, without conceding the value of that broad scientific insight, it is just possible that Rolf's attention to snow is more important for its virtue: the soul of science is fragile and easily crushed by its utility. Heralding the wonder of mundane phenomena is a scientist's sacred and too-often neglected responsibility. Rolf tended this responsibility in his communion with snow.

In any case, Allen never promised Rolf an opportunity to test any specific hypothesis—an over-credited, human-designed incubator of scientific knowledge. Rather, Allen offered a method free from the preconceived constraints of a hypothesis: clear your mind, open your eyes, and let nature show you something. The ponderous peril of this method is that it fails regularly, even when expertly employed, because one still needs the gift to see what is being shown.[16] What Rolf saw was a series of three severe winters, whose inherent unpredictability prevented them from being readily shackled to a hypothesis. Allen knew that Rolf had the ability to see, and the ability to be patiently faithful that there would be something to see. Confident about the future of researching wolves on Isle Royale, Allen retired the year before Rolf's PhD thesis was published.

Rolf Peterson and Don Murray would have gone on to become quite the team, except shortly before the winter study of 1979, Murray announced that he would retire after having been with the project for two decades. Hearsay has Murray quitting in protest of women participating for the first time in winter study,[17] but he had also become less able to comfortably keep up with the physical demands of working in a winter wilderness. By the time I met Murray in the 1990s, he was completely blind from a rare congenital disease. Yet, he was still able to describe aerial views of Isle Royale—dark basalt ridges rising from forests perforated with sparkling fields of snow on frozen lakes—in enough detail to suggest he'd flown just yesterday.

Bush pilots with the requisite skill are as uncommon as they are vital for wolf research. Pilot and plane are an extension of a biologist's sense of sight and the means of bodily movement. The project would have ended were it not for Don E. Glaser, who replaced Murray and turned out to be no less capable as a pilot. A decade into his tenure, Donnee would teach me all that I know about how to see the world from 500 feet over the tree tops.

While Rolf searched for meaning in blankets of snow, the wolves and moose carried on with their own lives. Let's circle back a few years to examine how they had been getting on. We already viewed aspects of their daily lives in the first chapters of this book, *Why Wolves?* and *Thoughts of a Moose*, and we did so with considerable intimacy.[18] But this time I'll dial the lens back for a broader view.

By 1970, the moose population had doubled, by which time it might have been possible to cling to the idea that nature's balance had merely shifted. By 1980, moose abundance had fallen to half its maximum. Wolf abundance increased for the first 10 years of Rolf's tenure, save one year. The wolves' social order fissured. Two packs became three and three packs became five.

Wait a moment. I've not prepared us as I should. We risk unconscionable callousness of the kind that resides with the aphorism *a single death is a tragedy; a hundred is a statistic.* When I say that moose abundance had fallen to half its maximum, that's no statistic. We have to read that for what is—hundreds of tragedies. Or triumphs, depending on the perspective.

In this passage we'll recall about 20 years of history in the lives of wolves and moose on Isle Royale. Absorbing that span and apprehending its insight precludes conveying all the joy and sorrow with words. Yet, it is possible with some effort for you to bring that sense of elation and anguish to the words.

But why? Why credit a sanitized account of life and then risk cold detachment in the process? Allow me to answer with an analogy. We can submerge ourselves in the excruciating immediacy of blood and love found in *A Tale of Two Cities* or *The Scarlet Pimpernel*. In doing so, we would know that the Reign of Terror really was a reign of terror. Those books will not, however, reveal why Robespierre ordered thousands of frivolous executions, why the sans-culottes so readily resorted to violence, or the long-term consequences of the terror. Those revelations require temporarily dissolving the fog of intimacy so that we might see details of a grander scale. The same is true for the wolves and moose on Isle Royale.

We should probably start over. By 1970, the moose population had doubled, by which time it might have been possible to cling to the idea that nature's balance had merely shifted. By 1980, moose abundance had fallen to half its maximum. Wolf abundance increased for the first 10 years of Rolf's tenure, save one year. The wolves' social order fissured. Two packs became three and three packs became five.

By 1980, the wolf population swelled to 50 sets of paws. While there is no possible meaning for the phrase "balance of nature" to match that circumstance, it was not complete chaos. The packs respected territorial boundaries defended by their neighbors, as much as they would have under less-crowded conditions. Rolf predicted that "future trends in the wolf population will depend on how the newly-established packs fare, and also whether or not additional pairs form."[19] The prognostication is iron clad in being a veritable tautology. Superimposed on that certainty was the implication that further increases in wolf abundance were plausible.

The pairs and loners would all perish within a year. The five packs of which the population was composed that year were in utter disarray.

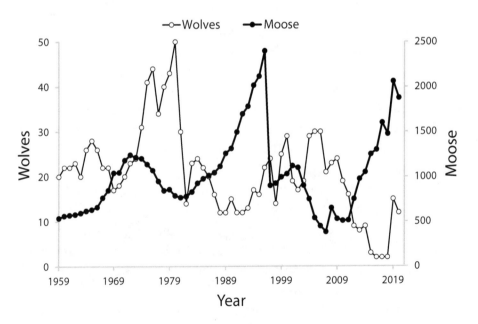

Fluctuations in abundance of wolves and moose on Isle Royale, 1959–2020

Middle Pack and Southwest Pack were short on food, adjusted their territorial claims, and fought over the same parcel of land. East Pack expanded its territory, enveloping nearly every square inch of land upon which Harvey Lake Pack was living. Northeast Pack was wedged into a few square miles at the island's northeast end, a forsaken spit of exposed bedrock and stunted fir trees. In 12 months, abundance had dropped by 40%, from 50 to 30 wolves.

By all appearances, the wolves' catastrophe resulted from a severe shortage of food that had been brewing over the past few years. Severe winters had left moose in a vulnerable state during several key years of the previous decade. With ample prey, some savvy alpha wolves took advantage and raised large litters of pups.

In those years, the alpha wolves tended to grant subordinate pack mates—pups from previous years—unlimited access to moose carcasses. As a result, each of these adult children could expect to renew their leases on life for another year. The normal situation is for alpha wolves to physically, sometimes violently, deny subordinate wolves access to feed on a carcass. Sometimes the rationing is imposed by a stronger sibling. The stomach pangs brought on by sibling rivalry, or parent-offspring conflict, motivates some adult children to leave the natal pack in search of better, self-sufficient prospects. Otherwise it's malnutrition. Either way, these wolves tend to live shorter lives.

More offspring and longer lives are the mechanisms of every population increase that has ever occurred. But too much of a good thing (for the wolves) led to the eventual and inevitable decline in moose abundance.

Wolves regularly travel down human summer hiking trails, where they incidentally deposit scats. Feeling obliged, Rolf collected these scats throughout the 1970s and '80s, picked them apart with forceps, sorting each scat into little piles of brown debris, and read them like tea leaves. Scats collected during the summer of 1980 were a harbinger. In a typical summer, wolf scats are filled almost entirely with the hair of calves and beaver. The hair of snowshoe hares is rare because only the most unfortunate hare ends up down the gullet of a wolf. Not

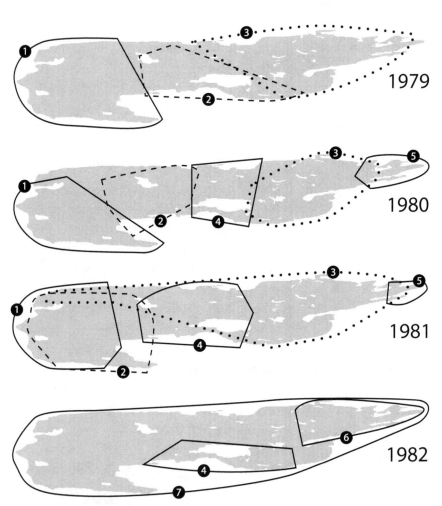

Territorial boundaries of wolf packs in Isle Royale changed dramatically from 1979 to 1982. Adapted from Peterson and Page 1988

so during the summer of 1980, when hare had become a regular part of a starvation diet. Hare hairs were present in almost 20% of the scats.

The winter of 1981 was far too weak to succor wolves during that year of need. With too little snow to rob the moose of their strength, wolves struggled to secure every meal. When wolf abundance declined from 50 to 30 in just 12 months, Rolf classified the event as a "natural outcome" of circumstances observed during the previous decade. The episode was dramatic but no more than a resetting to a more stable balance.

A year later Middle Pack dissolved, and Southwest Pack vanished. Northeast Pack, also missing. Harvey Lake Pack was reduced to a pair. A handful of survivors from the lost packs scattered and lived as lone wolves. Three pairs of wolves did not defend territories so much as they managed to kill a few moose and cling to their carcasses like castaways clutching a life raft. The "Gang of Four" was the only social group who had it together. They took their territory to be the entire island.

Many wolves starved and died quietly. Neighbors killed each other in bloody territorial disputes fought over food. The population crumbled to just 14 wolves. The survivors were shell-shocked but relieved from much of the intense competition that had resulted from so many wolves.

Then the research coffers went hollow.

However, allies of the wolf-moose research with political connections in Washington, DC, made a case to the National Park Service, and the agency found money that could be diverted, contingent on approval from the higher-ups. The highest higher-up was G. Ray Arnett, assistant secretary of the interior and a bold hater of wolves.[20] He heard about the diversion, forbade it, canceled the project, and issued an order to cease and desist. That news came while Rolf was conducting winter study in February 1983. The local park authority, charged with executing the order, did not disagree with Rolf's suggestion that the dictate clearly applied to park employees but maybe not exactly to Rolf and his civilian crew, Pilot Glaser and Doug W. Smith.[21] So Rolf, suffering from some congenital defect rendering his auditory nerves unable either to hear or interpret an order to cease and desist, kept on flying and counting.

Eventually, Arnett relented, rescinding the order and allowing the transfer of funds. Colleagues had advised him to find some other way to express his hatred of wolves. Within two years, Arnett resigned and was elected as executive vice president of the National Rifle Association.[22]

During that winter of the rescinded cancellation, Rolf found that the wolf population increased by an incredible 64% to 23 wolves. The population was composed of three well-formed packs with well-defined, well-respected, nonoverlapping territories. A return to normalcy might have been inferred from the population counts of 1983 and 1984, which indicated 23 and 24 wolves, respectively. Except that 10 or 12 of the 24 wolves in 1984 were pups. Those numbers and the laws of arithmetic meant that about 40% of the adults had died in the past year. Such a high rate of mortality is not quotidian, and its cause was not so easily explained.

The changes were so fast—50 to 14 wolves in just a couple years. Yet, the unfolding of a population's life also feels insufferably slow. Make all the observations you wish throughout the year, and, in the end, you add only one data point for the wolves and one for moose. It can take a few years or longer to discern an emerging trend. The challenge is not unlike observing precisely when the tide has turned on a wave-pounded beach. Slow-paced development does, however, have its virtues. In particular, the time to recognize a trend and its antecedents. The abutment of these two senses of time—simultaneously fast and slow—was never more towering than during the 1980s.

In 1981, Rolf considered the collapse to be "the natural outcome of events during the past decade."[23] A year later, he believed "accumulating evidence suggests that we are witnessing a predator-prey cycle with an extended period of fluctuation."[24] The system was self-correcting like a financial market with an overvalued commodity. The wolves and moose may have been knocked off balance, but the feedback had an intense homing instinct. The consolation that ecology could offer is a belief that resiliency is born from the intimacy between predator and prey, an unstoppable urge to return to the past. Timelessness born, not of nostalgia, but from the immutable laws of nature manifest in the mathematics of Vito Volterra.

After counting 24 wolves in 1984, Rolf explained, "Our current idea [is] . . . that the wolf population should stabilize at the current level while the moose population continues to increase."[25] By 1985, with 22 wolves living in three packs, the population appeared to have emerged from its dark hour.

With that concern alleviated and the passage of time, a new pattern presented itself as most obvious. Moose abundance had been increasing for the past four to eight years. With the first such sustained increase during Rolf's tenure, he could not help but notice the impact of all those moose eating the forest. This is not a revolutionary notion. If wolves eating moose is interesting, then moose eating the forest is no less interesting and no less important. This was not an epiphany granted by 25 years of research. The *idea* was as patent in 1958 as it was in 1985.

But one cannot begin, straight out of the chute, by documenting everything that is plausibly, or even certainly, important. Making all those measurements would be a serious challenge but not unthinkable. The exorbitant obstacle is conceptual, not logistical. The impossible dream is to penetrate the meaning of all those quantifications. Survey too many phenomena, consider them too quickly, and the result is a torrent of information that drowns knowledge. Comprehending such a deluge of data is like trying to visualize a virtual reality by reading its machine code.

The unique gift of observing the happenings of a place for a long time is the time itself, because time is the inescapable cost for understanding. Not merely the passage of time but time for meditating on the early indicators of trend and time for allowing patterns to more fully emerge. Allen knew that three years would not be enough time for knowledge of the depth that he sought. After three years, Allen and

Mech had spotted only a sparkle on the surface. A decade is not enough time. It took that long for Allen to figure out how to reliably count moose. How much time does it take? It took 25 years to expand the focus from wolves eating moose to moose eating plants. It takes more time—in one place—than most have patience and endurance.

The appraisal of moose herbivory began in the mid-1980s, and it would take another decade for the insight to ripen. That waiting did not coincide with a lull in the dynamics of predation. In 1987, Harvey Lake Pack crumbled after early signs of failure in the previous year. The wolf population slipped to just 16 wolves living in two packs. The precise nature of the mortality was unknown, but kill rates were low, and limited food supply was taken to be the most likely cause.

In 1988, the population slumped to an all-time low of 12 wolves.[26] And the moose population continued to "spiral upward, impeded little by wolf predation."[27] An upward spiraling moose population tends to be composed of young moose in the prime of their lives. Prime-age moose are not merely more difficult for wolves to kill. They are deadly.

Sometimes a moose "gets away" by putting distance between herself and the wolves in pursuit. Often wolves surround a moose, they stand off, and wolves abandon the effort because they cannot kill the moose without serious injury to themselves. Not being killed by the crushing blow of a hoof to the skull takes precedence over aching hunger and starving pups. While wolves lived amidst many moose during the mid and late 1980s, they were not awash in food. And while the difficulty of capturing prey may have dragged the wolves down, it was an inadequate explanation for the lowest-documented abundance of Isle Royale wolves.

Between 1981 and 1988, some 80 to 90 wolves died. Clues for the cause of the turmoil might have been found with any chance to necropsy a carcass. That fortuity arises from just one of two circumstances, the first of which is to retrieve the carcass of a wolf whose death is observed from the airplane. That opportunity is rare because we fly, on average, about 100 hours each winter, representing 1% of all the hours in a year—it's a wonder we glean any substance at all, let alone detect the moment that a wolf dies.[28]

The other occasion for necropsying a wolf is to stumble upon its bones while hiking in the forest. When a wolf dies, the carcass is quickly picked at by ever-present ravens, whose clamorous requiem calls on the two or three foxes within earshot. The foxes pull the bones apart and scatter them throughout the forest within days. Whatever soft flesh might remain is quickly requisitioned by maggots.

Between 1981 and 1988, the remains of only seven wolves were recovered. They confessed frustratingly little. One carcass was too far gone to determine cause of death. Two were "old, malnourished males with extremely heavy tooth wear."[29] Four had died at the fangs of other wolves.

The most common causes of death for a mainland wolf are being shot, trapped, poisoned, or hit by a car. When humans are not the primary cause of death, then starvation and attacks by other wolves are the immediate, or *proximate*, causes of death for a large share of wolves. This is so even for a healthy wolf population. Rolf yearned for carcasses to necropsy for insight about more *ultimate* causes of death. That desire went unfulfilled.

Disease and genetic impoverishment were suspected, but suspicion does not bear the weight of scientific scrutiny. In 1987, Rolf was asked by park authorities to prepare a proposal with plans to live-capture and radio-collar wolves and, more importantly, to obtain blood samples that would be screened for disease and genetics. Then, park authorities rejected the proposal Rolf created at their request.

No wolf is interested—to understate the matter—in being live-captured, anesthetized, and handled for a cause they do not understand. Handling wild animals in the name of science is generally a compromise of their interests and should be reserved for cases when the value in doing so outweighs infringement upon the animal. But this was not the park's concern. Instead, they stated that "not handling wolves would contribute to maintaining the park as pristine . . . thereby preserv[ing] the mystique of wilderness."[30] Although the park considered collaring wolves to be an "uncomfortable" proposition, they eventually acceded to soliciting reviews from scientific experts and subsequently accepted the proposal.

In reliving Isle Royale during the 1970s and '80s, I've made a crucial omission. What I've left aside was unknown to anyone until the 1990s. I excluded this critical detail until now because important aspects of anyone's life are routinely discovered after the time when first knowing would have meant a great deal. This climacteric was ultimately illuminated by having handled wolves, so I'll share it now.

Return with me to the Fourth of July weekend of 1981 when friends from Duluth and Chicago boated to Isle Royale in the company of their two pet dogs to reminisce, drink, and laugh. After a trip or two ashore for the dogs to relieve their bowels, they returned home. At home, the dog from Duluth began to vomit blood. Then came an uncontrollable and bloody river of diarrhea. The dog died four days after visiting Isle Royale. The Chicago dog is thought to have died under similar circumstances.

Felids from across the planet—cats, lions, bobcats, all the cats— have for generations dealt with feline distemper. The disease is often fatal and caused by a particularly small kind of virus called a parvovirus, because *parvus* means "small" in Latin. These little viruses are too simple to be counted among the living and too complicated to be credibly classified as inanimate. The parvovirus that causes feline distemper is a wadded string of DNA stuffed into a protective capsid, formed by 54 copies of a protein arranged by geometry to form an icosahedron. The surface of this 20-sided sphere is embedded with six copies of a second protein in such a manner as to create a shell accented with canyons, dimples, and spikes.

Without this capsid, the parvovirus would be quickly digested by enzymes of the body it aims to infect. The capsid also has adhesive regions, allowing it to stick to receptors of the host cell that are normally expecting to bring life-sustaining iron into the cell. The virus cannot hijack the host cell's machinery without first adhering to the cell surface. This business of adhesion is surprisingly specific. The parvovirus causing feline distemper can attach to the cell surfaces of any felid, and a few nonfelid species, but not canids.

Sometime in the 1970s, serendipity gifted a lineage of parvovirus with mutations at 17 of the 5,200 bases that code for its capsid. The first mutations did nothing, except to make the little virus less effective at brutalizing cats. The subsequent mutations, in concert with the prior mutations, restructured the capsid ever so slightly. Of all the possible mutations that could have had any affect at all, nearly all would have rendered the parvovirus inert and harmless. And these mutations had exactly that affect. Fortunately for cats, the mutant strain was unable to secure itself to cell walls. The jig would have been up for this line of parvovirus, except that it found its way into the body of a dog, where it thrived.[31]

This new thing—activated, if not animated—began infecting dogs by the scores, then by the thousands. Humans noticed the disease in isolated outbreaks during the summer of 1978. Over the next two to three years, it had spread throughout the world, infecting perhaps 80% of all dogs—wild and domestic. Millions died. That poor soul from Duluth is the first known case of hundreds of dogs to die from canine parvovirus (CPV) in northern Minnesota.[32]

Back on Isle Royale, one or more wolves caught a whiff of the feculence left by unfamiliar canids. Unable to refuse the impulse, they followed the odor to its source. Noses thrust to within inches, sharp and furious inhalations, and incensed excitement over invaders now gone (really gone). Between the inhalations and the excitement, some viral spores—real as they are invisible—wafted up from the surface, into the nose, down the throat, and onto tonsils and lymph nodes. A few days later, vomiting and diarrhea.

Pups were especially vulnerable with their altricial bodies unable to withstand all the retching. They died entirely spent and dehydrated. In July 1981, Rolf had determined that all three packs had produced pups, eight in total. None of them survived. No one knows how many adult wolves died of CPV on Isle Royale.[33]

The influence of CPV during the early 1980s was entirely unknown until four Isle Royale wolves—healthy by all physical appearance—were live-captured in 1988, along with another four in 1989. The blood drawn from these eight wolves was screened for the presence of antibodies to CPV (and a raft of other diseases). Antibodies in the blood would indicate having survived a past exposure to CPV. Two of the wolves tested positive, and two were marginally positive. Fecal samples from these eight wolves were also screened for antigens to CPV, the presence of which would indicate a wolf currently fighting a CPV infection. No antigens were found. Over the next 20 years, we drew blood from more than 30 wolves. None showed any sign of exposure to CPV. The disease had been a powerful exogenous force of the previous decade, and Rolf just barely caught its last reverberations.[34]

Samples of wolf blood were also sent to Robert Wayne of UCLA, a world authority in the genetics of small populations and populations of canids. From his analysis, he concluded that the wolf population had lost about half its genetic diversity since becoming established on Isle Royale in the late 1940s.

Other studies had demonstrated that genetic losses of that magnitude can raze a population. But those studies had, for the most part, been conducted on species like fruit flies or mice living in the artificial environment of laboratories. It was easy to raise doubts about the relevance of those studies to wolves in the wild. Most of these other studies were also conducted in highly controlled experimental settings where any ill effect could be pinned exclusively on genetic impoverishment, and the influence of other factors, such as food and disease, could be assuredly isolated. While it was clear that Isle Royale wolves had lost a great deal of genetic diversity, there appeared to be little prospect for understanding the *effect* of genetic deterioration on a population buffeted by disease and booms and busts in prey availability.[35]

The other critical conclusion of Wayne's analysis was that only a single female had contributed to extant diversity of mitochondrial DNA in the Isle Royale wolf population. That led, plain and straight, to the inference that the wolf population had been founded by a single pair of wolves and had been completely isolated ever since. Prior to this finding, the potential influence of immigration and emigration had been acknowledged,[36] but not afterwards. Genetic isolation seemed, by far, the most parsimonious impression from Wayne's findings, even though it would eventually prove misleading.[37]

With those retrospective insights derived from a few cc's of wolf blood, we can come back to the leading edge of the unfolding chronology. During the late 1980s and early 1990s, the wolf population lingered for the better part of a decade in the low teens. Rolf considered the situation "bleak." After seeing that abundance could plummet from 50 to 14 over a two-year period, Rolf expressed deep concern that the wolves could go extinct. That concern was shared by scientists, the media, and the general public. The park's default response would be to allow extinction.

By 1991, Rolf had developed an alternative view in three parts.[38] Part one: Rolf believed an essential purpose of a national park is the advancement of science and that the greatest scientific gains would be found in carefully observing the process of wolf extinction. But, in the same breath, Rolf also believed that wolves should then be reintroduced, in part, because that would also yield the greatest scientific gain.

Part two: The Park Service's guiding philosophy that humans not interfere with natural processes was fraught with contradiction and ambiguity, to the point of being unable to determine the best relationship between humans and Isle Royale's wolves. In particular, the Park Service's philosophy did not offer a basis for favoring extinction or predation. Extinction is a superbly natural process on small, remote islands, and predation is not only impeccably natural but it had also been disrupted by the "mayhem" of a human-introduced disease (CPV). Rolf voted for predation, but his point was that careful study of the philosophy offered no unambiguous basis for privileging one natural process over the other, though he raised the "sobering realization that deliberate management *by humans* [original emphasis] is the only hope for much of the world's wildlife, even in a place like Isle Royale."[39]

Part three: While science is important, "science is moot on the question of whether wolves belong aboard the Isle Royale ark." That question is decided, Rolf believed, by "think[ing] of our national parks as both laboratories and cathedrals" and knowing that a wilderness cathedral without the sacred imagery of wolves would be sacrilege.[40]

Rolf publicly asked the NPS to give his views and those of others the serious attention they deserved. For so asking, he was hauled into the office of park management, informed of their displeasure with his ideas, and reminded that future funding was contingent on their approval. Shortly afterward, park management proposed that winter study be conducted only in alternate years, rather than annually. Doing so would have undermined the scientific integrity of the research. The NPS never acted on this; the purpose of the threat seems only to have been to elicit submission.[41] My sense is that the park's opposition was a response to being threatened by the expression of a view it did not like, or it may have been bullying motivated by some personal grudge. The hostility may also have stood in defiance of common sense, or it may have been some combination of those possibilities.

By 1996, wolf reproduction was solid, mortality rates subsided, and wolf abundance had increased to 22, which is close to the long-term average. With some hesitancy, Rolf suggested that the "crisis" had passed. After more than a decade, Rolf reached the understanding that the best explanation for the wolves' foundering had been genetic impoverishment, exacerbated by the foul effects of disease and limited prey.

Wayne's genetic analysis also added fuel to a controversy among conservation professionals. Those with formal training in genetics tended to believe that loss of genetic diversity was a general threat to the conservation of small, isolated populations. Another group of conservation professionals—small in number but influential and tending not to have formal training in genetics—believed the threat was exaggerated.[42] With their recovery in the 1990s, the wolves of Isle Royale became a prime exemplar for a belief that the genetic health of small, isolated populations does not merit all the concern afforded by conservation geneticists.[43] Isle Royale wolves were used to serve that undue view for the next two decades.

Back in the 1970s, when Rolf first set himself to study the wolves of Isle Royale, he was impressed by winter's influence as an exogenous force occasionally capable of tipping the balance between wolves and moose. He even saw the great rise and calamitous fall of the wolf population as a "natural outcome" followed by a resetting to normal conditions.

But by the early 1990s, Rolf brooded the forlorn view that Isle Royale had become "a broken balance." Rolf saw balance, he saw that it would be regularly buffeted by external perturbations, such as severe winters and disease, and he saw resilience to those pressures. But only to a point, beyond which balance cannot recover on its own. In those cases, Rolf reckoned that we have a duty to examine our responsibility to restore the balance.

The Park Service responded to that duty with a formal promise, claiming extremely high priority to "convene a panel of . . . experts to identify and evaluate potential actions" in the event that extinction risk should ever again raise its ugly head.[44] They never did.

6 |

The Old Gray Guy

The wolf population stabilized by the mid-1990s, but not before the moose population began to swell. Predation's grip slipped, and the moose population tripled.[1] When a population's abundance sets a streak, it's high times among the population's individuals. Toward the end of that streak, when abundance is elevated, individuals fall on hard times.

These hard times struck moose as a rise in nutritional poverty. That flux washed over the wolves, as they are forever on the search for weakened moose. When times are good for individual moose, as they were in the late 1980s, about 40% of the moose killed by wolves show signs of malnourishment in the marrow of their bones.[2] In the years to follow, moose abundance rose like Icarus, individual moose melted, and 70% of the moose killed by wolves were malnourished. When moose abundance peaked in 1995, 90% of wolf-killed moose were malnourished. Wolves' work qualified less as predation and more as scavenging, though they were not quite patient enough to wait for their meal to die first.

The easy killing was certainly a boon for those wolves, but wolf abundance was too low, and the impact of wolf predation on moose abundance was flaccid. The relationships between populations had gone slack.

Malnourishment pressed itself onto the bones of moose during the great rise in moose abundance. Moose born during those years had brain cases with volumes that were 13% smaller, on average, compared to moose born in favorable years.[3] Malnourishment as a youngster predisposes moose to osteoarthritis later in life.[4] And so it was.

The forces of life rose up through individuals into their populations and then fell back from the population onto the individuals again like the burbling everyday chaos of a water fountain.

When the moose population exceeded 2,000 individuals—one of the highest densities of moose recorded anywhere in North America—Rolf reaffirmed his faith in snow and proffered like the oracle at Delphi that "winter and spring weather patterns might well dictate the course followed by the moose population."[5] He said so in March 1995.

The mother of all winters arrived early that fall. The island shuddered from the blast of another powerful, exogenous force. The most severe winter on record in this part of the world conspired with food shortage and a coincidental outbreak of ticks. The moose population collapsed. More than 1,500 moose died in the first five months of 1996, leaving about 500 survivors.

Back in the early 1980s, when the wolf population fell from 50 to 14 in two years, it was dramatic because 14 is so close to zero. The moose population in 1996 was in no danger of extinction. This collapse in moose abundance is staggering because so many died. Legions of sentient creatures starved beyond the point of no return and many more came close.

Some moose plummeted to their deaths after reaching too far for twigs that hung out over the edges of cliffs. We eulogized 17 of those deaths with records in our necropsy file. Others died with spruce between their teeth and cheek. Spruce is as edible to a moose as to you or me.

When moose could not contain the suffering, it spilled onto the forest. Balsam fir trees[6] that had in recent years grown to 12 feet in height held succulent twigs just beyond the nine-foot reach of a moose. These were the last morsels of food. Desperate moose grabbed the trunks of those trees, where the trunks are as much as a few centimeters thick, turned their heavy heads past the flexible limit of a young tree, and snap. Five years of patient reaching toward the sky was vanquished in a desperate moment. At one site, moose killed approximately one in every five balsam fir that were shorter than 12 feet tall.

An important part of each year's field work is searching for moose that have died so that we can perform necropsies. That is how we learn much of what we know about the moose population. In a typical year, discovering a dead moose takes some effort. Because moose hardly ever die on-trail, finding dead moose requires off-trail hiking—about 10 miles of off-trail hiking, on average, for each dead moose. 1996 was different. That spring, along just seven miles of park trail between Rock Harbor and Daisy Farm, 19 moose took their last step. As soon as your nose cleared the indecent odor of rotting flesh from the moose you'd just passed, the next dead moose began its assault on your senses. It was all-hands-on-deck for performing moose necropsies. The spring of 1996 on Isle Royale has the pitiful distinction of the highest density of moose carcasses ever observed. Anywhere.

The large supporting cast of scavengers—foxes, ravens, eagles, and wolves—were unable to take full advantage of the offering. Most carcasses were not visited by even one scavenger with a spine. Most of the scavenging that year went to a particularly well-endowed generation of maggots and carrion beetles.

The meteoric rise and crash of the moose population came and went without much doubt about why it happened. Canine parvovirus triggered or exacerbated the collapse of the wolf population, predation subsequently evaporated, and moose abundance increased until starvation conspired with a winter that piled snow up to the chest. You will recall, of course, the disease that started it all, didn't really start it all. It started when humans inadvertently brought the disease. Of the events that followed the disease's arrival, for which, if any, are humans culpable? Are we absolved because no one meant any harm? Are we off the hook because no one foresaw all that followed or because the final harms matured so far down the causal chain? Maybe. And maybe the question itself seems a tinge off-color. Without passing judgment on the answer, we may come to find by the end of this book that unanswered questions about culpability color the heart of our relationship with nature.

How that chain of events was influenced by the wolves' genetic isolation remained a mystery.[7] In 1994, Rolf had "hope[d] to resolve questions about the impact of genetic isolation [on the wolves] . . . by monitoring the next generation."[8]

Rolf might have meant the next generation of wolves, because it seemed too early to know that Leah and I would also be of the next generation. While I was still working toward a bachelor's degree, Rolf invited me to spend a couple of weeks at winter study in February 1994. That winter, day-time highs routinely failed to rise past −10°F. It was one of the coldest winters on record in this part of the world.

My mission, one day, was to snowshoe into a site where East Pack had killed a moose, perform a necropsy, and retrieve some of the bones. Don Glaser, the pilot, flew me to the closest inland lake, from which I would be retrieved by day's end.

I had never worked in such cold. I found the kill site—an accomplishment that Rolf took for granted. I collected the prescribed data and samples. Without rushing, I was prompt because the remaining daylight would be spent snowshoeing back to the pickup site, and I didn't fancy spending a night with a pile of frozen bones.

After loading my pack with 40 pounds or so of frozen moose parts— skull, mandible, and metatarsal bone, I started back along the snowshoe track I forged that morning. A little more than half way back, I stopped, set my pack down, and gnawed on a frozen-solid Snickers bar. After that short break, I reinstalled myself into the pack, got back into my snowshoe tracks, and resumed my trek.

A while later, Don flew overhead. He circled persistently and waved vigorously out the plane's window. I suspected some problem, but I had no idea what it might be. I picked up the pace. Before long, the problem was right before my eyes as I arrived to the kill site for a second time. After the frozen Snickers, I had obliviously walked down the outgoing trail to my right, not the homebound trail to my left. Returning to the pickup site before nightfall involved some athleticism. Later that evening in the sauna, Rolf asked—as is his style in mentoring young people—what I had learned.

My point is, while Rolf was reflecting on broken balance in the mid-1990s, I was just getting oriented. A year earlier, during the first few weeks of my first summer field season, I was assigned to help Brian McLaren, a PhD student. Our first task was a week-long mission to measure balsam fir trees as a means of understanding the impact of moose on the forest. On the first night of that trip, I found myself hungry and hypothermic beneath the clouds of an early spring storm, recessed in the depths of cedar-filled drainage. Loose debris of the forest stuck to my swamp-soaked clothing and skin. Thunder rumbled occasionally. Not the sharp kind of thunder that portends a severe, but quick, event. No, it was the duller kind that accompanies a long soaking. No tent, no food, no potable water, no raingear.

We were dressed in light clothing because it had been a nice day. We hadn't been thinking that the overnight temperature would slip to the mid-40s. We hadn't been thinking that we'd spend the night in these clothes. We hadn't been thinking.

I rehearsed the same set of thoughts throughout the night between periods of drunk-like, semislumber: "Don't look at my watch," I told myself. "The time will pass, but only if I don't look. In the morning, it'll be light again. Then we'll find our way to basecamp—in a snap. It's better to stay put until dawn. If we hike now, we'll easily walk right over the trail that leads to basecamp. Don't look at your watch." Neither Brian nor I had ever been so lost, or concerned about being lost, in our lives.

But we weren't really lost; we knew how to find our basecamp at Windigo. And we were on an island—walk in a straight line long enough and you'll come to an orienting, if not hospitable, shoreline. We were not lost, but our packs were—our packs, which had rain gear, food, water, and a tent.

Earlier that afternoon, we had found ourselves at a beaver dam. Something across the way caught our interest. We dropped our packs and crossed the dam. Once across, we saw up ahead and through the trees another drainage, just over a low drumlin, 100 meters off. For some unaccountable reason, we believed there was something over the drumlin worth seeing. Extending from the far side of the drumlin was another beaver dam crossing another drainage. A siren called our names and drew us to the other side. An hour later, we realized we didn't know where our packs were. And we soon realized the area contained three or four parallel drainages, each with three or four beaver dams.

We found our packs, but not until afternoon of the next day.

By 1994, social order among the wolves stabilized and the population consisted of three packs—imaginatively named East Pack, Middle Pack, and West Pack. The security provided by pack life was not enough to prevent wolf abundance from pitching and yawing. Seventeen wolves in 1994 surged to 24 in 1997. A year later, the number was pulled down to 14 wolves. Then shoved back up to 25 wolves in the following year.[9]

The trajectory of a population often drives the lives of its individuals. But sometimes, the life of an individual can drive the trajectory of a population and even entire food webs. An instance of the latter began in 1998 when a young wolf began his tenure as alpha male in Middle Pack. He didn't stand out right away because four of the six alphas were new that year.

Under new management, Middle Pack grew in number and demanded access to more moose. By 1999, Middle Pack squeezed West Pack territory tight up against the island's southwest shore. Middle Pack killed some West Pack wolves, starved others, and left no opportunity for the survivors to recover losses by raising new pups. West Pack was down to two wolves and then pressed into oblivion by 2000. Within a year of being guided by new leadership, Middle Pack grew from 4 to 10 wolves. A year later, the pack had grown to 12, by which time they controlled about three-quarters of Isle Royale.

With the obliteration of West Pack, the Middle Pack alpha turned his attention on competition residing to the northeast. Broad patterns in life can sometimes be traced to a single, specific event. February 16, 2000, is one of those dates for the wolves of Isle Royale. It was the kind of day that inspires questions like, what if Caesar hadn't crossed the Rubicon?

Middle Pack found a pair of wolves—dispersers from East Pack—engaged in the unauthorized killing and scavenging of moose in Middle Pack territory. Middle Pack caught and killed the male of that pair. Two days later, Middle Pack found the other wolf not far from Chippewa Harbor. A six-mile chase ensued. The next bits are from Rolf's field notes.

> We caught up to [Middle Pack] at water's edge just before noon, vigorously shaking themselves dry, while twenty feet out in Lake Superior, standing on a submerged rock in ten inches of water, stood a bedraggled wolf, cowering, its hindquarters almost underwater.
>
> . . . After several minutes of rolling in the snow to dry off, Middle Pack wolves either lay in the snow, watching the victim in the lake, or strutted stiffly back and forth along the shore in front of the hapless wolf. Suddenly, in quick succession, three wolves jumped into shallow water and leaped for the rock where their quarry stood quivering. Confronted by this snarling trio, it fought for its life, snapping furiously toward the lunging pack members. The lone wolf was forced backward into neck-deep water, but it retained its footing and held the attackers at bay; they retreated to shore to shake and roll again in the snow.
>
> . . . The pack led a series of attacks on the wolf on the rock, a dozen in all. The desperate defense of the lone wolf was effective, but the ordeal, including standing neck-deep in ice water, took its toll. . . . The loner adopted a new strategy—when attacked on its rock, it retreated into the lake and swam along the shore about thirty feet to another submerged rock, buying a few moments before the pack members jumped back to shore and ran down to the new location, where they renewed their attack.

This otherwise straightforward act of brutality took on a complex hue when "instead of pressing the attack," the alpha male of Middle

Pack "slowly wagged his raised tail . . . circled around to the wolf's side and regarded its hindquarters." The victim was a female in heat. The alpha male's arousal prompted the alpha female of Middle Pack to refocus the pack's attention on its original intent. The victim was back in the lake, attacked by all 11 wolves of Middle Pack. That bout lasted less than a minute. The pack retreated and the lone female rose to her feet.

> After forty-five minutes, the lone female, once again facing a violent attack, swam out into Lake Superior, heading for a rocky point about fifty yards down the shore. . . . I thought she might be able to reach shore and get away with an all-out run. But she could only crawl out of the water and stiffly walk a few steps before starting to shake her drenched fur.
>
> Within seconds, the Middle Pack arrived and knocked her over. The whole pack crowded around and bit her, shaking their heads as they held on. When the pack pulled back briefly, the female snapped at the wolves that surrounded her. The last wolf pressing the attack held the female's throat for several seconds, then left her lying motionless on the shoreline rocks. The Middle Pack retreated toward the forest edge, huddled excitedly, and then lay down in the bright afternoon sun, keeping an occasional eye on their victim. We thought the female was dead, but she raised her head briefly, which brought the pack running. They crowded around, vigorously biting her legs and throat. . . . At one o'clock, the female appeared to be dead at last.

Two and a half hours later, Middle Pack slipped away into the forest. Except for the trailing 12th member of Middle Pack, who had not participated in the killing. This wolf approached the carcass, full of an unease betrayed by a tail tucked deep between the legs. The nervous wolf reached in gently, muzzle first, and the dead wolf raised her head! The nervous wolf kept a respectful distance, but revealed himself to be a suitor as his muzzle moved toward her hindquarters. "His attention seemed to breathe life into her." By six o'clock, almost an hour after his arrival, they had both slipped into the forest.

Five days later, [we found] the wounded female and her suitor standing on a rocky knoll, about a half mile from the site of the attack. . . . [He] licked her wounded neck for several minutes. Judging from the number of icy beds nearby, the pair had occupied the site for a day or two.

. . . He had attended her neck wound and may have saved her life, but now he had become a pest, displaying an array of courtship behavior. The female had no evident interest. She whirled and snapped at his unwelcome advances. He prevented her from resting. Standing tall and wagging his tail nearly vertically in the air, the male moved constantly round and round the female, and she reluctantly moved in circles. . . .

Two hours later . . . the male was still making a nuisance of himself and the female continued to snap at him. . . . [That] was our last flight of the winter season.

A year later, a new pack had formed. They took their territory to be the land surrounding Chippewa Harbor. Three years later, we got a blood sample from the alpha female of Chippewa Harbor Pack. The DNA was a match to the blood left in the snow of the site where the female had been so brutally attacked and then cared for. The alpha female came to be known as Cinderella.

Middle Pack's failure to kill that wolf back in February 2000 may have been the only strategic error the alpha male of Middle Pack ever made in his plot to consolidate power. However, he could claim some of the success Chippewa Harbor Pack would enjoy over the years because their alpha male was his son.

In many populations, wolves often turn lighter gray, almost white, as they age. But we had never seen that on Isle Royale, not until the alpha male of Middle Pack grew into his later years. By that time, he had been leading Middle Pack for about six years, which is longer than most wolves are allowed to live. His large physique was intimidating. His brutish behavior was unquestioned. His reputation for killing moose and defending territory was storied. All this was known to us and the wolves. During the last couple of years of his life, we came to know him as the Old Gray Guy.

Looking in from the outside, the life of an ecologist may conjure adventures with wild animals in exotic locales. There can be truth in that impression. However, if that's all I made of my time in science, I would have lost my job long ago. A more complete picture involves a significant dose of accounting, math, and theory.[10] Accordingly, my first genuine contributions to advancing the understanding of Isle Royale wolves did not involve backpacks or snowshoes.

The Isle Royale wolf population is obviously small and plainly isolated from the mainland population of wolves. It has been known for most of a century that those conditions lead to losses in genetic diversity. Furthering that knowledge, even a Lilliputian advance, is likely to require quantifying precisely how much genetic diversity had been lost and how fast. One of my first contributions was to calculate, using the latest theories, that the wolf population on Isle Royale was likely losing 13% of their genetic diversity each generation (about four years).[11] That would mean they'd lost about 80% of their diversity since colonizing Isle Royale in the late 1940s.

Those quantifications offer very little on their own. But other similarly unassuming observations followed. Over the next 15 years, little insights accreted into something of broader value.

Some ecologists are motivated, I suppose, by the hope of discovering something grand. But I think it's more common for ecologists' motivation to be the simpler pleasure of the hard-earned, creative work of observing nature with their own eyes and molding that observation into a communal frame of ideas assembled by peers and predecessors. Leastwise, I've been motivated by that pleasure.

My motivations have also evolved in a manner that is more effectively shown than explained, as demonstrated by three contributions I had made earlier in my career. All three contributions involve understanding kill rate, that is, how frequently wolves kill moose. It is the rate at which predators acquire their food and the rate at which prey die from predation.[12] Kill rate meters the flow of energy and material from the prey population into the predator population.

You may recall from Chapter 4 that kill rate is also one of the four basic balancing rates that Volterra used in his two equations to describe fluctuations in the abundances of predator and prey. Being the only rate to appear in both equations, kill rate is *the* rate that directly connects predator to its prey and vice versa. That is the theoretician's contribution to the frame of ideas on kill rate.

Empiricists desire to observe the kill rate as directly as possible. They are prepared to see kill rate vary considerably from pack-to-pack and from year to year. Empiricists yearn to document that variation. Why the yearning? They don't even know why; it's their instinct.

The intricate dynamics of kill rate consist of the forces that cause kill rate to fluctuate from year to year. The consequences of those fluctuations on the well-being of wolves and moose feed back into that intricacy like an Escher painting. Inferring those causes and consequences from what begins as rudimentary observation is the passion of an empiricist. Theoreticians wait with bated breath so that they might anticipate whether the observations align with the theory.

We estimate kill rate from field observations in a most straightforward manner. Take, for example, the 2003 winter field season. We observed that seven wolves of Middle Pack killed 10 moose. They made those kills over a 44-day period. Our real interest is the *per capita* kill rate, which is the food available to each wolf, if each got an equal share.[13] The per capita kill rate for Middle Pack during the winter of 2003 was 10 divided by 7 divided by 44, which turns out to be 0.99 kills per wolf per month. Knowing that the edible portion of an average wolf-killed moose is 230 kilograms gives 7.5 kilograms, or 16.5 pounds. of meat per wolf per day.

Unfortunately, without context, that number means nothing. Does that rate siphon off more than the moose population can stand to lose, or not enough to sustain the wolves? Can't tell. Not with just that single value.

Building context requires time. By about 2000, I had nearly 100 estimates of kill rate to work with. That's roughly one estimate for each of three packs for each year over the previous three decades.[14] It continues to be largest database of kill rate ever observed for a large predator.

Context is also provided by hypotheses developed in advance of analyzing the numbers. Hypotheses provide the most robust basis for concluding, after performing an analysis, either "yeah, that's pretty much what I expected," or "wow—I didn't see that coming."

The first hypothesis I considered was that kill rate should be greater when prey are more abundant (because they are easier to find) and lower when predators are more abundant (because packs spend more time interfering with their neighbors' affairs and less time hunting when wolves are more abundant). I also expected kill rate to be higher when there were more calves (easier to kill) and when the winter was more severe (same reason). Aspects of those hypotheses are commonsensical, but there were hypertechnical considerations that left room for doubt.[15] Moreover, most empirical evaluations of kill-rate hypotheses had been limited to laboratory experiments involving invertebrates. This was one of the first assessments to focus on wild populations of mammals as they killed and were killed over the course of decades.

Sure enough, I observed all the hypothesized patterns.[16] But that was not the interesting result. Remember the fluctuation pie from Chapter 2, where we took all the fluctuations observed over the years in a phenomenon like kill rate, baked it in a pie according to some statistical recipes, and then figured how large a share of fluctuation pie can be predicted by various causes of fluctuation? Of the fluctuations contained in those kill rates measured each year over the previous few decades, no more than 40% of the fluctuations can be explained by the causes promulgated by the hypotheses. Hold that result, we'll come back to it.

The second kill rate hypothesis that I evaluated is that higher kill rates should be associated with more growth in the wolf population. The basis for thinking so is simpler than the first hypothesis. More food should mean fewer deaths and more pups. And this is exactly the pattern I found. But again, not the interesting result. The interesting and less expected result is that kill rate predicts only about 20% of all the fluctuations in growth rate from one year to the next for the wolf population. It's not surprising that other factors cause wolf abundance to fluctuate. But it is surprising that food explains so little. What could these other influences be? Bear in mind that other forces, such as winter severity, are indirectly represented in that 20% figure insomuch as they influence wolves' acquisition of food.

Those results affected me for not conforming with what I'd been told is the overarching purpose of science, namely to predict nature's behavior as a means of learning to manipulate and get the most out of nature. If that's the purpose of science, then good science would have high levels of prediction. But I had applied sophisticated analysis to some of the world's best data, and its predictive ability was meager.

What if the understood purpose of science needs updating? Maybe a better purpose would aim to understand the extent to which nature is complicated beyond hopes for predicting her behavior, as a means of learning to tread more lightly.

A second scientific contribution that shaped my understanding of ecology rose from the countless necropsies we've conducted on wolf-killed moose. Opportunities to perform necropsies accompany the discovery of moose carcasses in the course of observing the kill rate. After discovering a wolf-killed moose from the airplane, we wait however many days it takes for the wolves to finish feeding and move on from the site, typically four to eight days. Then we head in for the necropsy.

When wolves begin feeding on a carcass, while the flesh is still warm, they tear into the abdomen and pull out the rumen, a large medicine ball of smooth muscle filled with ongoing fermentation, as the microbes do not yet know that their host is dead. After the rumen is free and clear, wolves devour the liver and kidneys. Then they stick their heads into the massive cavity of the chest and tear out the lungs. These are the richest bits and, invariably, the first to be consumed.

These choicest morsels are also typically the exclusive pleasure of the alpha wolves.

Then the large muscles are consumed, hundreds of pounds of red meat. As the meat disappears, the wolves gradually disarticulate the skeleton to access smaller shreds of muscle. The front legs essentially fall off as larger muscles are consumed. The carcass is divided, most commonly into three large portions: neck and skull, chest, and the pelvis and back legs. That degree of disarticulation is routine but effortful. To disarticulate the frozen vertebral column of a moose with a purpose-built knife is not easy. Wolves perform the task with teeth and a pair of temporalis muscles. These are the muscles you feel at your temple when you chew. For context, the temporalis muscle of an 80-pound wolf is meatier than the bicep of an 80-pound human.

The rear legs are more difficult to remove as they are held fast to the hip socket with crisscrossing muscles and ligaments. Each finer degree of disarticulation yields fewer muscle fibers or sinewy gristle. Sometimes wolves don't even make the effort to disarticulate the rear legs. Excising the tiny bits of meat in every vertebral nook requires the devoted work of incisors. Sometimes some of the smaller bones are eaten and larger bones are broken to extract the marrow. Less often the hide is consumed—every square inch.

A moose necropsy involves recording answers to a series of questions of this ilk: How many bones are left? Have the legs, skull, and pelvis been disarticulated from the vertebral column? What portion of the legs are still covered in hide? The answers are used to calculate an index of carcass utilization that runs from zero to one. Zero describes an entirely uneaten carcass. One describes a carcass that was stripped clean. I found a way to reliably convert an index score into a close approximation of the percent of the carcass that had been consumed.[17]

Hundreds of necropsies conducted over a couple of decades indicate that a pack of wolves typically consumes between 90% and 95% of the edible portions of a carcass.[18] Of the food wolves do not eat, much becomes a banquet for foxes, ravens, and more. Who can diminish the value of the joy experienced by a band of chickadees who happen onto that mountain of leftovers? They'll pick strings of meat and tallow for weeks.

To contextualize that 5% to 10% of a moose carcass that wolves "waste," know that Americans waste approximately 40% of the food that we produce. Most of us are directly responsible for some of these losses by simply tossing food in the trash bin at home.[19]

If killing a moose is so difficult, why don't wolves eat every edible portion of a carcass? There is more to the question than is first apparent. Leaving food behind is widespread and has been studied in various species of zooplankton, spiders, predaceous mites, insects, shrews, weasels, marsupials, canids, bears, and, of course, humans. Ecologists have honored the behavior by christening it with jargon: partial prey consumption.

Scientists and children alike have a capacity for being attracted to questions that evade others. Why are leaves green? Why is water wet? Why do we age as we grow older? When such a question does, on occasion, catch the fleeting attention of an adult, it is usually dismissed in a moment with the response, how else would it be? But children and scientists give these questions their due attention. Accordingly, a small cadre of behavioral ecologists ask, why is partial prey consumption so common?

In their fact-guided meditation on that question, behavioral ecologists have prepared two possible explanations. The first posits a physiological constraint. Meaning? Prepare to be underwhelmed: the first possible explanation for not eating all that one has killed is simply because one is full. One does not have the time to digest all that one has captured.

An alternative possibility to this first explanation is easier to appreciate after reflecting on creatures whose dining habits should not be described to the squeamish. The creatures in question are spiders, who are said, by means of exquisite jargon, to practice extra-oral digestion. They grab their living, breathing prey, wrap it in silk, and use two horrifying fangs to inject the prey with "saliva" that begins digesting the poor, paralyzed soul from within its own skin. After the marinating agents have done their work, the spider sucks out the thick broth. You can imagine all that sucking would get tiring. The first slurps are easy and satisfying. But the last bits are stuck in the furthest crevices of the prey's exoskeleton, so small and difficult that it's hardly worth the effort.

The second possible explanation offered for partial prey consumption is that when prey are relatively scarce, it obviously pays to eat all that you kill. However, when prey are easier to catch, it pays to leave behind the least-choice parts. It may take more effort than it is worth to chew and digest the last few bits of low-quality scraps. When prey are really easy to catch, it might make sense to eat only the best parts. Partial prey consumption may be an intricate, albeit counterintuitive, behavioral adaptation shaped by natural selection. Something that ecologists call an "optimal foraging strategy."

A critical test for distinguishing the first, more obvious, explanation from the second, seemingly fanciful, explanation is to assess whether carcass utilization is greatest when food is most difficult to come by (in other words, when kill rates are the lowest). If it is, then there is a good chance the behavior represents an optimal foraging strategy.[20] These two ideas have been thoroughly tested for only two species, both of which were spiders. One spider species seemed limited by physiological constraint and the other seemed to be exhibiting an optimal foraging strategy.

For the wolves of Isle Royale, we found carcass utilization to be greatest when kill rates were lowest; the behavior appears to be an optimal feeding strategy shaped by natural selection. Humans exhibit analogous behavior. In particular, humans tend to throw away more food when food is easier to acquire due to, for example, their affluence.[21]

When humans hate wolves, they often decry wolves' gluttonous behavior. Thoughts of that ilk fuel the tumultuous relationship between humans and wolves. If wolves are gluttonous, then aren't humans four to eight times as gluttonous?

Again, I wondered about the understood purpose of science—to predict how nature behaves as a means of learning to manipulate and get the most out of nature. How did my work on partial prey consumption fit in with that purpose? *Perhaps an underrated purpose of ecology is the opportunity to reflect on what nature teaches us about ourselves.*

A third episode along my path to becoming an academic ecologist. Recall the estimate of kill rate that we calculated together—16.5 pounds of food per wolf per day. You would be right to think, "Wow! That's a large pile of meat." Incredulity subsides in recognizing that wolves are not allowed to eat all that they can kill. Much is lost to scavengers, especially ravens. A moose carcass in the forest becomes a community center for 15 to 20 ravens. If the carcass is not frozen, and if ravens are not dissuaded from hopping around on the ground by the presence of too many wolves, then each raven can take a few pounds of meat each day. They don't eat it all, not right away. They hide much of it in trees to be eaten later. Unchecked, ravens alone could consume much of a carcass in a fortnight.

Wolves' counter response is simple and sufficiently effective. They consume the carcass as fast as they can. They do so by means of two adaptations. First, wolves have impressively large stomachs. A medium-framed wolf can consume nearly 20 pounds of meat in one sitting. To apprehend what a feat that is, bear in mind that such a wolf is less than half the weight of a human who might order the king's portion at the steakhouse.

The second adaptation is best appreciated by recognizing that most carnivores lead reclusive, solitary lives. Think of Kipling's Shere Khan. By comparison to most species of carnivore, wolves lead conspicuously social lives.

Here come the ecologists' child-like questions. Why? Why are wolves and a few other carnivore species so intensely social but most are not? Their curiosity is amped by another observation known to ecologists since the 1980s. Among group-living carnivores, the socialites living in larger groups each tend to get less food. That is, the per capita kill rate tends to decline with increasing group size. The pattern has been observed in lions, African wild dogs, and wolves, including wolves on Isle Royale. The patterns beg the question, why don't individuals who find themselves in large packs break off into smaller groups so that each individual gets more food?

The quick (and inadequate) answer is that tackling and killing large prey, like moose, requires the teamwork of a large pack. This explanation falls short because pairs of wolves (even those living on Isle Royale) have demonstrated themselves more than capable of killing moose. And those wolves tend to capture more food, on a per wolf basis, than wolves living in larger groups.[22]

A richer answer lies with scavengers. A pair of wolves requires considerable time to consume a 1,000-pound moose, with each passing day, they lose more to scavengers. Wolves living in larger packs each get more food because they lose less food to scavenging ravens. They do this by eating a moose so quickly that ravens have little time to scavenge. The details that account for these losses are mathematically complicated and depend on some reliable foraging theory. While wolves in larger packs must share food among their brothers, sisters, parents, and offspring, that sharing ends up being not so costly as losing so much food to scavengers.[23] So ravens have something to do with explaining why wolves live such intensely social lives—a trait that is otherwise rare among carnivores.

Interesting. But how in the world could that result possibly help manipulate nature to "ease man's estate," which has been taken to be a fundamental purpose of science since Francis Bacon said so 400 years ago?

It can't. Nevertheless, that wolves live in packs because of ravens is cited, by peers and aficionados alike, as often as any piece of knowledge to come from the wolves of Isle Royale. Knowledge of this kind can serve only one purpose. That is to make you go "Wow!" Wow, that's so beautifully complicated . . . Wow, look how magnificently nuanced . . . Wow, how astonishingly connected.

While held in a state of compassionate wonder about nature, it would seem awfully difficult to intentionally mistreat her. How can one do anything but care for nature while astonished by her beauty, complexity, and interconnectedness?

What if science consists of two kinds of knowledge? The first kind helps us do things in the world—helps us conserve nature, restore damage we've caused nature, and live sustainably. But the knowledge enabling us to do good can also enable the most disgraceful endeavors—to live unsustainably and treat others cruelly. Attitudes determine whether we use knowledge for right or wrong, good or bad.

The second kind of knowledge sets our attitudes to compassion for nature's living things. In these late days, this second kind of knowledge may be more important. But it's not enough to merely acknowledge its importance. It may be necessary to set as a central purpose of science the advancement of wonderment-generating knowledge.

Cinderella and her Prince Charming rose to power in 2001 and still held power in 2006, by which time wolf abundance on Isle Royale had risen to 30 wolves—more numerous than at any other point in the previous 25 years. Interpack strife intensified with hunger pangs and fear of starvation.

Cinderella's Chippewa Harbor Pack held the island's prime real estate, with more moose in their territory than either Middle Pack or East Pack. East Pack broke the Lord's 10th commandment when they invaded Chippewa Harbor Pack territory on several occasions, appropriating kills on two occasions. Here's what happened on that second occasion:

January 31, 2006, 4:55pm. After killing a calf yesterday late afternoon, Chippewa Harbor Pack (CHP) was passing time lightly. Three wolves occupied the east end of Chickenbone Lake—one sprawled out on the ice while two pups play-wrestled with enthusiasm. The alpha male and a subordinate wolf chewed bones and hide at the calf carcass in the forest a half mile to the north. The alpha female and two other subordinate wolves were bedded between the kill site and Chickenbone Lake.

This was the third moose CHP had killed in eight days, so the wolves' bellies were full. But the pack had been engaging risky behavior. Just a few days earlier, these wolves had made a kill in country occasionally patrolled by Middle Pack, their neighbors to the southwest. Now, CHP was pushing its luck on the border of East Pack territory.

East Pack (EP) had not made a kill in eight days. This winter has not been easy for EP trying to feed nine hungry mouths in territory depleted of moose. Now they were on the move. Last night they traveled north to the Minong Ridge. By 0930 they had made Robinson Bay. After sleeping away most of the midday, they continued southwest toward the territorial border with CHP. By 1710, the EP had reached McCargo Cove. They traveled deliberatively, mostly in single file. With no apparent hesitation, EP cut south and uphill into the forest at a creek that drained into the south side of McCargo Cove.

They crested a hill and crossed an open swamp—still single-filed. Their travel had all the appearances of hunting—purposefully moving through more difficult and hilly terrain covered by deep snow and dense vegetation. But I don't think they are after moose. CHP was only 600 meters away and appeared to be unaware that EP was so close.

EP crossed the swamp and quickened their pace. They passed through another thick strip of forest separating two open swamps. After punching through the forest and on to the second swamp, EP was less than 400 meters from two of the CHP wolves (including the alpha male) at the carcass.

How and when did EP know that CHP was just ahead? EP probably smelled CHP; the wind was to their advantage, though very light. And EP certainly heard the ravens that attended the carcass. In the seconds prior to mayhem, CHP wolves continued to be unaware.

From the south side of the swamp, just out of sight, EP rallied, wagged tails, and even howled. Inexplicably, CHP did not respond. EP quickly circled around, outside the swamp's perimeter, to ambush the kill site from the most strategic direction. They tore through the forest edge onto the swamp with snow flying from the paws of eight wolves barreling down on the two CHP wolves at the moose carcass. The chase lasted seconds and covered no more than 30 meters. Under a thick clump of trees, the lead wolf from EP tackled the alpha male of CHP. All of EP surrounded and mauled the prostrate wolf. In less than three minutes, another lifeless carcass lay in the bloodied snow.

Meanwhile, a half mile away, two CHP pups on Chickenbone Lake continued play-wrestling while the third slept. After roughly 10 minutes, a CHP wolf ran out of the forest and onto Chickenbone Lake, tackled the sleeping wolf and aggressively wrestled with him or her for almost a minute, trying to communicate the tragedy that has just taken place. Several more minutes later, the remaining three wolves of CHP emerged onto Chickenbone Lake. At the gathering, they showed no sign of retreating to a safer location. Did the alpha female even know what had happened to her mate?

After 15 more minutes, the wolves of CHP got up and ran, seven-abreast, westward down Chickenbone Lake. They ran for about half a mile away from the brutal spectacle they had not seen, but surely intuited. They stopped—some lay down, others looked back. Were they waiting for the alpha male?

During this entire time, EP ran back and forth, crazed with furor, between the lifeless wolf and the lifeless moose. EP spent the next 24 hours at this site. CHP, still missing their alpha male, moved that night to a much safer portion of their territory just north of Lake Richie, where they spent the next two days, mostly resting.

It is horrific, but not unusual, for wolves to kill one of their kind. In a typical year on Isle Royale, one to three wolves are likely killed by one of their own. An alpha wolf typically kills two to four wolves in his or her lifetime.

After Prince Charming was killed, Chippewa Harbor Pack spent what was left of winter avoiding the portion of their territory that abuts East Pack's territory.

A year later, in January 2007, during our first research flight of the season, we tuned the telemetry receiver to the frequency that emits from Cinderella's radio collar. My heart sank. The beeps chirped at twice the normal frequency—twice, rather than once, for every second. Technically, that means the collar had been motionless for the past six hours. In truth, it meant that she was likely dead. That same day we landed the plane on Angleworm Lake, and I hiked north to the source of the mortality signal. She was likely killed by East Pack, as we were walking on land that they claimed. The condition of the carcass suggested she died after it had gotten cold, maybe mid-December, about 10 months after her mate perished. That she was recently killed by East Pack was further suggested by East Pack having visited the kill site on several occasions throughout the winter.

Cinderella was killed somewhere between her 9th and 12th birthday. She had seen more than twice the birthdays that most wolves ever see. She led the pack she founded for just over seven years. She was an experienced mother, having raised seven litters of pups. Life took its tribute. Both upper canines were worn to soft rounded stubs from years of battling moose. At the time of her death, she had a severely infected and loose upper canine tooth. She contributed to the pack's killing of moose in those last years of her life by pure tenacity. Cinderella had lived an eventful and worthwhile life.

In spite of genuine similarities, this is not a fairy tale. Cinderella and Prince Charming were neither helpless nor innocent. They had killed East Pack's alpha male in the winter of 2001–02. They had killed his predecessor the previous winter.

I don't know, maybe questions of deserving don't apply to nonhumans. In any case, Chippewa Harbor Pack did not fail with the deaths of their founders. They carried on—to the very end.

When poets say that life in its raw form is red in tooth and claw, this is the kind of behavior to which they refer. Events such as these inspire political theorists to say that uncivilized life has no arts, no letters—only continual fear of violent death, because that life is nasty, brutish, and short.

Cinderella was not the only fatality that year. The Old Gray Guy was sighted for the last time—just a glimpse on a park trail in September 2006—and was never seen again. Altogether, 12 wolves died between the winter field seasons of 2006 and 2007. This is almost twice the number of deaths that occur under average circumstances.[24] Only four pups survived to see their first winter, including two of Cinderella's pups. Under average circumstances, we would have counted about eight pups. The population tumbled from 30 wolves to 21 between the winters of 2006 and 2007.

In the spring of 2007, we live-captured and radio-collared six wolves. Blood tests indicated that two of the six had been exposed to canine parvovirus. Parvo had faded beyond the wolves' memory—not seen for the past four generations of wolves, or 17 years as we observe time. It took us a few years to realize that this viral shock coincided with the population's final unraveling. Two years later we live-captured another five wolves. Two tested positive for past exposure to canine parvovirus. That was the last time we live-captured wolves in this population.

We did not understand the longer-term trends into which we were temporally embedded. We had reason to believe that erratic kill rates were involved with the rocky rise in wolf abundance around the turn of the century and their more recent decline. We also began to wonder, with considerable circumspection, whether important themes in the population's trajectory were entwined with some hot summers and increasing tick burdens of the early 2000s. We assembled a hypothesis where the stress of heat and ticks had weakened moose, giving wolves the upper hand. The wolves took advantage. Predation rates increased. Moose declined. Too much of a good thing and the wolves left themselves with too few prey, resulting in the wolves collapsing.

If weather and ticks were only a trigger and not a sustained force, then the dynamics would self-correct, as prescribed by the balance of nature. Wolf abundance would decline. Moose would increase. You know how it is supposed to play out.

Each constituent element of the hypothesis is more than reasonable: moose overheat,[25] ticks likely enjoy the heat,[26] and wolves benefit from compromised moose. The hypothesis is true in the sense that everything is connected to everything else. While this kind of pantheism is robust as a metaphysical truth, it cannot be a causal explanation for every occurrence in nature. Soon we would discover that something else lay at the root of the moose decline.

The roots of this discovery require us to go back in time a few years to when Cinderella had just risen to power. That year, 2001, we found the skeletal remains of a wolf of whom we know very little, except that he or she died between the base of a cliff and the edge of a pond a few miles northeast of Lake Desor. The wolf died sometime in the late 1990s. The anonymous wolf is memorialized as specimen #3529, and his or her bones sit in a box on a shelf lined with several dozen boxes, each containing the bones of an Isle Royale wolf. This wolf is memorable because one vertebra from the lower neck—the seventh one down from the skull—is strikingly misshapen.[27] We took images and trolled for insight by sharing them with colleagues. It took a few years, but eventually we found Jannikke Räikkönen, a vertebrate anatomist from the Stockholm Museum of Natural History. She informed us that the vertebra—C7 to those in the know—was a unilateral intrasegmental transitional vertebra, with one side of the vertebra resembling a C6 with a transverse foramen. While few would fail to notice a malformation as gross as the one that caught our attention, Jannikke wondered if we had overlooked malformities in other wolves that would be too subtle to be detected by the untrained eye but no less important.

Over the decades, we had collected the skeletal remains of about 36 different wolves. Jannikke classified each vertebra from each wolf according to whether it had any of several classifiable malformations. She found many.

One of the most common malformations was lumbosacral transitional vertebrae, or LSTV. LSTV are unsure about their purpose in life and look to be a hybrid between lumbar vertebrae and sacral vertebrae.[28] The equivocation might be harmless were it not for critical nerves passing through these bones. Those nerves can be pinched as they navigate through the confused shape. LSTV also happen to be common among inbred lines of domestic dogs. Some dogs with LSTV suffer from any of the following: pain, partial paralysis, deficits in placement of the paws while walking, deficits in voluntary movement of the tail, loss of muscle tone causing weakness of the hind limbs, flaccidity of the tail, and incontinence.

LSTV occur in one of every 100 wolves in large outbred populations. That incidence jumps 10-fold among Scandinavian wolves, who are

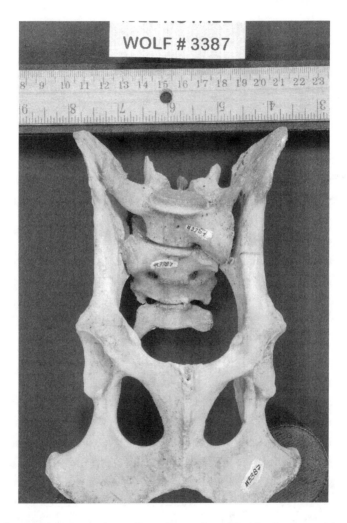

The pelvis and lower vertebrae of an Isle Royale wolf with a lumbosacral transitional vertebra (LSTV). J. J. Henderson

allowed to flirt with inbreeding.[29] On Isle Royale, where inbreeding runs without a leash, LSTV occurred at 30 times the rate observed in wolf populations with good genetic health.

Of the Isle Royale wolves that Jannikke had inspected, we knew the year of birth for most. That information allowed us to create a graph on which each wolf gets one symbol. Each symbol's position— left or right—lines up with that wolf's year of birth, which is indicated by the horizontal axis. If a wolf has normal bones, then it is represented by a zero and its symbol is a diamond, placed along the bottom of a graph. A wolf that has malformed vertebrae is represented by a one and its symbol is a circle, placed along the top of the graph.

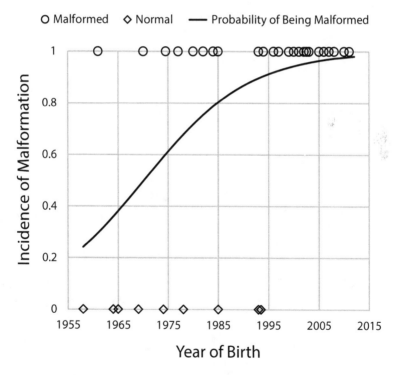

The incidence of vertebral bone malformations in Isle Royale wolves increased over time. Adapted from Räikkönen et al. 2009

The last wolf with normal vertebrae was born in 1993. We've analyzed more than 30 wolves born after 1993. They have all been malformed. To better see the pattern, we use a statistical technique known as logistic regression to show the probability any particular wolf would have a malformation, given the year of birth. That probability steadily rose over the years.[30]

The graph was offered to, and taken by, our scientific colleagues as the first indication of inbreeding depression[31] for Isle Royale wolves. The evidence for which we'd been on the lookout had been sitting all the while in shelved boxes. This discovery is an important example of a simple idea. The absence of evidence is not always reliable evidence of absence.[32]

That was the first of two breakthroughs in understanding the genetic history of wolves on Isle Royale. The second begins with appreciating that wolves have low digestive efficiency, especially during the three-day feeding binge that follows having killed a 900-pound moose. As much as 90% of what passes the throat also makes it past the bowels. While wolves miraculously transform the lifeless body of a moose into their own bodies and life energy, the greatest portion of a moose is transformed into wolf turds.

No indecency about it. Scats are a gift to ravens, delighted to gullet a little processed meat. What the ravens miss will be occupied in the spring by tenant carrion beetles.

That's not all. As a scat passes into the world, cells are sloughed off the wolf's intestines and onto the scat. Each cell contains a complete copy of the DNA that encodes everything that makes that wolf that wolf. For this DNA, we are grateful.

Scats seem to have something for every occasion. Another time I can tell you what we can do with a moose scat. But not now, I need to take us in a different direction.

After wolves kill a moose, after they feed and poop for as long as they like, after they leave the kill site, we snowshoe in and perform a necropsy. Toward the end of the necropsy, we go on scat patrol, pry-ing the rock-hard scat from the icy snow. A productive patrol will yield 30 to 60 scats. A productive season will yield 400 scats. With such a load, we are all but assured to collect at least one scat from each wolf.[33]

We began collecting scat with systematic devotion in the late 1990s. By 2007, we had accumulated several chest freezers full of fecal samples and secured enough funding to hire Dr. Jennifer Adams to analyze the DNA in a decade's worth of poop.

From each poop sample, Jen discerned a microsatellite profile (like a "DNA fingerprint") and turned those profiles into a pedigree, indicating who was related to whom. This pedigree building was not motivated or guided by any preformed hypothesis, and we risked learning only that each wolf has a parent and that most have several siblings.

But we got lucky. One day in 2010, Jen came to my office, fluttering with excitement. She said she had checked and double-checked the biochemical analyses. Jen effervesced telegraphic details of calculating and recalculating the probabilities of obtaining such a result by chance. She consulted colleagues with more experience on these particular matters. Jen told Leah before me—hoping for insight about how I tended to distinguish the incredible from the incredulous.

I interrupted Jen, "What are you talking about? What is this result that you are trying to annihilate with scrutiny?"

The tizzifying result was that a male wolf, represented in the database as M93, had a microsatellite profile indicating that he had parents—alright. They were Canadian. He had lived on Isle Royale for nearly a decade but was not from Isle Royale. He first appeared in the Isle Royale population in 1998. The previous year, this wolf apparently crossed an ice bridge connecting Canada to Isle Royale.

We have always been interested in connecting the genetic profile of as many wolves as possible to their identities in tooth and paw, especially the alphas. The connection is easy to make for the few wolves that we live-capture in the process of radio-collaring. We have blood samples from those wolves and recognize them on sight by their collars. Unfortunately, M93's profile was not represented in our meager collection of blood samples.

M93's scats were collected from the Middle Pack territory during the Old Gray Guy's tenure as alpha. M93 also appeared to be the father of most of the wolves who dropped scats in Middle Pack territory. Intriguing. But we needed assurance.

It turns out that the connection we sought was made quite a few years earlier. Upon being informed of Jen's genetic result, Rolf recalled a time when he watched, from the airplane, Middle Pack march across Siskiwit Lake. They were led by their alpha male, the Old Gray Guy. Before the pack left the lake and disappeared into the forest, the Old Gray Guy squatted and left one on a rock near shore—a signpost for other wolves, a snack for a raven, or an irresistible souvenir for Rolf. Don landed the plane. Rolf collected a portion of the scat. He labeled the bag "alpha male, Middle Pack, February 2003, Siskiwit Lake," with a black marker in the less-than-perfect penmanship you'd expect to be scribed onto a bag resting on your lap at 10°F.

Seven years later, and with that recollection, Jen dug through freezers looking for that bag. If the DNA in that sample matched the profile of M93 . . . Jen confirmed that it did. M93 was the Old Gray Guy.

We had known quite a bit about the Old Gray Guy, and Jen taught us a great deal more. He fathered 34 offspring. More, by a wide margin, than any other Isle Royale wolf. Several of his offspring rose to have prestigious lives of their own. One of his sons was Prince Charming, who cofounded Chippewa Harbor Pack in 2000. By 2003, East Pack was led by another son and daughter of the Old Gray Guy. One of the Old Gray Guy's first daughters became the alpha female of Middle Pack in 2002, when they became mates. Another son and daughter founded Paduka Pack in 2007.

I know. You're thinking, did he just rattle off three instances of incest? Yes. We'll come back to that. First, I need to firm up some context. The number of breeding wolves in a population is typically twice the number of packs, that is, one male and one female for each pack. In the years following the arrival of the Old Gray Guy, the population was composed of four packs and eight breeding wolves. Seven of those eight were either immediate offspring of the Old Gray Guy or the Old

Gray Guy himself. His DNA flooded the gene pool. All of this we learned long after it occurred. We had always known the Old Gray Guy's identity. But we were very late in learning his heritage.

In the barbarous dialect of geneticists, *ancestry* is uttered with special meaning. It is a number between zero and one, and every individual has its own ancestry number, which represents, very roughly, the portion of genes in a population that trace back to an individual who was born some years earlier. If, for example, an entire population traced back to 10 individuals, and each individual contributed equally to the gene pool, then those 10 individuals would each have an ancestry of 0.10, or 10%.

Phil Hedrick, one of the world's premier conservation geneticists, took an interest in the wolves of Isle Royale, and he calculated the ancestry of the Old Gray Guy. Within 10 years of his arrival, the Old Gray Guy's ancestry soared to 60%. In other words, by 2009, 60% of all the genes in all the wolves on Isle Royale traced back to the Old Gray Guy. That meteoric ancestry value illustrates the success of his lineage in spreading genes compared to the lineage composed of all the remaining, native Isle Royale wolves. The combined ancestry of the other few dozen wolves living on Isle Royale at about the same time was only 40%. No one had ever documented such a dramatic, dominating infusion of genes. Our ever-critical peers agreed when we called it a "genomic sweep." While the malformed bones were strong evidence of inbreeding depression, the ancestry values represented a gold standard.

Back in 1998, when the Old Gray Guy first arrived, he performed what geneticists refer to as a genetic rescue. Except genetic rescue is supposed to be an infusion of new genes followed by an increase in survival, reproduction, or abundance. Such improvements for Isle Royale wolves could be seen in the data by indulging eyes but not to the jaundiced eye of statistical inference. Even indulging eyes had to concede that the demographic response seemed lackluster.[34] The response was especially ho-hum when compared to other documented cases of genetic rescue that were few in number but dramatic, such as Florida panthers, Scandinavian wolves, and greater prairie chickens of Illinois.[35]

The equivocal response on Isle Royale made sense after we recognized that the benefits of genetic rescue began in 1998, just as the wolves were also entering a prolonged period of diminished food triggered by the severe winter of 1996. The two causes—genetic rescue and food shortage—worked in opposition to each other. If the winter of 1997 had been too warm for an ice bridge, if the Old Gray Guy had been killed by a trapper (a common fate of many mainland wolves) on his way to Isle Royale, if the Old Gray Guy hadn't been built of the tough stuff of which most alphas are made (most wolves are not), then the Isle Royale population may well have gone extinct in the late 1990s.

The only way for the Old Gray Guy to have been so successful was to limit the amount of breeding he did with wolves native to Isle Royale, who had become genetically inferior. That necessarily involved a fair bit of breeding within his own lineage. That is, inbreeding. That is, incest. The spectacular revitalization was also its own undoing. It began when the Old Gray Guy mated with his daughter, and it continued with brothers and sisters becoming alpha pairs and mates. The cost of inbreeding accumulated within the Old Gray Guy's lineage until it outweighed the benefits of the genetic rescue. That scale tipped in 2008. At which time, the wolves began what I expected to be their final plummet to extinction.

The Old Gray Guy's legacy lasted about a decade and a half, which is about three wolf generations. How had the population survived all those previous decades in isolation? We reasoned, with considerable help from Phil Hedrick, through it this way: if the population had been completely isolated prior to the Old Gray Guy, then the population should have lost about 80% of its genetic diversity. In truth, they had lost only a third that much. Well-established mathematical equations that describe these processes allow for the discrepancy to be explained only by three possibilities. The first two possibilities are a tad technical. First, the wolves could be minimizing the rate of inbreeding by selecting mates that are as unrelated as possible. That possibility is eliminated by rigorous inspection of the pedigree.[36] Second, the wolves with the least genetic diversity could have died younger than wolves with more diversity. That possibility was also dismissed by patterns of mortality in the population.

The third possibility is easy to understand. Genetic diversity can be restored by cryptic immigrants coming to Isle Royale without our knowing. The equations say that the genetic diversity we observed could be explained if something like two immigrants arrived every three generations. That nonnegotiable math demanded reconsideration of observations from years gone by.

There were those melanistic (black) wolves who came to Isle Royale in the late 1960s.[37] No melanistic pups were observed in the litters during those years, so it had been presumed that those immigrants never contributed to the gene pool. But only recently was it discovered that melanistic wolves do not always give birth to melanistic pups.[38] Hindsight increases the likelihood, to at least plausible, that those wolves did infuse fresh genes by reproducing with Isle Royale wolves. They did, after all, show every other (behavioral) sign of having done so. More than a decade later, a melanistic pup romped with litter mates in East Pack. Too much time had passed for that pup to be associated with the melanistic wolves of the 1960s. Retrospection hints that this pup was born to an immigrant wolf who was not melanistic, but carried genes for a melanistic coat.

Both of those cases involved melanistic wolves, which represent no more than 3% or 4% of wolves in the Great Lakes region. For the most part, gray wolves all look alike (to us), even the Old Gray Guy, whose immigrant status we detected only through sophisticated DNA sleuthing. If a gray wolf arrived to Isle Royale before the late 1990s, we'd be most unlikely to ever know about it. Cryptic immigration likely explains the genetic patterns we'd observed and the subsequent persistence of wolves on Isle Royale.[39]

Each immigrant arrives on ice that bridges the north shore of Isle Royale to the mainland. That channel of water is three times the area of Isle Royale and large enough to accommodate most of Rhode Island.

The bridge forms as large ice flows coalesce. Each flow jams into the next. Some are more than a mile across and weigh millions of pounds. They are fastened tight by pressure ridges, jumbled piles of ice chunks that form as the flows slowly crush into each other. The chunks are as large as automobiles and the pile can be several meters tall and many more meters across.

What a spell must come over the mind of a wolf who decides to cross. Every day she stands at the edge of the universe, looking past the shoreline of Lake Superior. Then one morning it is solid enough to walk on. Ice and ice and ice. The view from atop a pressure ridge is ice to the horizon. I don't even know if they know that the wrinkle in

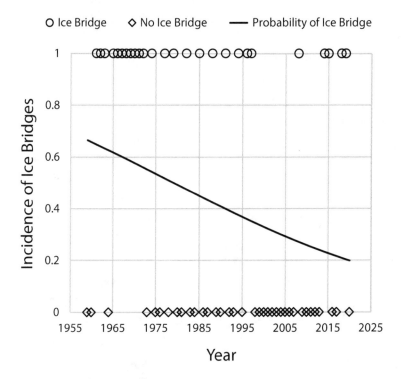

The frequency of ice bridges connecting Isle Royale to the mainland has decreased over time

the horizon is an island. I think they are just pulled by reckless enthusiasm for the uncharted.

During the 1960s, an ice bridge formed in about three of every four winters. We warmed our climate, and the frequency of ice bridges fell. By 2016, we expect an ice bridge just once every 8 or 10 years. By some forecasts, Lake Superior is expected to be largely ice-free by 2040.

For a few decades, and as late as 2008, we believed the wolf population was severely inbred due to its being completely isolated. We further believed the population had not been impaired by inbreeding. We could not have been more wrong on both counts.

The value of observing nature in one place for a long time might seem to be the gradual accretion of knowledge—painting one thin layer of new knowledge over the old, over and over, until it develops a shimmering patina of deep insight. Those have not been the lessons offered by Isle Royale. Rather, we observe intently, study hard, and consider all manner of explanations, saving only the unassailable ones. Later, the explanation that we had settled on with the aid of scrutinizing peers is upended.

Had we stopped watching at any point along the way, we would have been quite happy to carry thoughts that were quite wrong. The danger in paying attention for only a short while is developing a strong sense for an erroneous idea. It might be better to have not paid attention at all.

The patina never builds. Repeatedly, we strip away old layers and replace them with new layers.[40] The balance of nature becomes a broken balance. Connectedness and rejuvenation are mistaken for isolation and desolation. What we told ourselves was untouched by humans has fingerprints, fresh and old. What we accrete best are disproven ideas. Somehow, if not ironically, it brings us closer to some truth.

7 |

The Unraveling

The life of Frodo Baggins is inscrutable without knowing the hobbit community from which he came, or the transformation of the Fellowship he joined, or the unfolding of Middle Earth. The fate of every individual depends intimately on the course of the whole. The course of the whole is sometimes nudged by the life of an individual, but it is always composed by throngs of individuals who only seem inconsequential.

It's true for every life. The life of Alexander Hamilton, his family, and his nation melt into one. The lives of my grandparents are tiny daubs of light in grand impressions left by the Great Depression and World War II.

The lives of wild animals are similarly nested. The glimpse of a mouse skittering into a dark recess. We have no idea if it's the same mouse heard scratching in the wall yesterday or the child of the mouse that was here last year. Is the little mouse cruising down easy street or just getting by? Questions about how a creature came to be at this place and at this time rarely occur to us, unless that creature is a person.

Even when we are interested in such questions, wild animals are discreet. So, it's difficult to know how the small motifs of individual creatures flow into the larger movements that are the dynamics of an animal population or how those larger movements alter the motifs. Furthermore, the life of a population unfolds over generations, but the life of an individual unfolds day to day, season to season, and year to year. Adding to that challenge, balance-tipping events occur right beside the prosaic. We don't sort the two reliably in our own familiar lives, let alone among those whose lives are so different.

One of the great privileges of my life has been the opportunity to watch wolves in this way, bilaterally focused on individuals and wholes. Bilateral focus is also the best way to convey how the wolf population of Isle Royale unraveled, beginning at the point where we left them in Chapter 6. As with any multilayered story, a little context helps to understand forthcoming detail.

In January and February of 2009, Middle Pack's nine wolves governed most of Isle Royale. Paduka Pack had just begun to coalesce as a pack. They survived on the mercy of Middle Pack, whose territory enveloped Paduka Pack.

The eight wolves of Chippewa Harbor Pack were squeezed between Middle Pack and East Pack. The latter had been fading for a few years and disintegrated completely during the summer of 2009, shortly after the alpha female of East Pack died while giving birth.[1]

The wolf population declined from 30 to 19 between 2006 and 2010. In 2010, per capita kill rates were only 60% of the typical rate, and canine parvovirus seemed to linger for a second year. More than a third of the wolves died in 2010. That same year, the number of packs was cut from four to two. The number of reproductive opportunities declined concordantly. Vacant territories created a power vacuum.

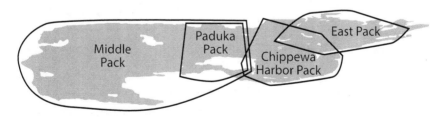

Territorial boundaries of wolf packs in Isle Royale during the winter of 2009

As the 2010 winter field season began, we became aware of a wolf whose life began to resemble the son of Lord Montague. This Romeo spent the winter dispersing from Chippewa Harbor Pack amidst the power vacuum and social confusion that accompanied the population decline. That he wore a radio collar allowed us an opportunity to imagine his experience with some vividness. What follows is scaffolding for that imagination—some notes from the field, from one day to the next and one week to the next, as they were recorded during the winter of 2010.

If you are less interested in the full breadth—which, like a Tolkien novel, requires you to keep track of which pack is where and doing what to whom—then you can read the following journal entries but just skim the passages of gray text without losing the thread of Romeo's life. He provides ample opportunity for exercising this merging of individual and collective. Nevertheless, the gray passages are valuable for providing context that is the inseparability between individuals and the collectives to which they belong. As you learned in Chapter 1, the telling of their lives is mixed with mine in the effort to follow how their lives unfolded.

January 16, 2010. We prepared the plane for work—outfitted the struts with telemetry antennae, checked and installed survival gear. We organized our field gear, photography equipment, and necropsy tools (an ax and plastic bags). We worked leisurely and let time pass lightly because clouds and wind would not have allowed us to fly, even if we had been ready.

January 17. First flight [of the season]! What a feeling to glide all morning above the trees and lakes. So much to absorb—expansive dark clouds rising over Lake Superior; the mottled colors of a landscape covered in alternating bands of conifers, deciduous forests, and lakes; and all the varied moods of Lake Superior.

We found three packs. Middle Pack wolves were traveling easily from Siskiwit Lake. Chippewa Harbor wolves traveled farther, from Moskey Basin to Harvey Lake. Paduka Pack slept much of the day on Beaver Lake.

From a site near McCargo Cove, we collected some remains of an old cow moose—the prescribed bones and 63 wolf scats. The moose was likely killed by Paduka Pack or Chippewa Harbor pack.

We saw tracks on Tobin Harbor that might have belonged to a pair of wolves—perhaps the core of a reconstituted East Pack? Too soon to say.

January 18. It was good that we woke before dark and took flight at first light. What began as thick, heavy clouds became snow. Within two hours we were back at the bunkhouse drinking cocoa.

In that short flight, we saw Middle Pack wolves, each curled up and asleep by the shoreline of Siskiwit Lake in nearly the same place we had seen them yesterday.

We also saw nine (!) wolves sleeping near the remains of a dead moose just NW of Harvey Lake. Yesterday, we saw wolves sleeping on Beaver Lake—thinking they were Paduka Pack—and wolves traveling SW toward Harvey Lake—thinking they were Chippewa Harbor Pack wolves. We did think at the time that these two groups were a little close to each other, given they were competing packs, and we thought Chippewa Harbor Pack was a little deep into what had been Paduka Pack territory.

Now it seems that both groups had been Chippewa Harbor Pack wolves. We also now wonder if Paduka Pack even exists any more—last year they numbered only three wolves.

January 19. Yesterday we discovered an unexpected need for repairs to our plane, the *Flagship*. The US Forest Service flew us a mechanic to make temporary repairs. Then Don flew our dear *Flagship* back to the mainland. Hopefully, that won't cost too much time.

Grounded, Rolf and I felt a bit without purpose.

The frozen harbor we land on had become a sea of snow drifts that were hard as cement. These are pretty tough on the plane's landing gear. So Rolf and I prepared a runway on the ice—clear of drifts and marked with spruce branches.

January 23. For each of the past few days we found evidence of a lone wolf hanging around the Washington Harbor area. He or she has been digging through the snow to eat half-rotten and now frozen apples from this past fall.

Went skiing today—it was warm enough that, on my return trip, my ski tracks had been invaded by snow fleas.[2]

January 25. Breakfasts have been tough. There seems nothing to talk about—no observations to review from the previous day and no day's plan to rehearse. To inspire conversation, each evening I've been reading from Joshua Slocum's *Sailing Alone Around the World.* At breakfast, I report on Joshua's progress. This morning he was almost taken by the biggest wave of the trip and made the difficult passage through the Straits of Magellan—marauding "savages" and extreme winds.

Snow poured from the sky all day. An angry north wind dragged the temperature down by 30 degrees in 12 hours.

January 28. By day's end the *Flagship* had passed her test flight.

January 29. We gave Don a great big hug, tossed his travel bag on the ice, refitted the plane with our research equipment, and refueled.

Don and Rolf took off for our first glimpse of Isle Royale in well over a week. It began snowing, and the flight was short. We only learned that Middle Pack had been spending some time just a few miles north of our bunkhouse.

January 30. By 0500 I was down on the harbor and fired-up the generator to begin warming the plane's engine. At 0600 I tugged gently on the corners of sleeping bags to rouse the crew from slumber. I organized flight gear while oatmeal burbled in the pot beneath the light of Rolf's headlamp.

It's calm, clear, and below zero. Two pairs of long underwear seem right for the day. Wool sweater, fleece jacket, and now it's time to lace up the mukluks. It's useful to all get dressed at the same time, otherwise someone overheats or cools down prematurely, depending on whether you wait inside or outside for the others to get dressed.

Predawn darkness and the snowmobile's roar steals my sight and hearing. The only conscious sensations are frozen air rushing past my face and seeping down my neck, and my clenched fists and braced legs as I hang, like an Arctic charioteer, from the trailer that drags behind the snowmobile—from the bunkhouse to the plane on the harbor.

Gently latch the plane door open and raise the flaps up from their overnight resting position. Now the installation begins: flight map, telemetry receiver, and two-way radio each into the front pocket; position tracker attached from the overhead fuselage braces; hand bag with binos, pencils, and camera accessories on the far side of the seat; precisely arrange the four-point seat belt so that I'll be able to put it on as I install myself and my bulky clothes. The portion of the cockpit where I sit is only about 24 inches left and right and about 24 inches fore and aft. Everything has a place, and every place has something.

Now the rear compartment . . . Slide the large pack with survival gear and necropsy gear—careful not to catch its buckles and straps on the tangle of electrical cords hanging from their sockets on the overhead panels flanking each side of the seat; put the midflight candy bars and water bottles where they won't get buried at the bottom of the storage compartment; slide the 300 mm lens and camera right behind me, and the 70–200 mm lens/camera beside it, where I can reach them during the flight.

It's all installed. Is there any part of the ceremony that I've forgotten? Ordeal is not the right word because it is not burdensome. It is not a rote ritual because we're paying too close attention. It is more like a thoughtful ceremony. Every action is rehearsed and has a purpose. I know what I can do with my mitts on, and I know which actions require bare hands. It's done exactly the same each time, as to not forget anything. Forgetfulness can spoil chances to make observations, but at worst, it can lead to disaster.

Moments before starting the engine, unplug the red 1 kW generator that had been warming the engine since 0400; remove the orange, tattered engine cover; stuff it in the blue pillow sack. I toss a mitt on the ice—the exact same spot as last time—and kneel on it as I stuff the engine cover into the storage pod below the fuselage and lock down the pod's door.

The starter whines, the cylinders fire, the prop turns, and the *Flagship* is alive. As the rpm's warm the oil, we stand on the ice evaluating the cloud forms and movement and discuss possible priorities for the flight. Don kills the engine, we get in. Right foot on the step, right hand on the overhead brace, knock the snow from the left foot and thread it through the jungle of telemetry cables and heater hoses on the floor boards. Once situated, power up the location tracker, connect the telemetry receiver to its power source, the intercom system, and the co-ax cables leading to the wing-mounted antennae.

It is all done, every step, with deliberateness and awareness—nothing can be forgotten. By 0830, we're sailing over the island, looking for wolves.

I flew with Don in the morning, and Rolf flew with Don in the afternoon. Here is a passage from Rolf's notes from the day:

In the morning, John had seen the Chippewa Harbor Pack with six wolves lounging near an old kill near Lake Richie. By early afternoon they were on the move, heading west overland, walking easily on top of the crusted snow. An hour later, we were circling the signals of two collared wolves, trying to spot them in thick cedar cover. Finally, six wolves came running out of the lowland, crossed a ridge, and plunged into another thick swamp, running full tilt.

We guessed they were chasing a moose, but when we next found them a half mile away, still running at top speed, we considered a wolf was the likely quarry. . . . We broadened the circles and found the wolf who was being chased.

As the gap closed, the chase was led by a year-old collared male wolf. Finding no other wolves in his company, he gave up the chase and turned back to the main group. They gathered in a rally centered on the whitish alpha male.

Meanwhile, the wolf being chased did not slow down, but ran out onto the rough ice of Todd Harbor, then turned and ran rapidly SW along the shore for another mile before slowing down and frequently checking its back-trail. The chase had occurred in a part of the island that has been claimed by the Paduka Pack during the past two winters, but also an area where Chippewa Harbor Pack regularly makes forays.

Back at the bunkhouse, we downloaded the photos and found that the wolf being chased was a large wolf, probably a male, getting gray around the head from aging. This wolf was noticeably grayer than the Paduka male last year, but overall the two wolves shared many similar features. . . . The wolf running for its life could easily be the Paduka male, now alone, but there is no definitive proof just yet.

Rolf and Don landed at sunset. I met them on the harbor to help put the *Flagship* to bed, and we returned to the bunkhouse as it grew dark. That's the kind of day we like.

January 31. Hungry for time in the air, we woke early. Gusty north winds had a different idea. We waited and waited, until we couldn't. By 1500, enthusiasm caused us to imagine the wind was waning. It wasn't.

Don arm-wrestled the control stick for much of the flight, matching every push and every shove as the wind bullied the *Flagship*. After several decades of flying, Don's motions are instinctive; they have to be. But instinct doesn't make the work any less tiresome. The most basic maneuver, circling, is a choreographed production. Don pushes the throttle as the *Flagship* turns downwind, and we race downwind, too fast to see anything. Easing back, the plane leans into the wind and the ground speed lessens considerably. Each circle is planned so that what we wish to observe is beneath the plane on the upwind portion of the circle. In the back, I occasionally get lifted off my seat. To write even a single letter is a chore. I memorize much of what I see.

We pinpointed the locations from where the wolves' radio collars were emitting their signals. Between wind and thick swamps, we never saw Middle Pack or Chippewa Harbor Pack; we only heard the beeps of their radio collars.

Despite bumps in the air and thick spruce on the ground, we saw (just west of Mount Franklin) a lone male wolf challenge a cow for her otherwise defenseless calf. We caught glimpses of the encounter just once for every six or seven of the circles we made. This wolf had been living with the pack to which he was born, Chippewa Harbor Pack. I expect this wolf may have an interesting tale to tell. Because he wears a radio collar, we'll be able to hear his story. We'll call him Romeo.

But this winter we haven't seen Romeo with the pack. He's been trying to make it on his own, and in doing so he faces two great challenges: killing moose by himself and finding a mate. And mating season is fast approaching; it commences shortly after our Valentine's Day. Romeo has been spending most of his time in territory that East Pack once held.

His continued presence in that territory reinforces our thinking that East Pack did not survive the summer.

February 1. Middle Pack spent most of the day milling about an area near Grace Creek overlook. They gnawed on the hide of a calf—it was all that remained of the calf that they killed sometime last week. Chippewa Harbor Pack spent the day hidden in a thick stand of spruce, about 200 meters from a site NE of Hatchet Lake, where they had been feeding on a dead moose.

Very little remained of either moose, and today was the first that we knew of either of these kills. They were probably made last week when we hadn't been able to fly.

The cow and calf survived—likely suffering no more than an awful memory. By morning Romeo ended up nine miles from where he began the night. He laid down where the Greenstone Ridge arches her back, black basalt rising almost vertically from the swamps that extend SW from Chickenbone Lake.

Today Romeo slept with an empty stomach, and perhaps with anxiousness for finding a mate. About 40 meters from Romeo, another smaller (probably female) wolf had also curled up to sleep on the same edge where swamp and rock meet. Is one of these wolves unsure about the prospects? Is the smaller wolf no prospect at all—perhaps a sister, visiting her brother for the day, and soon to return to Chippewa Harbor Pack?

February 2. Middle Pack left the scattered remains of their last meal for a swamp up the Washington Creek, about three miles NE of basecamp. Here they stood watch over an adult moose they had wounded during the night. The wolves engaged the moose just enough to keep it from feeding or lying down. This went on all day—the moose unable to escape, and the wolves unable to kill without risking mortal injury.

Most of Chippewa Harbor Pack remained hidden in the thick spruce near the moose carcass they'd been at the previous day. We began to wonder if the pack wasn't feeding from this carcass so much as guarding it. Three days ago, they chased a wolf from this general area out onto Pickett Bay. Perhaps that wolf—

who might be the alpha male of Paduka Pack—made the kill and Chippewa Harbor Pack chased it off the carcass.

For the second time in as many days, Romeo attacked another a cow and a calf. This time near Brady cove, and this time both moose left blood in the snow.

February 3. Chippewa Harbor Pack traveled across their territory—another day without a new meal. Middle Pack spent another day before the moose they'd wounded. But the moose was lying down. When a moose knows it's going to die, it doesn't lie down. If the moose was unable to stand, the wolves would have killed it. Did the moose know he or she wasn't at risk, despite seven lingering wolves? From photographs, we learned one of the wolves may have been injured by the wounded moose. By nightfall, the wolves left the moose.

Romeo left the cow and calf he had wounded the night before and roamed the drainages north of the Greenstone.

We also noticed the tracks of a couple of wolves near Hatchet Lake and wondered if they could belong to Paduka Pack.

February 4. Throughout the night and into the morning, most of Middle Pack traveled south across the Grace Creek drainage, over Red Oak Ridge, and on through the swamps south of Feldtmann Lake. Unable to kill the moose they'd wounded five days ago, they searched for an easier prospect. One of the Middle Pack wolves, perhaps thinking little of such a prospect, remained with the moose into which they'd already invested much. He or she ate blood-soaked snow.

Without food to unite their interest, or deep snows to restrict their walking, the wolves of Chippewa Harbor Pack had various ideas for how to spend the day. The alpha pair and one subordinate traveled across the barren expanse of ice that Lake Siskiwit becomes each winter. They announced their day's effort with scent marks and ended up at Lake Richie. The two collared wolves of the pack traveled together from near Lake Harvey to Chickenbone Lake, another wolf sauntered across Livermore Lake, and two others stayed at Harvey Lake.

While the three Chippewa Harbor Pack wolves had been crossing Siskiwit Lake, Romeo stepped onto the same barren field of ice about a half mile to the SE. He was walking toward a fifth, smaller (presumably female), wolf who had also just stepped onto the frozen lake about a mile to the SW. Intently focused on each other, both were unaware of the Chippewa Harbor Pack wolves patrolling their territory just to the NW, despite the close proximity. And the Chippewa Harbor Pack wolves hadn't noticed the tryst. The pair continued to approach each other with an interest that seemed to alternate between playfulness and restraint.

While a human could easily see a wolf standing on ice from even a mile away, wolves often pass other wolves at such distances without seeing one another. While the eyes of a wolf do well in low light, they have relatively poor distance vision. But scent is an entirely different matter. Wolves know a world of smells that is entirely beyond our imagination.

A north wind betrayed the Chippewa Harbor Pack wolves. In an instant, Romeo turned his body and all his olfactory attention toward the Chippewa Harbor Pack wolves. A moment later, he ran from the ice and into the forest. Without ever meeting on the ice, Romeo's friend did the same. The Chippewa Harbor Pack wolves continued on, never realizing the frustrated encounter.

After leaving one final scent mark at the NE end of Lake Siskiwit, Chippewa Harbor Pack continued off the lake, over a portage used by humans in the summer, and onto the next lake. Half an hour later the scene was safe. Romeo and his friend returned to the ice, laid down about 30 meters from each other and slept a short while. Then they disappeared into the forest—we don't know if they left together or separately.

February 5. East winds don't hit the shores of Isle Royale until they've passed over 100 miles of Lake Superior. When they call out to the lake, it vaporizes. Thick, churning clouds form and fly just 1,000 feet or so above the lake, and then pour over the island. Beautiful, but not so good for flying.

Middle Pack returned only briefly to check on their wounded moose and continued NE on toward Lake Desor. The two collared wolves of Chippewa Harbor Pack were just north of Lake Whittlesey. The clouds were gradually lowering toward the ground. Without learning any more about the wolves, we began flying home.

In the afternoon, we helped Don perform routine maintenance on the *Flagship* until the feeling had left our fingertips.

February 6. Two days ago Middle Pack left their wounded moose, cruised the forests for 18 miles, and within 24 hours they'd returned to that poor moose. The wolves returned only to pass by, as though they were merely checking a casserole dish in the oven. Yesterday and today, Middle Pack passed time with another 18 miles, this time to the NE, and then back again to the poor moose. On this cruise, Middle Pack showed us a moose they had killed a few weeks ago—one that we had not previously known about.

After being dissociated for a couple of days, nine wolves from Chippewa Harbor Pack reunited near Sargent Lake. Romeo had been nearby with another wolf, but he left by himself when all but Romeo regrouped under the leadership of CHP's alpha pair.

Within 45 minutes of flying this morning, we had seen all the pack wolves that we knew of. Also, the sun was occasionally shining, and it had snowed about an inch last night. These are conditions to scour the island for tracks of lone wolves we haven't yet discovered. We searched lakes, beaver ponds, and far reaches we don't visit as often.

The effort turned up tracks of lone wolves at three sites (Siskiwit Lake, Hatchet Lake, and Harvey Lake). Then it turned cloudy, making it impossible to detect the faint traces of lone wolf tracks laid in just an inch of snow.[3]

February 7. Light snow began falling at 0500. We slept in a bit. Snow fell through the day. What a relief. We'd flown each of the past 10 days, and now a chance to rest. Moreover, it has only snowed once in the past four or five weeks. The island is a tangled mess of tracks—wolves, moose, fox, and otter tracks all mixed together. It's all very confusing to sort out. A fresh snow would wipe away all the old tracks like an eraser clears a chalkboard.

We inspected photographs from the past few days of flying, and realized the alpha female of Middle Pack has survived to see her 12th winter. She is distinguished by her whitish coat and a nonfunctioning radio collar. She is one of the oldest wolves to ever live on Isle Royale. Her posture makes us wonder if she'd been injured by the moose Middle Pack attacked eight days ago, or perhaps arthritic vertebrae have stiffened her spine. Injury would explain why they haven't killed that moose—why they wait so patiently for her to die. Big changes are in store for Middle Pack if this old lady's best moose-killing days have passed.

February 8. Snow. For the second day, we slept in . . . a few chores around the compound. We finished the day with a sauna.

February 9. The wind blew too hard to fly the *Flagship*. However, the exchange flight, with Leah on board, arrived from Ely. She'll spend most of the next couple of weeks processing the samples that we've been collecting—balsam fir, moose pellets and urine, and wolf scats. Hundreds of samples in all. We collect them, but she leads the effort to transmute these samples into data, like an alchemist working lead into gold.

February 10. Trees rocked, loose snow formed miniature snow tornadoes on the harbor, and the *Flagship* tugged at the ropes that held her to the ice. The wind blew all day. I worked on a scientific manuscript, Rolf reviewed field notes, and Don counted his toes (practicing for the moose census).

The snow is only about a foot deep but rock-hard from the rain we had several weeks ago. In most places, wolves and people walk right on top, but not moose. Moose bleed from dragging their shins, day after day, through those crusts.

February 11. Hundreds of thousands of years ago, Gaia's skin groaned and left a series of north-south fault lines across the island. For several thousand years, these gashes have been filled with swamps. Yesterday, Middle Pack killed and ate a calf at the edge of a swamp lying in one of these earth fractures. While they ate, the moose they'd wounded 12 days ago finally died at the edge of another swamp. After having gone without food for almost two weeks, each wolf in Middle Pack will soon have consumed about 80 pounds of meat in a four-day period—if they can eat it before the ravens do.

Don dropped me at Hatchet Lake. I hiked up the creek to conduct a necropsy. The bones were scattered throughout a cedar swamp, where fallen trees and upturned roots formed a confused tangle. I climbed and crawled to find the few bones that I could.

Although Chippewa Harbor Pack had defended this kill site, we wonder if they stole it from Paduka Pack—if they even exist. Because that site was quite old, I found only five wolf scats. Perhaps the DNA in these scats will tell us whether Paduka wolves had been here.

Chippewa Harbor Pack toured their territory during the past couple of days. They visited an old kill site, killed a moose in a small stand of conifers at the base of Minong Ridge, and wounded another moose south of Angleworm Lake. As they fed, Romeo played the role of prodigal son, managed to ingratiate himself to the pack he's otherwise abandoned, and fed from the same carcass that fed his parents and siblings.

February 12. The air was still and clouds obscured the sun—perfect conditions for counting moose.

To cover each moose survey plot thoroughly, we typically fly nine overlapping circles that allow us to see most of the ground in the plot from more than one angle. We've demarcated 91 plots; each is one square kilometer in size. Some are perfect rectangles, others have irregular boundaries defined by creeks, shorelines, or ridge tops. Together the 91 plots represent about 20% of the island's area. If the weather is favorable, it takes us about three weeks to count the 91 plots. Regardless of the weather, the moose counting takes about 900 circles of the plane.

I know which way is north for about the first three circles. Because the plane is banked on a 40° angle for most of the survey, I experience only the ground right below the plane and only out as far as the wing tip allows. A continuous strip of trees, snow, and rock passes beneath the plane's strut like a conveyor belt. The engine fades into a deafening white noise. The proprioception is not flying but floating with a nearly imperceptible centripetal tug on your trunk. The only conscious stimulus is visual.

A casual glance down below will not suffice. For the entire survey, your eyes scan continuously close and far, forward and backward, scanning quickly past open areas where moose would be easy to spot and dwelling on thick patches of conifer trees. Many moose are observable beneath a thick tangle of spruces or cedars only for a second—not a metaphorical second, but a literal second—as the plane passes over the tangle from just the right angle.

If a moose can be seen, we must see it. The accuracy of the estimated population size depends on it. In past years, we've tested ourselves with radio-collared moose; depending on how thick the habitat is, we have about a 75% chance of seeing each moose. That sightability factor is taken into account when we estimate abundance from these counts. But the correction factor is accurate only if you stay hungry to see the next moose, as if it were the last moose on the planet.

Mesmerized by the roar of the engine and the trees slipping past your field of view, your eyes are vulnerable to glazing over. The only way is to be greedy and gluttonous for the sight of another moose. You can fly for what seems to be quite some time without seeing a moose. Without the reward of seeing a moose, the mind is tempted to wander through the corridors of your memory thinking of the most irrelevant things. Every neuron and every synapse needs to be tasked with looking for moose. Counting moose is a mentally exhausting kind of meditation.

Today we spent about three and a half hours counting moose on 15 survey plots. After all 91 plots are counted, we subject the numbers to a random stratified sampling algorithm based on the negative binominal distribution, the sightability correction factor, and the Akaike information criterion—it's a fancy way of extrapolating from the moose we count to an estimate of the total moose abundance.

February 13. Flew for half the day, until the snow started to fly. We still managed to pick up a moose kill near Harvey Lake, and see that Middle Pack had stayed at their kill site, Chippewa Harbor Pack killed a moose near Lane Cove, and Romeo slept at the site of Chippewa Harbor Pack's last kill.

February 14. Six inches of warm, heavy snow. That'll change a few things. It is liable to send the wolves to the shorelines where the traveling is easier. It will give a break to the moose living away from the shorelines. The snow will also make it easier for us to detect the tracks of wolves we don't know about, especially in the area where Paduka Pack used to live.

Spent much of the day writing scientific manuscripts. Then some time at the bone pot.[4] Even when it's cold, it's best to dress lightly. Fishing around in the hot water warms the hands. Steam, pouring from the pot, warms the chest. The bones we collect still have quite a bit of flesh on them, which needs to be removed before the bones can be studied and archived.

At the bone pot, every type of bone has its own recipe. Drill small holes in the hooves—it's the only way to attach an ID tag. The mandibles get warmed just a bit with the incisors sticking out of the water. The idea is to only thaw the frozen ligaments that hold each tooth to its socket. Otherwise they're impossible to pull. If the incisors are left in the pot too long, the rings fade. Then it's tough to see the annual cementum rings that correspond to the moose's age. After the incisors are pulled, the mandibles and other bones can be hard-boiled. Skulls take the longest, metatarsals the least time.

But don't over boil a calf skull—it'll be reduced to a pile of soft fragments. And drill two larger holes in the mandible and metatarsals to let the marrow drain out. Otherwise the marrow diffuses through the bone, leaving it a greasy mess to handle.

I hope you understand that these are only principles and this is just an overview of the knowledge you would receive from some time at the bone pot. It is fetid and foul, but it is also a cauldron of knowledge.

After a couple of hours of boiling, it's time to pull the skulls for a little detail work. The posterior portion of the skull is a wild landscape of nooks and crannies, to which the tissue holds tight. Care is required to always know where that knife blade is and where it'll go should it slip—this is not the time or place to cut yourself. This is also a good time to work the inside of the cranium with a knife—otherwise the brain won't come out. Then back in the pot for a little finishing work.

Pull a bone out of the pot with a missing tag—stop everything. Figure out which moose it is and retag it. The small bones get put in a strainer and then into the pot. Otherwise it's tough to fish out a small bone amidst all the other gunk that accumulates at the bottom of the pot.

After boiling, the bones are inventoried. Then they wait until summer when they'll spend a few months in the open air and sun. By fall, they'll be ready to sit on a museum shelf waiting for their stories to be told.

February 15. Wind and snow. Spent some of the day writing, and then some time at the boiling pot, processing moose bones.

February 16. Ditto.

February 17. Wind.

February 18. It has been five days since we last saw the wolves. We woke early, hoping for a good long day, and left basecamp at 0800.

After some difficulty locating any wolves, we realized the right telemetry antenna was not working at all. Sometimes the switchboxes go bad. We changed that. Sometimes the cables go bad. We changed them. Occasionally the antenna itself goes bad. After we spent more than an hour of fiddling with co-ax cables, hose clamps, and electrical tape, the wind began to howl.

Don fought the gusts and Rolf recorded notes while winds pushed and pulled on the *Flagship*. Both Middle Pack and Chippewa Harbor Pack had left the comfort and safety of kill sites deep inside their territories. Romeo traveled the island, looking for a mate.

February 19. The air was still, and the sun was filtered by a high, thin overcast. We counted moose on 24 plots today. It won't be long until we're finished counting moose.

For each moose, we record their activity (standing or bedded) and age (adult or calf). Standing moose are, for the most part, foraging. During the first two daylight hours of the day, about 80% of the moose we see are standing. By midday that drops to less than 40%, as most moose are bedded, chewing their "cud" from the morning's bout of foraging. And by day's end, more than two-thirds of the moose are foraging again.

Such a crude and simple view for such a complicated creature. Each of the 50 or so moose we counted in the past two days has its own life. One was a middle-aged bull, maybe six years old, and at the peak of his life. He mated with maybe three or four cows this fall, and he still had time to put on enough fat in the late fall to easily make it through winter. He'd seen wolves earlier this winter, but they didn't even bother. We counted him; we just don't know which one he was.

Another moose was an old, tired cow, maybe 13 or 14 years old. She had a calf this year. Wolves took it in late summer. But she'd successfully raised two other calves when she was younger. In the past year, she developed periodontal disease . . . thousands of miserable twigs to chew each day. She may not make it to taste the tender shoots of bluebead lily that rise each spring. We counted her too; we just don't know which one she was, though wolves certainly know.

Moose are like all the people we brush past in our own lives, people whose names we never learn. They all have a story. With some knowledge you can imagine those stories, and they are true—we just don't know who is living which life.

Middle Pack killed a moose last night. Except for a few hours of rest on McCargo Cove, Chippewa Harbor Pack continued traveling. Romeo did the same.

February 20. Teeth, hooves, blood, bruises, adrenaline, exhaustion. Romeo killed a moose. Very likely, this is the first moose he'd ever killed. He'd seen his parents, the alpha pair of Chippewa Harbor Pack, do it many times. He would have even helped his parents kill moose. He'd wounded moose a couple of times this winter but never killed one. His pride was heightened because he killed this moose with the help of a girlfriend. By early morning they slept with full bellies while a dozen ravens celebrated the accomplishment with a feast of their own.

It was all good for Romeo, save one minor issue. The moose they killed was less than two miles from where Middle Pack had been resting on the hillside overlooking the frozen remains of their own moose carcass. It is difficult to imagine Middle Pack not discovering what Romeo and his friend had done in their territory.

Chippewa Harbor Pack wandered their territory until they found a moose they could kill about a mile north of Daisy Farm campground.

By day's end, everyone—save a loner wolf or two—was at a kill site, feeding or bedded.

February 21. Middle Pack left their unfinished carcass, heading straight for Romeo's kill site. They deliberately swung around the kill site and made their final approach from the north. They charged the site. Tails flailed the air, paws pounded the ground, and the urine of an alpha male spattered the snow. Romeo and his friend, however, slid out to the northeast just a few minutes before Middle Pack arrived.

They were not running, each in their own direction, or as fast as they could, like wolves who thought they might die. In an hour they only put about two miles between themselves and Middle Pack, when they could have easily made six miles if they had needed. Meanwhile, Middle Pack ran north and down onto the chaotic ice of Lake Superior's shoreline. The alpha male was frenetic. He led the pack southwest along the shoreline (while Romeo and friend eased on to the northeast). Then Middle Pack's alpha male turned his crew around, and they traveled northeast along a low shelf just above the lake, while Romeo and friend were on the next higher ridge and a couple of miles up the shore. Middle Pack's pursuit was completely frustrated when the shelf they were on was squeezed by open water of Lake Superior on the left and a cliff on the right. They returned to the scene of the crime and satisfied themselves with feeding on the calf that Romeo had killed.

Romeo lay down to rest just about four miles northeast of his kill site. His friend, less comfortable with the nearby threat of Middle Pack, continued on to the northeast. We would not see her for at least another four days.

Chippewa Harbor Pack fed on their carcass all day, and most of the pack wandered out on to Moskey Basin to rest and socialize.

As a reminder of all that we were missing, we saw the tracks of two lone wolves (each traveling separately) near the southwest end of Siskiwit Lake. The lives of these wolves were, no doubt, as fierce as Romeo's.

February 22. Fog cloaked the island all morning. Not until about 1300 did we take off from Windigo. Middle Pack continued to guard the calf that

Romeo had killed. Chippewa Harbor Pack continued to feed on their carcass. In the past 24 hours, Romeo covered many miles. We watched him travel by himself from Lake Mason to Moskey Basin. With the sun sagging low and the *Flagship's* fuel gauges dipping toward empty, we headed home. We headed home as Romeo was soon to discover the tracks of his parents and siblings leading from Moskey Basin straight to where they were presently feeding on another carcass. Romeo's big adventure seemed like a distant memory by now.

February 23. Wind and snow.

February 24. Snow.

February 25. Middle Pack finally gave up guarding the calf that Romeo had killed. They traveled NE to within a mile of Florence Bay, where they discovered a depression in the snow on top of a beaver dam. The subordinate wolves gathered, tails wagging, jostling each other for a closer look. They watched and waited for the alpha male, who was bringing up the rear. After applying his nose to the simple depression in the snow, he marched to the nearest tuft of brown grass poking out of the snow. He raised his leg, urinated, and then tore at the snow with his paws. Middle Pack had found a site where Romeo had bedded before he had completely exited from Middle Pack territory a few days earlier.

At that moment, Romeo was 16 miles farther to the northeast. Alone, but safe within his parent's territory. He fed from the carcass of a moose his parents had killed a few days earlier. While he fed, Romeo's parents and siblings had traveled more than 10 miles to the southwest, passing through an old kill site near Hatchet Lake. Tracks on Hatchet Lake suggest that two of the Chippewa Harbor Pack wolves mated on the ice. We caught up with them shortly after they left Hatchet Lake—up over Mount Siskiwit and half way to Lake Siskiwit.

More loner tracks just a mile northeast of basecamp.

February 26. Near the edge of a spruce stand, about a mile southwest of Little Todd Harbor: The wolves of Middle Pack approached slowly, the moose eventually stood up. From our view, the moose showed no sign of alarm. The wolves never approached too close. The moose gradually maneuvered to a safer position, with a downed spruced to his rear.

The wolves kept a safe distance—25 yards or so. For an hour, we circled above in the *Flagship*. The moose milled around. The wolves did the same. The wind increased, and we could no longer comfortably continue circling the scene. We flew home.

February 27. Wind.

February 28. More wind, no flying. Rolf and I learned a bit from reviewing images and comparing notes of all the behaviors we'd seen in the previous couple of weeks. From that review we realize that a younger female is successfully challenging the alpha female of Chippewa Harbor Pack for the attention of the alpha male and control of the other females in the pack.

March 1. Sometime in the past day or two, while we were grounded, Chippewa Harbor Pack abandoned their kill site just north of Daisy Farm. I hiked in. All that remained of a 900-pound moose was a jawbone, one shoulder blade, and the bones of one leg. Not even a skull could be found.

Without food to hold his interest, Romeo left his family (Chippewa Harbor Pack) again. We watched him tour through Hidden Lake, Duncan Bay, and onto Robinson Bay—searching for food and a mate—probably not in that order.

Romping, chasing, tumbling, and tackling. Paws, teeth, and fur flailing in every direction. For 50 minutes, the wolves of Chippewa Harbor Pack socialized on Moskey Basin. Some had fun. Others were put in their place. Like never before, the new alpha female of Chippewa Harbor Pack asserted herself with other females in the pack. The old alpha female continued becoming familiar with her new role.

After an intense period of socializing, Chippewa Harbor Pack left Moskey Basin, cruised through forests to Lake Mason, returned north to Lake Richie, through the swamps toward Angleworm Lake, and back to a place that never escaped their thoughts. Among these ridges and conifers, they wounded a moose on February 11. The pack had also passed through the area on the 18th and 19th. Today we found what their interest in the place was. One of the wolves stood victoriously on a moose carcass.

March 2. Chippewa Harbor Pack fed and slept all day at their kill site. For access to the fresh carcass, Romeo tried once again to ingratiate himself to his natal pack. By nightfall, Romeo left the area.

About four miles west, Rolf and Don found him just a few hundred yards from another lone wolf. Turns out Romeo found a kill made by another wolf (see tomorrow's entry), probably an old-looking loner sitting on top of a nearby ridge (this could be a remaining male from the old Paduka Pack).

Throughout the day we noticed more than the usual number of wolf tracks—the tracks of loners, pairs, and trios. They cut across several lakes on the island's east end. Some probably belonged to members of Chippewa Harbor Pack, most we don't understand. We followed most until they faded into nothing. One set took us to a lone wolf crossing the vast expanse of ice that covers Lake Siskiwit. So many questions. Does this wolf belong to a pack, or has she been living alone? Have we seen her before? If we could only know just a tiny bit of what she knows.

March 3. For a second day, the wolves of Chippewa Harbor Pack slept beneath a brilliant sun and perfect skies near their kill site.

The alpha female of Middle Pack has never seen a winter like this one—not in the 11 or so years that she's been alive. They travel almost nowhere, and they eat almost nothing. For most of the past 20 days, they've been at or near the carcass of a moose just a few miles up Washington Creek. They first wounded that moose on January 30.

This morning was different. The day began with the alpha pair mating. Then they traveled south to Lily Lake, and then on into the Big Siskiwit Swamp. The wolves were headed into country that holds plenty of moose. Perhaps they will have a fresh meal tonight.

Romeo found his friend and a meal. While Romeo and his friend fed from the carcass of a half-eaten moose at the base of Mount Siskiwit, another pair of wolves perched on a rocky ledge overlooking the kill site.

Very likely one or both had killed the moose, but they were now denied access to the site. We have no idea how long this pair of wolves has been together, but the behaviors we observed gave the impression they had just met. If we only knew what stories they had to tell.

March 4. With each passing day the sun hangs in the sky a little longer. Each afternoon, the snow becomes soft and wet. Snow fleas fill every footstep. By morning the snow is crusted again. Middle Pack traveled to the south shore over the nighttime crust. They slept all day with a grand view of Lake Superior, rather than travel through soft snow. Chippewa Harbor Pack also avoided the lousy travel conditions and slept for another day at their kill site.

Romeo's girlfriend led him 20 miles to the southwest, straight into Middle Pack territory. Romeo was willing until they came to tracks Middle Pack had left on the south shore. At the time, Middle Pack was just one and a half miles down the shore (10 minutes away, by a wolf's pace). For 45 minutes, Romeo and his girlfriend inspected the tracks and tested each other. He was unwilling to continue on. They would try to kill him. She was unwilling to leave him. Very possibly, Romeo's friend is a daughter from Middle Pack. Since we first saw Romeo with his new friend, Middle Pack has been missing a wolf. They settled on a night in the thick cedar swamps nearby, but no closer to Middle Pack.

March 5. During the night, Romeo and his friend returned northeast to more comfortable country. Tracks on Hay Bay suggest that they mated.[5] They continued on to feed on the carcass they'd co-opted a couple of days earlier at the base of Mount Siskiwit. They have each other, and they have learned how to get food. Now all they need is to eke out and defend a territory.

March 6. For two days, Middle Pack traveled at night and rested during the day. They traveled the south shore, around Cumberland Point, up Washington Harbor, farther up the creek, and back to the kill site where they've spent much of the winter.

For two days, the three senior wolves of Chippewa Harbor Pack slept at their kill site. Unable to pass the time idly, three subordinates from that pack went on an overnight trip. Often the alpha pair of a pack will leave the subordinates at a site while the alphas go off for a day or two to patrol the territory. It is unusual for a group of subordinates to make such a trip while the leadership stays put.

A warm, hazy wind blew all day. Winter's end isn't far off.

March 7. Sometime this morning a thought occurred to the alphas of Middle Pack. It was time to check on that moose they'd wounded nine days ago near Little Todd Harbor. With ravens waiting on their arrival, Middle Pack found their moose, dead beneath a lone spruce tree just a few yards from where they'd wounded it. If past behavior is any indication, they'll spend much of the next few weeks at this site.

We were surprised to find Romeo traveling alone. Where was his friend? As he headed southwest across Mud Lake, he stopped every few moments to look over his shoulder. What was his concern?

We continued on up the island, wondering about Romeo and listening for the telemetry signals of Chippewa Harbor Pack. Within a few moments, the first faint signals registered. The signal took us straight to where Romeo and his friend had been feeding.

Chippewa Harbor Pack was bound to discover the site. For the past three days, they had been sleeping at their kill site just about two miles to the northeast. As a son of Chippewa Harbor Pack, Romeo wouldn't have been in mortal danger. However, they would kill his friend if they could. Was Romeo alone because they'd killed her? Did they both escape death by splitting up? Perhaps she and Romeo will be reunited by the afternoon.

From Mud Lake, Romeo went on to Hay Bay and then up the Little Siskiwit River.

We thought he'd bed down and lay low for a while. However, when he got to the river's source, he reversed course and walked down the river. Near the end of the river, he turned again and headed up the river for a second time. Then he bedded down. Did he know his friend was alive? Was he laying a track that would be easy for her to find?

March 8. Chippewa Harbor Pack left the creek bottom where they struck fear at Romeo and his friend. Near Little Todd Harbor, Middle Pack rested about half a mile east of their feeding site. Another half mile farther to the east, Romeo was bedded. Thick cover prevented us from seeing any of the wolves. We only heard signals. We can only suppose that Romeo's friend had returned to Middle Pack, her natal pack. Why else would Romeo take on the risk of being so close to Middle Pack?

The snow is going fast, and the lakes are turning to slush. The morning forecast caught us by surprise—several days of rain. Very likely, today is the only day this week we would be able make it to the mainland. After a short flight to find Romeo and the packs, we decided to pull the plug. To leave by 1500, we'd have to dismantle the operation quickly—tear down the bone boiling pot, strip the *Flagship* of her research gear, pack food into boxes, decommission the water hole, and secure the compound.

Like so many episodes in life, winter study often ends suddenly and with no definitive conclusion. But it's not an ending, just a pause. In a couple of months, we would be back for spring field work.

In the interlude between winter field seasons,[6] I did not believe Romeo would survive to see another winter. In part because most dispersing wolves—excepting those who return to their natal pack—die in the process of dispersing. We know what Shakespeare would have done with Romeo, and I did think Romeo would die. But I also thought he should survive—knowing that I have no authority whatsoever on how it is that fate should conduct her affairs.

The following winter Romeo did return home, and he received more than a prodigal son's welcome. Because we really did not expect it, it took us a couple of weeks of observation during January 2011 to figure it out that Romeo had become the alpha male of Chippewa Harbor Pack.

The pack seemed to be doing well under the leadership of Romeo and his new mate—also serving for the first time in her life as an alpha. They killed enough moose to keep everyone alive and to expand the family by two.

The only neighbors left were Middle Pack, who faired not as well over the past year. They had been led by a seasoned alpha female. We know nothing about the circumstances of her death. We only know she was not present in 2011 after leading the pack for seven years. She had been an important wolf—with significant impact on both the demography and genes of the entire population. In her lifetime, she had only two mates, with whom she produced 27 or 28 offspring. Calling to mind, and surpassing, the depths of incest practiced by the royalty of old Europe and even the Greek gods of Olympus, her mates had been her son and, before that, her father—the Old Gray Guy.

Romeo's style of neighborly relations with Middle Pack would receive a nod from Machiavelli.

February 19, 2011. Don dropped Rolf at Mud Lake. Middle Pack's attraction to the place was our attraction. We couldn't see a moose carcass from the *Flagship*'s privileged vantage, but some things can only be seen from the ground. Sure enough, Rolf found a palate, a mandible, and a few other bone fragments from a calf that Middle Pack had killed about a week ago.

After an hour of searching every square mile of Chippewa Harbor Pack territory, we hadn't found them. We were a bit surprised—they are usually pretty easy to find. Don and I picked up their tracks where we'd last seen them a couple of days before. They took us to Intermediate Lake and then SW down the length of Siskiwit Lake. Don spotted a couple of wolves up ahead, sleeping on the lake. They were Middle Pack wolves, sleeping right where Chippewa Harbor Pack tracks turned north from Siskiwit Lake. That's not easy to explain. Middle Pack usually rests in discrete locations, far from the edge of their territory and in the forest, away from shore—not on the ice in the tracks of Chippewa Harbor Pack.

And where is Chippewa Harbor Pack? We turned north, gained altitude, and heard a telemetry signal. We soon saw Chippewa Harbor Pack, on the march—all nine of them. They were four miles farther west than we'd ever seen them before and were pushing farther into Middle Pack territory while Middle Pack slept.

An hour later, Chippewa Harbor wolves marched across Lake Desor. Another hour and they were a few more miles west. They traveled across ridge tops, where the traveling is easy but moose are generally absent.

We took this to mean they were not interested in hunting moose. They were fixing to claim territory. They encountered no evidence of Middle Pack wolves, which had remained along the southern shore of the island since we arrived.

No wolf in Chippewa Harbor Pack has ever been detected this far southwest before, except one—Romeo. He was leading this expedition.

When the *Flagship* landed at sunset, Chippewa Harbor Pack was about six miles from the bunkhouse and heading our way. They could be here in an hour or two. As they trotted into the darkness, did Romeo recall memories from last winter? Or was his mind focused on the boldness of the moment?

After putting the *Flagship* to bed for the evening—unloading, refueling, engine cover, and struts tied to the ice—we took our telemetry equipment to the hill behind the bunkhouse. No signal. An hour later, at 2030, faint beeps from the two strongest collars. At 2200, the telemetry receiver sounded off right inside the living room. The signals of all three collars were so strong they penetrated right through the walls of the bunkhouse. They were on the harbor, no more than a third of a mile from the bunkhouse.

From the porch we listened. A long, low howl. Then another. The sound waves echoed as they rose above the frozen harbor. For several minutes we only heard the breath of this single wolf. Darkness transformed the sound into synesthesia. If you listened, you could see the air rush from his lungs, through his throat, and past his teeth. It was Romeo's voice. Without any warning, a roisterous choir erupted—howls, yips, and wails in nine different voices. The sound was intimidating even if you weren't the intended audience. Chippewa Harbor Pack wanted everyone to know they were here in Washington Harbor.

February 20. Chippewa Harbor Pack spent the night in the harbor. But by midmorning we were unable to hear any more telemetry signals. They'd moved on. Snow and wind kept us from flying to find out where they went. Did they return home, the same way they came? Did they continue to tour Middle Pack's territory? The snow is as hard as cement now, and wolves won't be leaving any tracks. We may never know.

February 21. The wind has exceeded 20 knots for 78 of the past 96 hours. Maddening.

February 22. This was not the day we'd expected. Wind and clouds rather than calm and sun. We flew anyway.

Chippewa Harbor Pack walked onto the ice of Moskey Basin. We wondered where they'd been and what they'd done since we last saw them march through to the far side of Middle Pack's territory three days ago. They carried the narrow waists of wolves with empty bellies. They haven't killed a moose in about a week, though we may have missed one.

After observing Chippewa Harbor Pack for some time, we began the tedious search for Middle Pack. It always takes so long, and we often never even find them. Searching in flat light is worse because we have little chance of seeing tracks, even if they are right before our eyes. It's routine now. Start at Mud Lake, turn one-mile S-turns heading SW parallel to the south shore, continue on all the way around Siskiwit Bay to Houghton Point then to Long Point.

A few minutes into our search, we were stunned. The collar was transmitting two beeps per second, twice the normal rate. Mortality mode. Beneath us on the ground, an acre of wilderness—every square foot had been trampled by the feet of wolves. Typical of scenes where wolves kill another. Chippewa Harbor Pack killed the alpha male of Middle Pack.

Don wrestled the wind to fly circle after circle over the site. After many passes, we pinned the source of the telemetry signal to the south side of a small clearing about a mile north of Hay Bay. We saw a raven fly from the ground. That's the spot. We couldn't see a carcass, but it would be there.

When a wolf dies, ravens and foxes take the carcass in days. If we are to recover anything of the carcass, it'll have to be retrieved immediately. I marked the spot on a map and planned the route I'd take overland from Hay Bay, where Don would set me down.

This wolf died at the base of some shrubby cedars on the edge of that small clearing. I stood for some time. Being close to a wolf is usually sad because it is almost never by their choosing or on their terms. The last time I touched the alpha male of Middle Pack was in April 2009 when I fit him with the radio collar that told us today that his time had passed.

His flesh hadn't frozen, but rigor mortis had set in to his legs. Ravens and foxes had eaten all of his internal organs and chewed half his ribs. He'd probably died 24 to 48 hours ago. We'll know more about the injuries he sustained when we necropsy him. Except for a few scent marks on nearby bushes and the sea of wolf prints that surrounded me, there were not many other clues.

In 2007, after his father (the previous alpha male) died, this wolf had become alpha male at the young age of two years. An early and promising start, but now at age six, he is dead. If his upper canines—broken and worn down to little more than nubs—are any indication, his four years as alpha have not been easy.

For now, I had to figure out how to fit him onto the sled. A pint of bloody water pooled where his organs had been. As I realized this would be a bit messy, I also realized that for my hands all I had were the chopper mitts I was wearing. With all the mindfulness due to the corpse of any creature, I slipped a knife between his right rib cage and scapula to remove his front leg. I rolled him over and removed the other, wrapped them in plastic, and put them in the sled. I could curl his neck and upper spine just enough that the rest of him would fit in the sled.

I attached myself at the waist to the sled, and we began the procession home. This was the last mile of Isle Royale wilderness that he would ever cover. Maneuvering the sled through cedars and alders and over downfalls and ledges, it was all about momentum. If I paused unprepared with the sled on an obstacle with center of gravity too far aft, the 60-pound load could pull me off my feet and on my back. If the CG was just forward enough, and if my snow shoes were planted firmly, then the energy of the momentum would be absorbed by my face in the snow. The sled halted in both manners on several occasions. The trip home certainly counted as exercise.

At dinner we discussed the implications. The surviving members of Middle Pack will likely begin living the life of dispersing wolves. They face a significant risk of starving to death over the next year. Isle Royale is now inhabited by a single pack, Chippewa Harbor Pack. It's been four decades since the population was reduced to a single pack. Predation pressure on the moose could be cut in half over the next year. More calves will survive. Older senescent moose will be able to live a year or two longer. Beaked hazels, red-osier dogwoods, and aspen saplings will be browsed back. And the balsam fir trees—dominant source of food for moose in the winter—if they had the awareness, they would have trembled the day the alpha male of Middle Pack died. It's a new world.

February 23. I spend the day with the alpha male of Middle Pack. Massive contusions across the back, back of the head, and over the rib cage. He also suffered some arthritis in the spine. This is not the first time we've seen arthritis in such a young wolf on Isle Royale.

February 24. The wind granted us just two hours over the island. For the past two days, Chippewa Harbor Pack traveled around the east end of the island. We spent most of the flight looking for remnants of Middle Pack or other loners but found little more than the track of perhaps just one wolf on the far SW end of Isle Royale.

Two of the three remaining Middle Pack wolves died within the year, and the pack dissipated like morning mist.

Romeo was inbred but also among the least inbred wolves alive at the time. He looked to be the kind of wolf who could lead the population through another generation.

One day at the start of the next winter, Romeo, his young son, and another pack-mate came to a pit in the earth. Fallen logs and jagged rocks protruding from the pit walls provided something on which to crawl down. The pit is an artifact of 19th-century copper mining. For many decades it had been filled with water to within about 15 feet of the earth's surface. Because it was early winter, the water had frozen. Wolves, being wolves, explored. They climbed down the walls. The ice was thick enough to support the weight of one wolf, and even a second wolf, but not three. They all drowned.

When natural explanations for the fate of an individual fell short, Thucydides—the father of historical explanation—invoked the influence of the gods. These gods were neither omniscient nor all powerful. But they did interfere with the lives of mortals, often inadvertently. Today, so many eons after the *Götterdämmerung,* misfortune still flows from the fickle and vain actions of those acting as if omniscient and all powerful.

Chippewa Harbor Pack would never recover from the drownings. Four other wolves from the pack died that year. No pups were born to the pack. For the first time since we began observing Isle Royale wolves, no pups were born anywhere in the population. A few months after the death of Romeo, we counted just nine wolves across the entire island.

The mortar that bonds a pack together includes the ever-present danger of their neighbors. With Middle Pack and the threat they represented gone, the social bonds of Chippewa Harbor Pack started to crack. One male dispersed from the pack and found the sole survivor of Middle Pack, a female wolf. She allowed a relationship to develop, even though he had participated in killing her father the previous year. This new couple was followed by his brother—a collared wolf that we called Pip. Pip's brother allowed the tag-along, so long as Pip didn't interfere with the new relationship. We called these wolves the Trio.

Another wolf fled from Chippewa Harbor Pack. In January 2013, we did not understand why this wolf—who we'd begun to call Isabelle— had not moved from a small patch of thick spruce on the island's west end for a 10-day period. It was the beginning of the winter field season, and we didn't know what she had been doing in the weeks before our arrival. Maybe she had been injured? Or maybe she was content to scavenge some old moose carcass? All clues were protected by the forest canopy. Were it not for her radio collar, we would not have known of her presence at all.

By February she had left that patch of spruce, and we began to understand. On February 2, we watched Isabelle running for her life, chased by two wolves, north along the shoreline. A half-mile lead was reduced to nothing in just a few minutes. At Rainbow Cove, the smaller of the two wolves tackled Isabelle.

February 2, 2013. They both tumbled into Lake Superior. It was 5°F, and they rolled in chest-deep water. A third, larger wolf stopped at the water's edge, ready to attack, but hesitated, apparently less enthusiastic to dive in.

Isabelle and her assailant were a seething pile of swiping paws, flailing tails, and striking teeth as they rolled in the breaking surf. Isabelle got in a few good bites and dodged several others. Then the third wolf dove in to join the attack against Isabelle. She took several bites on the back, belly, and near her neck. This was no scrappy dog fight. These wolves lunged and sank their teeth into Isabelle with the same fury and power they use to bring down a 900-pound moose. Any well-placed bite could tear a gaping wound, deliver considerable damage to internal organs, or cause severe internal bleeding. They were trying to kill her.

Unable to take any more hits, she slipped lose and began swimming. She didn't stop until she was about 30 meters offshore. Waves were building in the 20 knot winds. She'd take a mouthful of water with some of the larger waves. She was stuck between a watery death and the bone-breaking jaws of two violent wolves. After about 10 minutes, she couldn't take the cold water any longer.

While Isabelle was swimming, Pip finally showed up on the beach, about four minutes after the attack began. With Pip's arrival, we realized that the attacking wolves were part of the Trio to which Pip belonged.

As Isabelle came to shore, when her feet touched bottom, she curled her lips, bared her teeth, and waved her head back and forth, threatening counterattack to any advance. She fought, but in her condition, there was no way she could fend off two wolves. The two assailants kept Isabelle in water up to her knees and often up to her chest. They attacked, retreated to dry ground for a moment or two, and attacked again. Occasionally a breaking wave would overtake all three of them. This cycle of attack-retreat-attack went on for the next 20 minutes. Each attack was terrifying.

Pip never participated. He only watched. It's complicated. Pip and Isabelle are brother and sister from Chippewa Harbor Pack. This spring they fed from the same carcasses. The male assailant is also a brother (probably half-brother) to Isabelle, though this may be the first time they've ever met.

Thirty minutes after it all began, the wolves gave Isabelle a break, letting her alone at the water's edge. With her assailants just 20 meters away, Isabelle stood on guard for a few moments, then sat, and once or twice she laid down. But she never set her head down. She did not give in to the cold, pain, or fatigue. They could have killed her, but they did not.

Higher on the beach, Pip's companions rolled in the snow, scratched vigorously at the cobblestones, rubbed chests against each other. Even Pip joined the celebrations. All the while, waves occasionally washed over Isabelle's hind end. A couple of times she tried to walk away, along the water's edge. Each time the female charged—forcing Isabelle to stay on the edge of the water and on the edge of life itself. This went on for another 30 minutes.

Then the Trio left the beach all together. A few minutes later Isabelle limped away into the nearby swamp. And then night fell.

Isabelle survived the night. And she survived the next night. No one traveled far from the attack site. But the Trio also refrained from further attack. Six days after the beach scene, we traced the thin ribbon of wolf tracks through the snow that brought us to the Trio on Siskiwit Lake. They had been on the track of a loner but were interrupted by the call of a long nap in the bright sun. Why the Trio did not exercise their ruthlessness with this wolf, I do not know.

Isabelle began the day sprawled on a sun-soaked knob, a couple of miles from where she'd been the previous two days. Her posture was still unusual, as it had been for the past several days. She laid on her sternum with rear legs extended and head flat on the snow. We watch her rise and sit with her rear legs splayed. She looked over her shoulder for a while, then plopped back onto the snow and resumed that awkward position. There was no blood in her snowy bed or any injuries visible from our vantage point.

Sometime in the morning darkness, the Trio began traveling west along the south shore from Long Point. Not wanting for food, they left behind a largely intact moose carcass. Their entire path showed a double track with the male and female walking side-by-side the entire way. They will breed soon.

By midday they found the place where Isabelle had been recuperating from her previous beating. They crossed Washington Harbor and continued their sweep by sticking to the outer shore until reaching the steep cliffs of the north shore, where they were forced inland.

The Trio was just a mile or two from Isabelle and heading directly toward her. They had the advantage of approaching from a ridge top that ran to just above Isabelle's position. But the wind carried Isabelle's scent away. Consequently, the Trio passed right by, missing Isabelle by a quarter mile. Oblivious, Isabelle continued sleeping.

Suddenly the Trio reversed course, running. We caught only glimpses through the thick forest cover. They ran in one direction, then the other—but generally away from Isabelle's current location. The Trio had just encountered, for the first time, tracks Isabelle had made two weeks ago, when she spent about five days localized in this area. For an hour or two, the Trio frenetically sorted the tangle of tracks—discerning the fresh from the old. One set led out of the jumble and straight to Isabelle. They incited themselves upon those tracks and chased her down. They attacked. On the north shore, about three miles from Isabelle's morning bed, the female of the Trio had isolated Isabelle on a rocky promontory glazed in ice 20 feet above the waters of Lake Superior.[7]

The female of the Trio laid just behind shoreline trees with her sight focused on Isabelle's every move. Her presumed mate licked her back, suggesting that she'd sustained injuries from Isabelle. Pip was curled up in the background, far enough to be uninvolved with the mayhem surrounding his sister, but close enough to the other female, hoping for an unlikely opportunity to gain her favor.

Isabelle sat in bloodied snow. She had open wounds on both rear flanks and her right shoulder, in addition to torn flesh on her left rear foot from an injury of several days earlier. It would be another long night for Isabelle.

The next nine days unfolded this way:

February 9. We flew straight to the promontory. Two ravens sat on the blood-soaked snow where Isabelle had been. Another raven flew circles above. An eagle sat in a nearby tree. We didn't see Isabelle, but her collar was not in mortality mode, and the signal said she was about half a mile northeast of the promontory. The Trio, minus Pip, slept on the ridge top about two miles to the south. Wind and clouds prevented us from learning any more.

February 10. The weather was too poor to do anything except fly once around the harbor. So we did. That was enough to hear that all of the wolves were in about the same location as yesterday. And, if her collar is telling the truth, Isabelle is still alive.

February 11. Ten and a half inches of wet heavy snow fell. Thirty knot winds.

February 12. Only a short flight on account of poor weather. The Trio was still located about three miles from the bunkhouse and less than a mile from where they'd last confronted Isabelle. Now, their interest is held by the carcass of a moose. They haven't even finished the moose they left at Long Point.

Isabelle moved to an area north of Lake Desor—country not often visited by either the Trio or the remnants of Chippewa Harbor Pack. We're grateful to learn as much as we do in these short flights.

February 13. Isabelle moved farther northeast. We didn't see her with our eyes, but the telemetry signal said she was just northeast of Little Todd Harbor.

The Trio left the kill site. They headed toward their previous kill site by Long Point and away from Isabelle.

We counted moose on another 12 survey plots. In doing so, we found another set of twins. That brings the total up to nine sets of twins—really quite extraordinary.

February 14. Wind and snow. Rolf hiked from the bunkhouse to the carcass that the Trio had abandoned yesterday. Upon arriving to the site, he first noticed the wolves had eaten the hide and muscles of the moose's abdomen, exposing the rumen. Yet the rumen was still interred within the carcass.[8] The carcass was partially snow-covered. Wolves had eaten only a small portion of it—the left hind quarter and the portion of the right hind quarter that wasn't frozen solid to the ground.

This moose wasn't killed. It died here, and the wolves had scavenged it. If this moose had died before the cold of winter, the immediate surroundings would have been covered with maggot casings. That was not the case. This moose likely died in November or December. His antlers were grossly misshapen, covered with an unusually deep-grooved surface and still mantled in velvet. He was an old, decrepit bull, no longer able to balance the hormones that control antler growth.

The Trio has had four moose in the past month. A little arithmetic . . . four moose divided by three wolves divided by 30 days. That's 1.35—not the highest rate of prey acquisition we've ever seen, but it's darn near. Even more impressive is that in two of four cases the wolves appeared to have scavenged moose after they had died on their own. The moose near Feldtmann Lake was an old bull with arthritis that was as severe as it gets. This latest moose was old enough that it couldn't produce normal antlers. Normally these moose are killed by wolves before they die from senescence-related health issues. But now moose are numerous enough, and wolves rare enough, that the oldest, most decrepit moose are increasingly likely to die before wolves kill them.

February 15. Weather allowed us to fly but kept us from learning much. Isabelle continued traveling northeast—deeper into no-man's-land. The Trio returned to the carcass they'd been scavenging.

February 16. Morning snow gave way to afternoon clearing. Rolf and Don counted moose on another 10 survey plots. The wind and snow do not share our interest for getting those plots counted. The weather makes us nervous about finishing the moose count.

The Trio scavenge from their carcass near Hugginin Cove. Isabelle was a few miles farther northeast than she had been yesterday—now just about a mile west of McCargo Cove. The terrain here is high and rugged. Thin soils prevent too many trees from taking root. It's open country with only a scattering of trees. We saw tracks of running moose and a running wolf. Near a few small conifers were an alert cow and her calf, bedded with its rear end backed up against one of the conifers. A small patch of blood stained the snow. Isabelle had gotten her teeth in the calf—but perhaps only once. It is quite an accomplishment for a lone wolf to kill a calf that is protected by its mother. But Isabelle is an accomplished wolf.

February 17. Isabelle, the cow, and the calf had rearranged their positions slightly. The cow and calf were bedded beneath a different set of trees. We never saw Isabelle, but the telemetry signal indicated that she was bedded beneath some other nearby conifer tree.

Isabelle did not, apparently, consider returning to her natal pack to be an option. If she had, she would have done so before trying to kill moose. Only Isabelle and the pack know whether she has rejected the pack or the pack has rejected her.

The calf didn't appear to have sustained any more injuries. His or her safety is greatly favored by keeping his or her rear end protected by a thick conifer tree that limits Isabelle to attacking only from the front. Because the trees are so sparse in this country, the moose cannot easily leave, and Isabelle—being just one wolf—cannot easily attack. Isabelle has time.

Don and I counted moose on more survey plots until we were chased home by a snow squall.

Drawing blood is significant, but still a long way from completing the deed. The standoff could have gone on for a few more hours or days. Whenever it ended, either Isabelle or the moose would have experienced the intense relief known only to those who have narrowly escaped with their lives. We do not know who received which fate because that was the last flight of the winter field season, and life carries on whether we pay attention or not.

With so much in the world against her, I expected the next time we heard from Isabelle it would be the mortality signal on her collar.

A year later, static. From the cockpit of the plane, nothing but harsh static across the entire island. We presumed Isabelle had been killed and the radio collar destroyed in the process. On a flight back to base-camp to refuel, Don suggested that we fly out to Houghton Point. I was dubious: it's out of the way, takes time to get there, and we often find nothing. But I agreed.

Houghton Point is on the far side of Siskiwit Bay, which was locked in ice. Bare black ice, pancake ice, frost-covered ice, fields of jagged ice chucks—some chunks would measure 20 feet across—heaped onto each other. The bay was filled with every kind of ice formation and hues, ranging from pure white, calming light blue, dark foreboding shades of blue, and black. It goes on and on for more than 20 square miles.

Don set our course across the bay to Houghton Point. It's pretty rare to find anything of interest so far from shore. So, I had been looking down on the ice but not seriously. I merely gazed. Then, some set of neurons in my melon, some set not under the control of my conscious mind, jolted and hauled my consciousness from mesmerization. A wolf! A lone wolf traveling on the ice. And she was wearing a radio collar. It was Isabelle. I checked the telemetry receiver—static. Isabelle's radio collar was dead, but she wasn't. She figured out how to kill and not be killed. While those are two of the three requirements for a good wolf life, the prospects for her realizing the third requirement are nil. She will never find a satisfying mate—not on this island. Every surviving male is a brother or half-brother.

The next time I saw Isabelle, I held her skull in my hands. Rolf had prepared the skull—using a warm bath and sharp blade to carefully remove the soft tissues from the bone. Both lower canine teeth had been broken off and the stubs heavily worn. Her incisors were infected. One lower premolar was also broken off and heavily worn. Her teeth were older than her years.

An ice bridge had formed in late January 2014, a week or two after we last saw Isabelle alive. She had never before seen one. Instinct whispered the suggestion. She started walking. About a week later, she was shot and killed at the edge of Lake Superior on an Indian reservation.

During the same winter that Isabelle left the world, the mated pair of the Trio was raising three pups. The new pack still tolerated Pip's presence.

With the birth and survival of three pups, the Trio had become a full-fledged pack by January 2014. We followed them throughout that winter, including making detailed observations on the last flight of that winter season, March 1, 2014. They were never seen again. All six of those wolves disappeared without a trace.

The ice bridge connecting Isle Royale to the mainland persisted for another three days. Then the wind took it out. I'm not sure which is more likely, for the entire pack to have died in a year or for the pack to have left. I do hope they fared better than Isabelle.

After the diaspora from Chippewa Harbor Pack, all that remained of the pack and the population was an alpha female (Romeo's widow), a son, and a daughter. A year later, the alpha female died and the half-siblings mated. To tell the whole truth, they were closer than half-siblings. They were also father and daughter.[9] They mated and gave birth to a severely inbred pup who never looked quite right. He or she died sometime between 10 and 21 months of age. As of March 2018, this brother and sister were all that remained of the wolves of Isle Royale.

8 |

Sense of Place

By March 2011, wolves had become too few in number to counterbalance the growth of the moose population. The wolf population was functionally extinct. By 2016, moose abundance doubled. Animal populations grow like investment accounts. That is, by rates. Moose showed every sign of continuing to grow at 22% per year. That's more than double the rate an investor would expect on a good stock mutual fund. Isle Royale would be overly endowed with moose in just a handful of years.

I expected the forest would soon take a hit and was concerned that moose would do long-term damage that would not be easily undone. Because a growing moose population has considerable momentum, preventing that damage would require acting before the damage began.

The health of ecosystems inhabited by large ungulates—like moose—depends on the presence of a functioning population of predators. That's a maxim from the field of science devoted to protecting ecosystem health.

I advised the National Park Service that Isle Royale's ecosystem health would soon be at risk. I expressed that concern in March 2009, when I first observed early signs of teetering, two years prior to the collapse of predation. Our annual reports to the NPS anticipated much that would come to pass. From the first moment, the NPS looked recalcitrant to even acknowledge the concern and conveyed an impression of prejudging the appropriateness of doing nothing. If the wolves go extinct, so be it. That course of action is demanded by wilderness policy. Nature will be allowed to take its course—whatever that may be. There will be no intervention. End of discussion. I was left with the unambiguous impression that I should keep other thoughts to myself, or I'd have hell to pay.

My concern included the significant role that humans played in bringing wolves to functional extinction—disease, mine-pit drownings, and most importantly, climate change. Climate warming had caused a decline in the frequency of ice bridges that form during the winter between Isle Royale and the mainland. During the 1960s, ice bridges formed in 8 of every 10 winters. At that rate, a wolf or two occasionally immigrated to Isle Royale.[1] The infusion of new genes carried by the immigrants held inbreeding depression at bay. By 2009, the frequency of ice bridges had declined to about one per decade and are expected to soon be among the stories we tell grandchildren about how it used to be.

Because climate change was involved, the question of what to do about the wolves on Isle Royale was really a much bigger question about *how should we respond to damage caused by climate change in our national parks?* Will we let climate change take whatever it likes without any resistance? Or would we fight to save America's Best Idea? The NPS's response to wolves on Isle Royale would set precedence.[2]

I advocated for genetic rescue. Bring one or two wolves to Isle Royale right away. The infusion of new genes would get the wolf population back on its paws before the moose population got carried away with itself. That would be the quickest, easiest, least intrusive, most natural way to restore wolf predation. Mimic what the Old Gray Guy had done in the mid-1990s.

The NPS appeared to be ignoring the concern. I drew what attention I could to the concern. I wrote technical articles, gave presentations, and wrote an editorial for the *New York Times*.[3] The first public response by the NPS came on April 9, 2014, with a press release:[4]

> For the past two years, park managers have discussed island and wolf management with wildlife managers and geneticists from across the US and Canada and have received input during public meetings and from Native American bands of the area.
>
> "Our decision on a way forward is supported by our review of the best available science, law and policy," [Superintendent] Green [of Isle Royale National Park] said. While the park will not bring wolves from outside to Isle Royale in the near term, Green said there is time to fully explore the consequences of such an action.

Journalists asked to see a copy of the review that led to this decision. A park representative said that there was no review or evaluation that could be shared—just a decision.[5]

Lack of transparency is poison to the lifeblood of democracy. Aspiration for democracy is what first vivified the NPS. The entrance to America's first national park, Yellowstone, is a triumphal archway, inscribed, not with some sacred principle of environmentalism, but with a celebratory aphorism of democracy. Letters standing two feet tall, in all caps, read "FOR THE BENEFIT AND ENJOYMENT OF THE PEOPLE."

Democracy depends not only on transparency but also on knowing—for better or worse—what the people think. Shortly before the April press release, the NPS created an email account to which any individual could express their views on the worsening condition of Isle Royale wolves. About 1,000 citizens responded. About 8 out of every 10 people who offered a comment expressed favor for actively conserving wolf predation by genetic rescue. Nearly 90% expressed favor for either genetic rescue right away or reintroduction of wolves should they go extinct.[6] Those insights were made public only with the assistance from the Freedom of Information Act.

The NPS began disseminating "educational" materials and delivered presentations so that we the people might offer informed views. The dominant theme of these materials was that extinction is a natural process on small islands. An important message was that the wolf and moose populations on Isle Royale are stable.[7] The materials included the statement that "even though wolves kill moose for their food, they do not make a large impact on regulation of the size of the total moose herd."

I had explained to the NPS what the science clearly showed: The wolf population was *not* stable. The moose population was *not* stable. Circumstances were changing quickly. If minimizing the risk of long-term damage mattered, then the time for a deliberate decision was now. And I reminded them of the judgment championed by conservationists, namely, the integrity, health, and resilience of ecosystems inhabited by large ungulates (like moose) depends vitally on the presence of a healthy, functioning population of predators (like wolves). The NPS was unmoved.

Without any public record of how the NPS arrived at its decision, the best one can do to understand their position is to note off-the-record statements from individuals within the NPS and voices they may have been listening to.

There was one wolf biologist whose voice harmonized with the predilection of park decision-makers. He believed that "despite the high level of inbreeding, the [Isle Royale] wolves seem to have behaved and functioned ecologically like any outbred population." And, "thus, potentially in 2013 or 2014, the [Isle Royale] wolf population could increase by over 60%, as it has done before." That last phrase, "as it had done before," refers to the natural tendency for all animal populations to fluctuate in abundance over time. This wolf biologist also wrote that "wolves have influenced [Isle Royale] vegetation via their predation on moose. The question of whether such effects are positive or negative is a matter of judgment.[8] . . . Suffice it to say here that any such concerns are premature at this time because [Isle Royale] still harbors a functioning wolf population that could well persist for many years with or without human intervention."[9]

These views belonged to Dave Mech, Durward Allen's former student, whose pioneering field work launched six decades of wolf research on Isle Royale. Mech expressed those views in 2013, two years after the wolf population ceased performing its ecological function.[10]

With respect to off-the-record statements from individuals within the NPS, some voices expressed concern for setting poor precedence: our parks are about to be drubbed with climate change. Conserving wolf predation on Isle Royale would create an expectation to act similarly in myriad other cases. Any effort to resist would be as influential as a "fart in a hurricane." Acting on Isle Royale would also "open the flood gates." Better to refrain than create so many difficult-to-meet and meaningless obligations.

Some within the NPS expressed base resentment for the attention that I brought to the wolves of Isle Royale. Allowing wolves to slip from Isle Royale would certainly solve that problem. Some thought it would be right to allow wolves to disappear and let humans stand in for the ecological role of wolves. Serious consideration, they thought, should be given to hunting moose on Isle Royale in the absence of wolf predation.

About a year prior to that NPS press release, Ted Gostomski, a science communicator for the NPS, objected to my case for restoring wolf predation. He believed that wolves are an ancillary feature of Isle Royale's wilderness character and ought not be protected from extinction.

> Wilderness is a subjective character made manifest in different ways to different people. . . . When the idea of creating a national park on Isle Royale was first catching on in the 1920s (about 20 years before wolves first arrived on the island), "wilderness was a much less exact word—a word ripe for interpretation, a word that, through the efforts of many individuals, became synonymous with Isle Royale." In other words, it was the place itself that defined wilderness. Given that these discussions occurred at least 40 years before the passing of the Wilderness Act (1964), it is fair to say (and it has been said) that Isle Royale helped to define what wilderness is, and it did so before wolves arrived. Wolves are part of Isle Royale's wilderness character now, but they are relative newcomers.[11]

Gostomski also questioned the wisdom of conserving predation because climate change might at some indefinite point in the future cause Isle Royale to become inhospitable to moose. But the centerpiece of Gostomski's argument was that wolf predation on Isle Royale should not be conserved or restored to demonstrate the NPS's conviction that wolf predation is not "too big to fail"—just like felonious bankers from the 2008 financial crisis.[12]

That same year, another perspective was offered by Tim Cochrane, the NPS superintendent of the nearby Grand Portage National Monument, former cultural resource manager of Isle Royale, and author of two books, one on the historic fishing culture of Isle Royale and another on the history of Ojibwe people on Isle Royale. Cochrane summarized his view in an essay entitled "Island Complications: Should We Retain Wolves on Isle Royale?" (italics added):[13]

> Out of this choppy history [of Isle Royale] it's difficult to conclude what is "*natural*." But it is the unchanging geographic situation of Isle Royale, its *remoteness* in Lake Superior, that has been and should continue to be the primary determining fact in the national park's management. Its character and *integrity* as a remote archipelago must be acknowledged and heeded. To supersede the *insular character* of Isle Royale by reintroducing wolves is arguably toying with its biological and *historical authenticity*—and, perhaps, with the most fundamental biological-given of island life, which is the screening Lake Superior has done through the millennia of which animals and plants make it there.

Historical authenticity, the remote insular character of Isle Royale, the subjective nature of wilderness character, integrity, and naturalness—those are the themes, I agree. And a right response to the collapse of the wolf population requires a right handling of those themes.

As we appraise these themes, know that there is a two-tiered interest. One interest is to know specifically how to best respond to the Isle Royale case. The other interest is to know, more broadly, how to best respond to dozens of other cases like Isle Royale, many just around the corner, in other parks, with other treasures. These cases all involve rising anthropogenic threats that begin outside of national parks and relentlessly creep in. The list of such threats includes climate change, the increasing incidence of invasive species and wildlife diseases, and pollutants that circle the globe before settling to the ground. "Island Complications" states, and I agree, that deciding how to care for such places depends on the particular history of a place.[14] Both tiers of interest profit by reaching deep into Isle Royale's genesis.

None of us witnessed that beginning, but we can generate a pixilated sense by first forming, as best we can, a mental image of the Grand Canyon. We can take a little prompting by Bill Bryson:

> Nothing prepares you for the Grand Canyon. No matter how many times you read about it or see it pictured, it still takes your breath away. Your mind, unable to deal with anything on this scale, just shuts down and for many long moments you are a human vacuum, without speech or breath, but just a deep, inexpressible awe that anything on this earth could be so vast, so beautiful, so silent. Even children are stilled by it. . . .
>
> The scale of the Grand Canyon is almost beyond comprehension. It is ten miles across, a mile deep, 180 miles long [and 17 million years old]. You could set the Empire State Building down in it and still be thousands of feet above it. Indeed, you could set the whole of Manhattan down inside it and you would still be so high above it that buses would be like ants and people would be invisible, and not a sound would reach you. The thing that gets you—that gets everyone—is the silence. The Grand Canyon just swallows sound. The sense of space and emptiness is overwhelming. Nothing happens out there. Down below you on the canyon floor, far, far away, is the thing that carved it: the Colorado River. It is 300 feet wide, but from the canyon's lip it looks thin and insignificant. It looks like an old shoelace. Everything is dwarfed by this mighty hole.[15]

And by a few words from the person who worked to establish the Grand Canyon as a national park, John Muir: "In the supreme flaming glory of sunset the whole canyon is transfigured, as if all the life and light of centuries of sunshine stored up and condensed in the rocks was now being poured forth as from one glorious fountain, flooding both earth and sky."[16]

Your conjuring of the Grand Canyon is sufficient if you can understand why George Wharton James wrote: "I have seen strong men fall upon their knees. I have seen women, driven up to the rim unexpectedly, lean away from the Canyon, the whole countenance an index of the terror felt within, gasp for breath, and though almost paralyzed by their dread of the indescribable abyss, refuse either to close their eyes or turn them away from it."[17]

Envisioning the formation of Isle Royale requires superimposing two scenes. While gazing at the Grand Canyon with our inner eye, summon a second vision of Kilauea volcano in Hawaii National Park. Again, we'll do well to get a little help from another word-master, Mark Twain, describing the 1840 eruption of Kilauea:

> At night the red glare was visible a hundred miles at sea; and at a distance of forty miles fine print could be read at midnight. . . . Countless columns of smoke rose up and blended together in a tumbled canopy that hid the heavens and glowed with a ruddy flush reflected from the fires below; here and there jets of lava sprung hundreds of feet into the air and burst into rocket-sprays that returned to earth in a crimson rain.[18]

The Reverend Titus Coan described the eruption this way:

> There had been activity such that no man had seen. It was reported that, for several days before the outburst, the whole vast floor of the crater [Halemaumau] was in a state of intense ebullition; the seething waves rolling, surging, and dashing against the adamantine walls, and shaking down large rocks into the fiery abyss below. The heat was such that no one dared to venture close for a clear look.[19]

Kilauea had erupted almost continuously, day and night, between 1983 and 2018. It produced enough lava to pave about 20 miles of road each day. Enough lava, in total, to cover the state of Vermont in an inch of lava.

I hope you are still holding that image of the Grand Canyon. From that colossal canyon, remove the mighty Colorado River with the power of your imagination. Replace it with an apocalyptic ribbon of fire, smoke, and molten rock pouring up from miles below the earth's surface. Now, in your mind's eye, stretch this fiery canyon from 180 miles long to more than 1,200 miles long. Widen it from 10 miles to more than 50. The Grand Canyon and Kilauea are, by comparison, not worth even a roadside pullout.

We summoned this conflagrant gash with our imaginations, but it is not imaginary. It formed 1.1 billion years ago as the North American continent rifted apart, and it is the cauldron from which Isle Royale's foundation emerged. The fracture ran from the present-day location of Detroit north by northwest, up past the lower peninsula of Michigan, then across the Lake Superior basin to Duluth and Ashland, then it turned south by southwest, running at an acute angle across the Mississippi River and across Iowa. The breach ended in eastern Kansas. Isle Royale would eventually form from the north edge of this rift, and the details of how that happened are worth knowing.

Lava flowed, up through this breach, episodically for 30 million years[20]—about as long as the Colorado River has been carving the Grand Canyon. With each episode, the basin was coated with another layer of lava. Some layers are just a few inches thick, while others are hundreds of feet deep. Some of the intervening periods were long enough for thin layers of nonvolcanic, sedimentary rock[21] to form. In all, enough magma spewed to bury the entire state of Arizona beneath two miles of lava.[22]

The rift was on its way to becoming an ocean, but then suddenly gave up that particular planetary urge. The earth stopped bleeding lava. The crack healed. That fissure is the largest midcontinent rift on our planet to have grown so prodigiously large and not become an ocean.

After the bleeding stopped, there was a brief 40 million years of geologic quietude. Then the earth heaved again. Under the strain of a

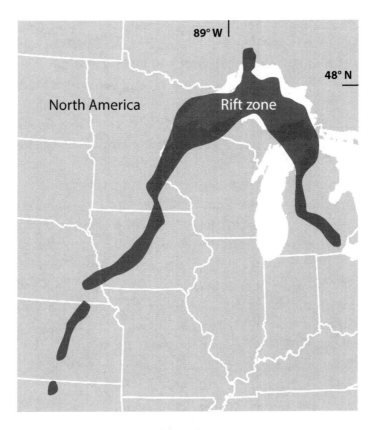

Location of a great rift zone in North America about 1.1 billion years ago

stretching continent, the north and south sides of the rift cracked like a dried-up scab. The now-solid layers of lava were no longer supported by the adjacent crust of the earth. The weight transferred to the earth's mantle, which is not quite solid and flows like asphalt. Thousands of square miles of the earth's crust slumped.

The earth sighed and the continent squeezed in on itself. The north and south edges of the great plains of solidified lava had nowhere to go, except upward. And so, they were shoved upward, producing two ridges that we recognize today as Keweenaw Peninsula on the south shore of Lake Superior and Isle Royale to the north. The basin between those ridges is filled with the waters of Lake Superior.

While some of the layers of this solidified flood are more than 100 feet thick, they are nevertheless exceedingly thin in comparison to the area they occupy. Very thin. If we could take that sheet of lava and scale it down to the length and width of a sheet of printer paper, that lava sheet would be about 5,000 times thinner than the paper. For a rock this broad and thin, the rules of Rock Paper Scissors do not apply. At this immense scale and extreme proportion, a rock can be cut as cleanly as a sheet of paper and bent just as easily.

Imagine those solidified layers of lava to be like a stack of *National Geographic* magazines lying flat on a table. Reach your fingers beneath the magazines' spines and lift gently. The spines, which had formed a single smooth, vertical surface, are now facing upward. But the edge is no longer smooth. The magazines have also slid past one another slightly and the upper convex corner of each spine is now a ridge, and the concave corner formed by the spines of two adjacent magazines are now valleys. This is the kind of lifting that conduced the long swamps, rocky ridges, and deep fjords that confer much to Isle Royale's

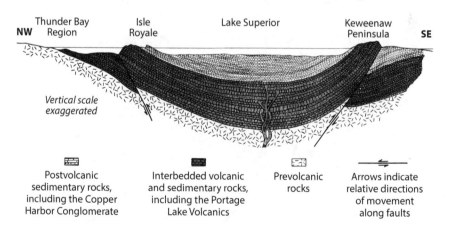

Layers of volcanic rock left by the rift zone and their relationship to Isle Royale.
US Geological Survey Bulletin 1309

character. This genesis of topography accounts for even finer detail, such as Isle Royale's north-facing slopes being steeper than south-facing slopes.

When the fiery rift became an immense, featureless valley of dark rock, the only form of life on the planet was bacteria. The sun shone, the wind blew, clouds rolled in, it rained, and the sun shone again. For eons there was nothing more. With the passage of more time, dust and dirt and mud slowly eroded into the valley from adjacent, massive sandstone cliffs that resided to the north and south.[23] For half a billion years, that is how it was—before plants and invertebrates evolved, taking residence in the basin and on the ridge that would become Isle Royale.

Standing near the water's edge on a bare and tilted boulder of basalt. That is one of my first memories of Isle Royale. If it were one shade of gray darker, it would be black. This particular behemoth is larger than a large suburban lot. It slinks into the head of Moskey Basin, a long tongue of Lake Superior that encroaches deep into the island. Standing behind me are spindly fir trees and lumpy spruce. They wrap around to the left and right and follow the shoreline to the horizon. The trees are restrained by the impenetrable basaltic shore. Standing there feels like being swallowed by Isle Royale, like being swallowed by solitude.

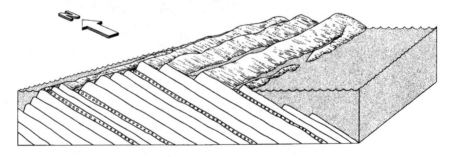

The ridge-and-valley topography of Isle Royale is a result of the layers of volcanic rock left by the rift zone. US Geological Survey Professional Paper 754-A

That scene is now solidly imprinted. When first commissioned to my store of memories, it was just scenery—a purely aesthetic experience. I had no idea what I was looking at. Only later did I learn that the rocks were venerable, even by rock standards. Now I know that boulder has experienced the last four pirouettes of the Milky Way Galaxy.

My early exposures to Moskey Basin were ignorant of its being formed by the groove between two layers of lava. Nor had I known that portions of the Greenstone Ridge Trail, which runs the spine of Isle Royale, are a solidified layer of lava—the thickest layer in the midcontinent rift, which took 1,000 years to cool from molten lava into solid rock. Most of the midcontinent rift is buried, up to six miles, beneath the earth's surface. I had not known that Isle Royale is one of the few places where it peeks above the surface.

Knowing the history of the topography that supports a landscape contributes mightily to a *sense of place*, which is formed when the natural history, culture, and geography commingle in our minds and form the stories—lyrical stories and scientific stories—that define a place. To develop a sense of place is to triangulate points of connection between a place and a person's experience, knowledge, and values. Sense of place enriches one's relationship with a place in the same way that your love for a person is enriched by learning about some faded episode in their history. Sure, this all may just be the oxytocin talking. If so, it would still contribute essential insight.

Whatever it is, it is certainly not just me talking. According to the great American writer Wallace Stegner, in his essay "The Sense of Place," Wendell Berry, the nature writer and poet said, if you don't know where you are, you don't know who you are. I would add, if I do not know who I am, then I do not know how to behave in the world. Wallace Stegner also says that sense of place rises from the

> kind of knowing that involves the senses, the memory, the history of a family or a tribe. He is talking about the knowledge of place that comes from working in it in all weathers, making a living from it, suffering from

its catastrophes, loving its mornings or evenings or hot noons, valuing it for the profound investment of labor and feeling that you, your parents and grandparents, your all-but-unknown ancestors have put into it. He is talking about the knowing that poets specialize in. . . . No place, not even a wild place, is a place until it has had that human attention that at its highest reach we will call poetry.

Stegner's sense of place is unquestionably anthropogenic, but it is not anthropocentric, as clearly expressed by the essay's conclusion: "Only in the act of submission is the sense of place realized."

Another 80 million years passed before the desolate ridges and valleys were colonized by trees and newly evolved amphibian-like creatures. The great rift that carries Isle Royale's parent rock floated across the earth's mantle, spending some time near the equator. So, for a while the rocks that support Isle Royale supported tropical forest. A hundred million years later, the first Isle Royale wild flower bloomed at about the same time that the dinosaurs first appeared. Thousands of generations of dinosaurs lived their lives on land that we now float over in the ferryboats that bring us to Isle Royale.

Two and a half million years ago, our human ancestors began to walk upright in Africa. At about that time, and very suddenly, the rocks of Isle Royale turned cold. A large portion of the northern hemisphere was subjugated by several waves of glaciers.[24] Glaciers flow like rivers and advance southward when they flow faster than the melting at the glacier's margin. When glaciers advance, they scour the earth with the full force of their weight, plucking into their icy claws every lithic object on their path—rocks, pebbles, boulders, dust, and dirt. Glaciers retreat when the margin melts faster than the rate of flow. As they retreat, all that earthen material horded by the glacier is released.

As the last wave of glaciers retreated north, just 12,000 years ago, the margin of a glacier hesitated over the southwest end of Isle Royale for a moment that lasted hundreds of years. Ice flowed forward and melted back in equal measure. The glacial conveyor belt dumped 2 trillion pounds—give or take—of rocks, pebbles, boulders, dust, and dirt on what is now the southwest portion of Isle Royale.

That material formed the thick, rich soils that now support a rich forest with a diverse mixture of deciduous and coniferous trees. Glacial deposits also settled into the valleys, taking some edge off the corrugated topography.

After that hesitation, copious melting. The glacial margin retreated rapidly and conveyed far less material on the island's northeast end. Consequently, that portion of Isle Royale is barely covered in a thin layer of soil. Bedrock pokes through everywhere, and the ridge-and-valley topography is cut in sharp relief—as the ankles of every Isle Royale hiker know.

Many valley bottoms have since been filled with swamps, beaver drainages, and small inland lakes—places moose find especially comforting throughout the summer. The soil and glacial-carved topography of eastern Isle Royale is particularly hospitable to balsam fir trees, which are the dominant forage for moose during the winter.

Consequently, the eastern portion of Isle Royale is home to considerably more moose than the western portion. Wolves arrange themselves accordingly. Because defending a territory is difficult work, wolves defend enough territory to contain all the moose they need and no more. As a result, the western portion of Isle Royale tends to support one pack with a large territory and the eastern portion more commonly supports two packs each with smaller territories. Life is gently shaped by geologic history.

Geologic history has also shaped humans' use of Isle Royale. Human families hunted mastodons on the shores of Lake Minong, which is what we call the lake that eventually transformed itself into Lake Superior. Lake Minong lapped the base of the thinning, retreating glaciers. With the recession, the waters of Lake Minong subsided some, and Isle Royale emerged like Venus on a half shell in plain view of the lakeside humans. We humans may well have been the first terrestrial vertebrates to arrive on Isle Royale. We likely brought insects stowed away in our luggage and seeds embedded in the mud that clung to our belongings.

Our ancestors first explored Isle Royale, I suppose, because humans are prone to explore. By this point in history, our ancestors were fully human, equipped with the same model of mind and body we come equipped with today. The only important differences between us and them stem from differences in our technologies and tools.

Our tools mean enough to who we think we are that we divide pre-history into long, monotonous blocks of time named for the most so-phisticated material from which we made tools. The Stone Age is, for example, sliced into finer periods, such as the Paleolithic, Mesolithic, and Neolithic, representing the early, middle, and late Stone Ages. Each period marks what seem today to be positively trivial advances in tool making.

The Bronze Age got rolling when we mastered enough metallurgy to create an alloy of copper and tin. The early Bronze Age is the Chal-colithic period, or more plainly the Age of Copper. A lump of copper can be pounded with relative ease using handheld stones to make high quality tools. The tough part is finding a lump of copper. Most copper on the planet is tied up in complicated minerals, many of which con-tain sulfur and are given erudite names such as chalcocite and chal-copyrite. That copper is impossible to extract from the minerals with-out modern chemistry.

Isle Royale was given a copper accent about 30 million years ago, after its foundation was fully formed and while it awaited glacial pol-ishing. The earth's periodic heaving and sighing created faults in Isle Royale's foundation. Mineral water seeped into the cracks and gaps left by the faulting. By some geologic accident, sulfur was absent from these mineral waters. But the solutions did contain tiny amounts of silver, minute traces of gold, and lots of copper. The water came and then went, leaving behind much of the metal that had been in solu-tion. Many of these cracks and gaps were filled with pure copper. Some of the copper dollops are smaller than a penny and others larger than an automobile. Some veins of copper run straight to the surface, and occasionally one finds a lump of copper just sitting on the forest floor of Isle Royale.

Copper is what humans found on early visits to Isle Royale. Our ancestors also figured out how to extract chunks of pure copper embedded in basalt by heating the rock with fire, pouring cold water on it, and smacking it with a stone hammer to separate the copper from the rock.

Copper tools had become state of the art in North America by about 6,000 years ago. The earliest evidence of humans on Isle Royale is from about the same time. That evidence is largely restricted to shallow pits and associated tailings from which the copper had been mined. That primitive mining carried on for about 1,000 years.

Isle Royale's copper is no curio. Isle Royale[25] is extraordinary for being among the few places on the planet where copper exists pure—ready for use, right out of the ground. Copper from Isle Royale (and Keweenaw Peninsula) also has an isotopic signature that distinguishes it from copper mined elsewhere in the world. That isotopic signature matches a copper knife unearthed at an archeological site in Illinois. The knife was found near a mastodon that died 6,000 years ago. Lake Superior copper has also been found among the artifacts of a grave site at Cahokia, a city that flourished about 1,000 years ago with as many as 20,000 people near the present-day site of St. Louis, Missouri. Other tools made of Lake Superior copper have been discovered at sites near the Gulf of Mexico and in the desert southwest.[26] The trade in Isle Royale's copper was an industry and a continent-wide commerce, probably the continent's first commerce in a nonrenewable resource.

By the time French explorers made first contact with North Americans, Isle Royale's copper had faded to legend. Jacques Cartier's second voyage to the New World (1535–36) took him up the St. Lawrence River, no farther than 100 miles past Quebec City, where he heard rumors about copper from the shores of a large lake several hundred miles to the west. It would be another 100 years before the French would make it to the Lake Superior region themselves. When they did, in the 1620s, they found some Ojibwe people, but not many. The Ojibwe people's numbers likely plummeted after the onslaught of small pox that raced ahead of European colonists. The Ojibwe people were using Isle Royale as a fishing ground, and they harvested maple syrup from the southwest portion of the island. Isle Royale was special because maple trees were rare in the forests of the nearby mainland. The Ojibwe people also harvested berries, but they did not mine. The cultural knowledge, and apparently the interest, to mine copper was lost.

European colonists made their first foray into commercial copper mining a century and a half later. In 1772, Alexander Henry attempted and failed to mine copper on the south shore of Lake Superior near the Ontonagon river. Of the miscarried experience, he wrote, "The copper ores of Lake Superior can never be profitably sought for but local consumption. The country must be cultivated and peopled before they can deserve notice."

Seventy years passed and Henry's conditions for copper mining were met. Benjamin Franklin negotiated with the British for a United States boundary that made Isle Royale part of the United States in 1783. Sometime later, the shores of western Lake Superior were well connected to the rest of the nation by rail, demand for copper had risen, and the region had been cultivated. Cultivated in the sense that the United States government "negotiated" the Treaty of La Pointe in 1842 and took Isle Royale from the Ojibwe people.[27] Copper mining on Isle Royale commenced the following year.

Over the next 12 years, 100 tons of copper were taken from four sites across the island—McCargo Cove, Snug Harbor, Todd Harbor, and in the rocks between Rock Harbor and Chippewa Harbor. The effort cost more than it produced and ended in economic failure. One of the open mine shafts associated with this effort would eventually claim the lives of three wolves in late 2011[28]—part of the population's final unraveling.

Mining technology and transportation improved to usher a second mining effort that lasted eight years, leaving marks at Island Mine, Conglomerate Bay, and Minong Ridge. This so-called Minong operation ended in 1881 with economic failure, sending 150 men and their families packing.

A third mining enterprise began in 1889 and was supported by a village of 135 people, established on the island's southwestern shore. Numerous holes were dug. No copper was ever found. The site of this village is now the place from which we base our winter field season.

Copper mining and deep geologic history are two of the primary hues in Isle Royale's polychromatic history. Any visit to Isle Royale and any contemplation of the island is enriched by knowing that history. The enrichment occurs by transforming a place from a lifeless stage into a living character with memories of the past and—I hesitate to press the keys, but cannot refrain—hopes for its future. I know this is the oxytocin talking, now. And that is the point of developing a sense of place.

We're almost ready to wrestle with how Isle Royale's history might guide a decision about whether to conserve wolf predation. Beforehand, there is one more hue in Isle Royale's history that deserves focus.

In 1903, Isle Royale supported five hotels that provided fishing and camping for a cadre of wealthy, white humans. The island provided refuge from the harried pace of modern, urban living. Well-to-do vacationers came to Isle Royale for its rugged isolation, craving the "insular character of Isle Royale." Recreation grew steadily. By the 1920s, there were six resorts, whose amenities included bowling, dancing, and golfing. Two hundred cottagers traipsed the shorelines of Isle Royale during peak season.[29] They came and went, along with their stuff, by any of the several ferryboats that serviced Isle Royale several times a week.

Isle Royale mirrored a nationwide countertrend in "wilderness" recreation that pretended to resist America's slow transformation into an urban nation. Widespread urbanization worried many because rural living is what imprinted Americans with their character. Rurality made Americans rugged and independent, and it fostered the American brand of democracy. Leastwise, that's the story some Americans told themselves.

When the nation was young, fewer than one in five people lived in cities. By 1920, about half did.[30] Crime and pollution rose alongside urban life. City life eroded physical and moral characters. The best-available therapy for this fall from grace was wilderness recreation. Again, that's the story some Americans told themselves.

That moralizing instigated tourism on Isle Royale.

Playful visitors relished what they took to be Isle Royale's untouched forests. Isle Royale's forest had, by and large, been spared the lumberjack's ax because white pine is rare on the island. Otherwise Isle Royale would likely have been razed along with the other forests of the Great Lakes region. There had been two recent logging operations on Isle Royale, but the economic difficulties of extracting timber left those operations short-lived and limited in scope.[31] Vacationers also seemed to have not acknowledged that portions of Isle Royale had been burned to find copper. That seemed long ago and those forests had grown back.

When a pulp mill was constructed on the south shore of Lake Superior, some believed the small spindly spruce of Isle Royale would become marketable. In 1922, Minnesota Forest Products Company purchased 66,500 acres of Isle Royale from the Island Copper Company.

That transaction wreaked outrage among cottagers and resort owners. The threat of logging launched a movement to protect Isle Royale. Cottagers wanted to protect their darling vacation spots. They also thought that garnering official protection for Isle Royale would raise the pecuniary value of their properties. Even the timber company was supportive, so long as they could maintain mineral rights.

The selling point offered to those unlikely to vacation on or profit from Isle Royale was that it was Michigan's last untouched bit of nature, the last remnant of Michigan's once-great forest. Isle Royale had wilderness value.

The movement to protect Isle Royale organized under the leadership of Albert Stoll, a well-connected journalist from the *Detroit News*. In 1923, Stoll and colleagues proposed that Isle Royale become a national park. They brought their proposal to Stephen Mather, director of the NPS, which at the time was a fledgling agency that oversaw 17 national parks.[32] While Mather wanted more national parks, Isle Royale did not meet the most basic criteria for beatification. Its scenery was not sufficiently beautiful. Mather rejected the proposal.

Less than two months later, Mather overturned his rejection and embraced the proposal. But Mather was not moved by the cottagers' concerns. His reversal was a chess move in the contest of national politics.

When Congress created the NPS in 1916, it charged the agency with an impossibly dualistic mission to "preserve unimpaired the natural and cultural resources and values of the national park system for the enjoyment, education, and inspiration of this and future generations." In plainer language, the NPS was to maximize enjoyment of nature without tarnishing it. Mather paid more attention to the "maximize enjoyment" portion of his mandate and less to the "without tarnishing it" portion. By his assessment, protection and valuation would follow only if enough citizens revel in these voluptuous landscapes for themselves.

The early years of the NPS also coincide with the time when Americans first fell in love with automobiles. By 1920, 1 in 10 Americans vacationed by automobile, and many of those car vacations were to wilderness areas. Mather did not follow this trend. He paved the road—literally—so that others could follow the trend that he did so much to create. Mather collaborated with colossi in industry and government to build more than 2,000 miles of road in national parks and hundreds of car campgrounds to go along. He also instigated an interstate highway system that would connect the western parks. The dust and noise of all those roads splitting magnificent landscapes were conspicuous. To many it was offensive and a betrayal of the NPS's mission.

Gifford Pinchot, director of the US Forest Service, took note. He was not offended by the roads; he was offended by the very idea of an NPS. Prior to the NPS's creation, Pinchot argued that care for national parks should be entrusted to the Forest Service.[33] As the nation's premier land management agency, they would do the job with exceptional efficiency. Pinchot was disturbed by the interagency competition that would be created between the Forest Service and a National Park Service. The Forest Service had already "lost" several parcels of land when they became national parks. That trend would surely continue with the creation of a National Park Service.[34]

Others were concerned that Pinchot's Forest Service was not sufficiently committed to the kind of protection that would be required in caring for national parks. That concern was fueled, for example, when Pinchot successfully argued against John Muir for construction of Hetch Hetchy dam in Yosemite National Park. Pinchot's management philosophy flowed from the idea that the entire universe is divided into two metaphysical categories, "man and natural resources." That is, men and stuff for men to use. The Forest Service is the agency whose mandate was built not on preserving nature but on the idea of using nature as much as desired without infringing on future generations' interest to use nature as much as they desire.

In the end, the National Park Service was created. Pinchot, believing that wilderness car parks were the NPS's Achilles' heel, developed his own plan for providing Americans with wilderness recreation—a plan that excluded roads and automobiles. The Forest Service would show the fledgling Park Service how to properly perform its mission. In developing this plan, the Forest Service leaned on the ideals of Aldo Leopold, who envisioned wilderness as "a continuous stretch of country in its natural state, open to lawful hunting and fishing, big enough to absorb a two-weeks' pack trip, and kept devoid of roads, artificial trails, cottages, or other works of man."[35]

The Forest Service's recreation plan was issued in 1922, and its centerpiece was a roadless wilderness recreation area in Superior National Forest, located in northern Minnesota just 50 miles from Isle Royale as the crow flies.

Mather realized that he might outflank Pinchot if he turned Isle Royale into a national park—one without roads. Within two months of rejecting the proposal to promote Isle Royale as a national park, Mather announced that Isle Royale would make the "finest water and trail park" and would serve as the Park Service's showcase for what a "primitive" park should be. At Mather's prompting, Congress authorized Isle Royale as a national park in 1931.[36]

Authorization meant that Isle Royale would become a national park after the private land on Isle Royale had been purchased. In 1935, Franklin D. Roosevelt broke precedent with an executive order that authorized the expenditure of $700,000 to purchase the remaining private land on Isle Royale. The establishment of Isle Royale as a national park was imminent. That same year, shortly before the lands were actually purchased, logging began on the southwest side of the island. No fuss was raised, and no one was bothered. The reason to protect Isle Royale as a park had shifted. Logging was no longer seen as a threat. The new threat was development—buildings and roads—that would ruin the island's wilderness character.[36]

Isle Royale become a national park by the confluence of two interests: some politically active locals protecting their cottages and competition between two alpha males in government. Both interests rode on a rising tide of wilderness recreation inspired by fear that Americans were losing their "manly" character.

History is a fickle guide—sometimes implying how to behave, other times how not to behave, and often not clear about which case is which. History also teaches that rich endings sometimes result from motivations that are celebrated as gallant by some and decried as hoggish by others. One's sense of place forms from *how* one uses the history of a place to guide today's relationship with a place.

In preparing for Isle Royale becoming a national park, the Park Service commissioned Adolph Murie to study the island's forest and wildlife. Murie had just earned a PhD from the University of Michigan studying deer mice in Glacier National Park. Most of Murie's accomplishments were ahead of him. He would study coyotes in Yellowstone National Park and be the first to study grizzly bears in Denali National Park. He also studied the relationship between wolves and Dall sheep in Denali. The watermark of his work ran counter to prevailing trends.

He endeavored to understand each species—not in isolation but in relationship to the entire ecosystem. This thinking is commonplace today but was novel in his day.

Murie's observations showed that wolves were not a nightmare in the dark heart of wilderness. Rather, he showed how wolves belonged and fortified the landscapes they inhabited. And they deserved to be treated appropriately. Those views made Murie unpopular in the Park Service.[37] They reluctantly eased efforts to eradicate predators from Yellowstone and Denali. Amnesty came too late for wolves in Yellowstone.

Notes from Murie's trip to Isle Royale reflect the broad scope of his study. These notes also reveal how he had been inculcated by a prominent feature of Isle Royale. In particular, he was fazed by extensive areas "in which the top of every tree is broken off, and there is little else [for moose] to eat except bark. No poplar reproduction was noted. The winter moose food is practically gone from the island."[38]

Murie had planned several reports from his observations on Isle Royale. But limited time and funding allowed for only one report. It focused on Murie's portentous unease with what moose had been doing to the place. *The Moose of Isle Royale* was published in 1934. While it is a collection of wide-ranging observations on the behavior, diet, and habitat of moose, all observations point in one direction:

> So far as the moose is concerned the overbrowsing of winter foods is most serious, for at this season the food supply is generally greatly restricted, but from the standpoint of conservation in general, summer overbrowsing may be as serious as winter overbrowsing. . . . For instance, a reduction of water plants may have a rather direct deleterious effect on the fish fauna. . . .
>
> To prevent further devastation of vegetation on Isle Royale it would seem highly advisable that control methods be initiated to reduce the moose population.[39]

Murie knew that reducing moose abundance would be difficult, and he thought it should be accomplished by any means necessary—hunting, culling, or predators. His attitude was rooted in his belief that the "land should not be teeming only with moose but teeming with all nature."[40]

Murie has been in good company to believe that unchecked moose devastate forests. Another respected biologist, George M. Wright, surveyed the flora and fauna of Yellowstone in the 1930s. After elk abundance surged in the wake of the NPS's extermination of wolves, Wright described the vegetation as "hammered" and "battered" and described the overall condition as "frightful."[41]

During the 1930s, a few luminaries saw that damage caused by removing predators would be as harmful as damage caused by roads and development. In the century that followed, the national park system became a showcase for the destructive effect that ungulates, living in the absence of predators, have on ecosystems. In western parks, overabundant elk have been trouble for Grand Teton, Rocky Mountain, Olympic, Mount Rainer, and Yellowstone National Parks. Damage caused by elk is the reason we reintroduced wolves to Yellowstone in 1995. In eastern parks, overabundant white-tailed deer have caused considerable damage, especially in Great Smoky Mountain, Shenandoah, and Catoctin Mountain National Parks.

In 1981, the National Parks Advisory Board chided the Park Service, "even the most superficial review of animal problems in the parks reveals that overpopulations are at the root of many difficulties."[42] In 1995, a respected group of wildlife biologists reflecting on the past century of management in the Park Service, which tended toward predator eradication more often than not, wrote: "One of the most wide-spread wildlife overpopulation concerns in the NPS is the four- to ten-fold increase in white-tailed deer populations in eastern US Parks. Some 49 park units in five NPS regions have identified 'possible resource imbalances' caused by high deer densities. It is a long-standing concern."[43]

Cottagers had wanted to protect Isle Royale as a means of protecting a privileged life style. But they knew such a parochial concern would not inspire broad support. Stoll and others adopted a more urgent and populist rationale: "The ax is after [the island], just as it has pursued and taken vast sweep upon sweep of Michigan's once almost measureless forest."[44] Isle Royale must be saved because it was "Michigan's last bit of untouched nature."[45] They extolled, "The heavily-timbered shoreline, hills and valleys of Isle Royale have never been molested by the lumberman's ax. . . . Preserve Isle Royale. In years to come, we shall thank the fates for the foresight of those who planned the acquisition of this property for all the people."[46]

Murie objected to both the cottagers' underlying rationale and Stoll's purported rationale. He hated that Great Lakes forests had been spoliated. But he pointed out that Isle Royale's ecosystem is importantly unrepresentative of the region. The forest is different. White pine, hemlock, tamarack, and jack pine are all common on the mainland but rare or absent on Isle Royale. The mammals were different. Chipmunks, woodchucks, field mice, flying squirrels, skunks, black bears, deer, and porcupines—again, all common species on the mainland and conspicuously absent on Isle Royale.

Murie wanted to protect and preserve Isle Royale, but he cared as much for the underlying reason. Knowing *why* protection and preservation is warranted speaks volumes about *how* to protect and preserve Isle Royale. Protecting a place for the wrong reason risks protecting it in the wrong way, which would be to not protect it. The intimate conjunction of why and how is a chink in the old adage that ends justify means.

The cottagers cared little for the brainy connection between why and how, even though little foresight is required to know that protecting Isle Royale as a national park would be an unlikely means to protecting their interest to maintain private cottages. Murie argued that the primary reason to protect Isle Royale was that "Isle Royale is practically uninhabited and untouched. The element of pure wilderness which it contains is rare and worthy of best care. True wilderness is more marvelous (and harder to retain) than the grandiose spectacular features of our outstanding parks. It alone labels Isle Royale as of park caliber."[47]

At the same time, Murie mused about introducing wolves to Isle Royale: "Were it known if and to what extent our larger predators . . . such as the timber wolf prey on moose, a possible solution to the overpopulation on the island would be the introduction of an effective predator."[48] Over the next decade, musings became conviction precipitated by his own pioneering research on the wolves and their prey in Denali. Murie advocated introducing wolves to Isle Royale, even though there was no evidence to suggest that they had ever before been present. He believed that introducing wolves would be vital to maintaining the vigor of Isle Royale's wilderness character.

Murie's position is noteworthy because he is an architect of the American understanding of wilderness. Murie was not alone. Aldo Leopold—also an architect of wilderness ideals—wrote the director of the Park Service in 1944 advocating for the introduction of wolves to Isle Royale. The director rejected the idea, concerned over "the possibility for adverse public reaction."[49]

Sigurd Olson was of like mind. He conducted pioneering research on the relationship between wolves and deer in northern Minnesota during the 1930s. Olson was also president of the Wilderness Society and the National Parks Association. He helped draft the Wilderness Act of 1964. His knowledge of wolves and wilderness led him in 1936 to chair a committee of the Ecological Society of America, whose task "was to promote the introduction of wolves to the newly established Isle Royale National Park."[50]

Murie, Leopold, and Olson all supported the introduction of a species to a wilderness area where there had been no evidence that the species had ever before lived. They knew the value of wilderness, and they knew the threats to wilderness. None saw any problem with humans bringing a species that was never known to have inhabited Isle Royale. It was intervention, it was unnatural, but there was no problem. These champions of wilderness understood that wilderness could be harmed as much by the absence of predators as by the building of roads and development.

The architects of wilderness seem to stand apart from a vocal minority of wilderness advocates of the 21st century who believe that restoring Isle Royale wolves today would affront the doctrines of wilderness. (More on those beliefs in Chapter 9.)

Laurits Krefting conducted additional research on the moose of Isle Royale in the 1940s. That experience led him, in 1951, to also propose wolf introduction as a solution to the adverse impact of too many moose. The same year another biologist familiar with the case, A. M. Stebler, recited the same concerns and offered the same solution. The appeals were rejected by the park superintendent who believed wolves were "vicious beasts" and feared a public relations disaster.[51]

The Park Service could not be persuaded by reason or science. But they were persuaded by the pizzazz of Lee Smits, another newspaper man from Detroit. He pressed hard for introducing wolves and led a private effort that resulted in the release of four captive-raised wolves on Isle Royale in 1952. The release was carried out even though it was known by 1952 that wolves had already colonized Isle Royale on their own.[52]

Three of the introduced wolves were killed or removed after they became a public nuisance and the other disappeared. In the meantime, wild wolves took root and flourished. Concerns about moose dissolved. Ten years later Durward Allen began what would become the longest continuous study of any predator-prey system in the world.

Recall that "Island Complications" told us that "out of this choppy history it's difficult to conclude what is 'natural'" and warned not to toy with Isle Royale's "biological and historical authenticity."[53]

It is a shame that histories are too often ignored, because they repay whatever attention they're offered. Particular histories of Isle Royale authentically shaped my life in particular ways. Other histories of Isle Royale have shaped the lives of others in other ways.

History does not dictate future choices. Thank goodness. A path can be charted, however, when history federates with a good moral compass. One of the compasses available to us points to what's natural. Let's see where that would take us.

9 |

All Natural

Because "Island Complications" is a prodigious essay, it is no surprise to find it reaches beyond history and strikes the heart of another fervid idea, whose jurisdiction includes the wolves of Isle Royale and expands well beyond:

> Most people who are familiar at all with Isle Royale assume that the national park's famous populations of wolves and moose are "natural" residents of the archipelago. . . . But a historical view . . . reveals a much more complicated situation. . . .
>
> All the large mammals on Isle Royale have changed in the 20th century. Coyotes and lynx have gone and wolves appeared. Woodland caribou were extirpated and moose arrived. . . . Red fox arrived circa 1925. Otter were missing for much of the 20th century but now are quite common. And a little earlier, in the late 1800s, beaver were nearly extirpated. This radical composition turnover may be an effect of island biogeography. One primary indication of island biogeography is that the island(s) being studied have only a subset of the animals and plants found on the nearest mainland. Island biogeography also routinely maps species turnover on islands, as species "wink out" and different ones "wink in."[1]

Then "Island Complications" admonishes the reader not to toy with "biological and historical authenticity." The passage alludes to a tangled knot of ideas that runs throughout much of the essay. Tracing one thread reveals a line of thought that seems to be that Isle Royale's essential, timeless nature is its insularity. On islands, extinction is a natural process. What's natural is good. Ergo, goodness is allowing the wolves of Isle Royale to go extinct. The loss of wolf predation should not be merely accepted but celebrated.

Logicians—they are the lot who study logic as a vocation. You may be unable to name a single living logician. They are rare for sure, but they exist, and they deserve our gratitude for their vocational sacrifice. They endure soul-scarring education to be worthy of (among other logician duties) abstracting the logic of specific, detail-filled arguments into generic expressions stripped of distracting minutiae. We'll perform one of their abstractions, and you'll want to thank the next logician you meet for his or her service.

Here it goes, alakazam. The generic form of that logic about extinction on islands is as follows: X is natural, what is natural is good and right, therefore X is good and right. That logic has a notorious sibling: Y is unnatural, what is unnatural is wrong and bad, therefore Y is wrong and bad.

The simple, stripped-down form facilitates seeing how a diversity of specific thoughts—really consequential thoughts—are bound by a single logic: Herbal medicines are good because they are all-natural. Vaccines are bad because they are unnatural. American Spirit cigarettes must be good because they advertise "100% additive-free tobacco" and that "it's only natural." Homosexuality must be wrong because it is unnatural. This is just a small sample of the absurd conclusions to arise from this particular logic.

Another responsibility of logicians is to maintain and guard an encyclopedia of logical fallacies. These are stripped-down forms of argumentation that are essentially defective but still trick people on a regular basis. You'll recognize some entries in this compendium by their common names: slippery slope, red herring, straw man, and cherry picking.[2] Because logicians also have a style to maintain, many entries are known by their formal Latin names. One of the most notorious members in this encyclopedia of poor reasoning is *Argumentum ad Naturam*, argument by appeal to nature, and it seems to have slipped into "Island Complications."

Eighty percent of Americans, according to recent polling, believe that foods labeled "natural" are free of antibiotics, GMOs, toxic pesticides, and artificial colors and flavors.[3] They are wrong, because marketers are unscrupulous *and* because nothing in that list is "unnatural." *Argumentum ad Naturam* employs twin subterfuges by presuming that the difference between natural and unnatural is meaningful and by presuming that a thing is good simply because it is natural or bad because it is unnatural. A thing might be natural and good, but being natural is not enough of a reason for its being good.

Having exposed the general peril of thinking "X is good because its natural" and "Y is bad because its unnatural," let's untangle the next knot, whose thread we find in observing the tendency for island species to "wink out" and "wink in."

Johann Forester was thoroughly immodest about his mission as ship's naturalist: "My object was nature in its greatest extent; the earth, the sea, the air, the organic and animated creation."[4] He would certainly see a great deal of nature, as the ship's captain, James Cook, was planning to sail around the world via the southern oceans in search of an undiscovered continent. The breadth of Forester's mission required attending all phenomena, from the obviously important to the assuredly obscure. Among his documentations are tallies of the species found on each of the scores of islands that he visited. After reflecting on those tallies, he discerned that "Islands only produce a greater or lesser number of species, as their circumference is more or less extensive."[5]

That odd observation is remembered today as the seed for island biogeography—the study of how the geography of an island affects its biology. To mothers who learn that their child aspires to become island biogeographers, patience and faith. Some really upstanding people have pursued this dream.

For example, Professor Darlington from Harvard documented, in 1957, a pattern like the one Forester described. He did so with more focus and precision, showing that the number of beetle species on Caribbean islands increased with the size of the island.[6] He observed similar patterns in the diversity of reptiles and amphibians among the same islands.

In 1962, an amateur bird watcher noticed the same pattern in the number of bird species in the East Indies, plants in the Galápagos, and terrestrial vertebrates on islands in Lake Michigan. The "amateur" was Frank Preston, whose day job was researching the physics of glass—work that led to the development of Corelle glassware. What struck Preston about his distinctive recreational pastime was the similarity of all those species-area relationships. Being facile with mathematics, he discovered that the patterns could be described by a simple equation. This equation is the $E = mc^2$ of ecology and essentially as simple: $S = bA^z$. Preston's equation and the data he collected are shown in the graphs below.

Regardless of where on the planet we find islands, and regardless of the kind of species we consider, the number of species on an island

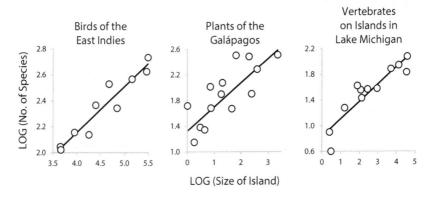

One of the most basic patterns in nature is for the diversity of species to increase with the size of the area being considered. Data from Preston 1962a

is so consistently related to the island's size that the number of species is pretty reliably predicted from that single equation. Whenever nature repeats herself in such a mathematically obedient manner, the Pavlovian reaction of a well-trained scientist is to nose past the curtain to discover whether some law of nature is pulling levers and pushing buttons.

A couple of years after Preston's discovery, two great ecologists of the 20th century, Robert MacArthur and Edward O. Wilson, developed the fundamental reasoning for the precise and repeatable pattern: extinction occurs more frequently on small islands and less frequently on larger islands, where populations can grow to larger sizes, thereby insulating themselves from extinction. The establishment of species (and reestablishment of extinct species) occurs more frequently on islands close to the mainland (or another island) and less frequently on more isolated islands. The balance between those two rates (extinction and colonization) determine the number of species on an island.

MacArthur and Wilson backed that intuitive reasoning with extraordinary mathematical deduction. They published their findings in a 1963 paper with the impenetrable title, *Equilibrium Theory of Insular Zoogeography*. The paper's modest conclusion is the archetype of stupefying erudition: "The main purpose of the paper [has been] to express the criteria and implication of the equilibrium condition, without extending them for the present beyond the Indo-Australian bird faunas."[7] The dynamic-duo realized the poverty of their writing shortly after the paper was published. They reworked the presentation, corralled more equations, redoubled the explanations, and published a revised treatment of the same topic. The effort paid a large dividend. Species-area relationships, like those depicted in the previous set of graphs, are now understood to be a universal pattern that goes far in explaining one of the most basic features of any ecosystem: why each ecosystem has as many species as it does, no more and no less. Many ecologists hail this species-area relationship as one of the few laws of nature that rule in ecology.

That would have been enough, but MacArthur and Wilson had more insight to convey. The treatise's finale is, at once, understated and grandiose: "The same principles apply, and will apply to an accelerating extent in the future, to formerly continuous habitats now being broken up by the encroachment of civilization, a process graphically illustrated by . . . maps of the changing woodland of Wisconsin."[8] That text is accompanied by simple maps of Cadiz township, in southern Wisconsin, showing the loss of forest and the increase in areas cleared of forest.

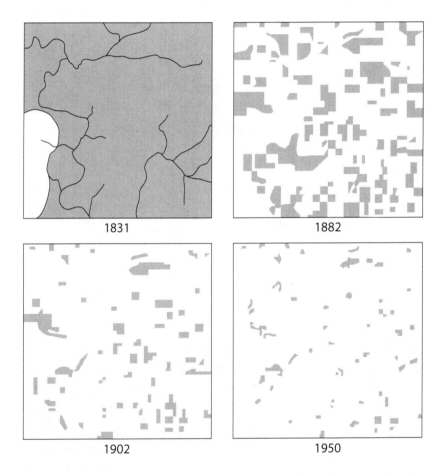

Maps of Cadiz Township over time, showing the loss of forest (*gray areas*) and the increase in areas cleared of forest (*white areas*). Adapted with permission from Curtis 1956

The new ending still leaves much to the reader's imagination, but plenty enough readers applied that imagination. The "islands" in island biogeography are not necessarily literal islands. They are a metaphor for caves, streams, mountain tops, and tide pools. MacArthur and Wilson were invoking any kind of habitat (say, forest) surrounded by some other kind of habitat (say, cleared fields) that would leave a (forest-dwelling) species as if on an island. The tragic, jarring realization is that humans were turning vast swaths of intact habitat into small, isolated islands.

Sixteen years later, in the summer of 1983, a young William Newmark packed his car, unfolded his map of western interstates, and toured the national parks. He was less interested in being awestruck by grand scenery than in meeting with park naturalists, looking for answers to two questions: What mammals live in this park today? What mammals lived here decades ago, when this place first became a park? The answers earned Newmark a PhD and a place in the history of conservation science for demonstrating one of the most basic principles of conservation. Most of the parks had lost a number of species. They went locally extinct, quite naturally, because we had turned parks into small, isolated islands—rich habitat surrounded by an ocean of land taken for human use.

Preston—the binocular-wielding, glass physicist—understood the repercussions a quarter of a century earlier: "If what we have said is correct, it is not possible to preserve in a state or national park, a complete replica on a small scale of the fauna and flora of a much larger area."[9]

The wisdom of island biogeography is *not* its explanation for why extinction is a natural process on an island. Island biogeography is a warning for the fate of life banished to islands. Because humans have generally neglected the warning, island biogeography's backup lesson is a prescription for averting annihilation of the life that humans exile to islands created by habitat fragmentation. To reduce the risk of extinction on an island, provide for an adequate stream of immigrants whose arrival is more likely to depend on assistance from humans.

"Island Complications" warns of peril if we "supersede the insular character of Isle Royale." Isle Royale is, for sure, marvelously well characterized by its island-ness. But island-ness is not Isle Royale's distinction—the planet is rather generously sprinkled with literal islands, and humans have battered the planet into a sea of habitat islands.

What sets Isle Royale apart is being inhabited by wolves freed from the cruelty of human persecution, moose who do not pass each autumn in fear of the hunter's bullet, and a forest unfettered from industrial logging. The distinction is not the presence of wolves, moose, and forest. Isle Royale's distinction is for being a place where humans have allowed herbivory and predation—the *relationships* that bind wolves, moose, and the forest—to manifest unencumbered by human imposition—freed from neck snares, bullets, and feller bunchers. Isle Royale is distinctive for being one of the last such forests.[10]

Is that distinction worth saving? It's not intrinsic to Isle Royale's basalt bedrock, sparkling harbors, or even its being an island. The distinction is not even part of Isle Royale's deep history. Humans foisted the distinction onto Isle Royale because humans persecute large predators in just about every forest that can be reached. That is, just about every forest on the planet.

The question remains, is that distinction worth saving? Saying yes is a weak gesture toward restitution for humans having overindulged at the expense of others.

Is extinction on islands a natural process? Yes. It's natural in the same way that it's natural for glaciers to melt in the warm sun. The proper questions, however, are *How did it get so warm?* and *How did Isle Royale become such a distinctive sanctuary for predation and herbivory?* This line of questioning leads to the realization that natural and unnatural are less meaningful categories than is often supposed. The concern is further illustrated by another relevant question: Did the wolves go extinct because Isle Royale is an island or because of disease, mine-shaft drownings, and climate change—all brought by humans? Finally, even when the word "natural" is brimming with genuine meaning, it is still an insufficient reason for goodness.

Argumentum ad Naturam go back from whence you came, with your neck beneath the logician's boot.

If "extinction on an island is good because it is natural" does not amount to much of a reason, what about "extinction of Isle Royale wolves is good because the wolves are unnatural"? Maybe that's a reason to celebrate the loss of predation on Isle Royale. "Island Complications" offers exactly that impression.

First some background. Wolves crossed an ice bridge to Isle Royale in 1948 or 1949 and established a population. The colonizers would have been a pair of wolves or an entire pack. If it had been a pair, there is every reason to believe they would have reproduced and become a pack within months of arriving. When a pair of wolves has food and space, that's what they do.

In January 1952, Lee Smits, a publicist from Detroit, convinced the NPS to release captive wolves onto Isle Royale. He convinced the park even though it was known that a wolf population had just established itself. The Park Service granted permission with uncharacteristic swiftness. In August 1952, four wolves from the Detroit zoo were released—Queenie, Lady, Big Jim, and a black wolf named Adolph. These wolves had never before seen the wild or been on their own. Having been raised and fed by humans, they were not afraid of humans and had no experience killing moose or anything bigger than perhaps a rabbit. The poor wolves did not know how to behave in the wild and became a nuisance. Within a month of being released, park staff resolved to remove the wolves. One wolf was trapped, two others were shot and killed. Big Jim was never seen again.

Of Big Jim, Durward Allen wrote in the 1970s: "Jim was left as the sole survivor in the wild, and for some years he was something of a legend on the island. As we see it now[11] . . . Jim would have been at a great social disadvantage in any attempt to associate with his own kind. It is unlikely that he made any contribution to the genetics of today's population."[12] That is the last time, to my knowledge, anyone wrote anything substantive or original about those four wolves. There was no concern for the exceedingly remote possibility that Big Jim contributed genes to the Isle Royale wolf population, until 2013 when "Island Complications" asked "could the founder [of the Isle Royale wolf population] have been one of the Detroit Zoo wolves, a female nicknamed 'Queenie'? Moreover, could another of the zoo wolves, a male called 'Big Jim,' have survived long enough to interbreed? . . . Any wolf reintroduction decision should ideally be informed by a determination of whether Queenie or Big Jim had a founding effect among Isle Royale's wolf population."[13] The relevance of being so ideally informed is the not so subtle insinuation: if the wolves of Isle Royale were marked by the imprint of humans—that is, if they are not natural—then wolf predation should not be restored. The loss should be greeted with relief.

The NPS reinforced that concern with "educational" materials designed to influence the public's view about whether to restore wolf predation on Isle Royale. According to these materials, "one or two wolves made it to Isle Royale"[14] in 1949. Because one wolf cannot establish a population, the implication is that it is at least plausible that the population's establishment depended on zoo wolves. To cinch that possibility, the NPS materials describe the fate of the four zoo wolves as one having been trapped and another shot and killed. Of the other two, they said, "The remaining two wolves were shot at (but not hit). These wolves were occasionally seen and presumably reverted to a 'wild-state' and augmented the population." As well, they stated, "Jim was one of the wolves known to survive and assimilate onto the island." The other wolf that survived is identified as Queenie. Another educational document from the NPS simplifies the story: "In 1952 one or two of four stocked wolves successfully assimilated into the population."[15]

The Park Service account allows only one of two possibilities: the wolves of Isle Royale exist because humans introduced wolves, or the wolves' gene pool was contaminated with genes of wolves brought by humans. Suppose it is so. Suppose that Queenie or Jim did contaminate the gene pool. Further suppose that circumstance to be offensive. Then, I guess the best you could hope for is to (I cannot believe I'm writing this) cleanse the population of its impure genes. To that queasy end, no genetic traces of the Detroit wolves can be detected today.[16] If they had been part of the population, their genetic imprint is either indistinguishable from the rest of the gene pool or it was removed by a process of mate selection and offspring survival that favored genes carried by the wild-born wolves.[17] That process is called natural selection. Perhaps the name of that process is comforting to those with undue affinity for what's-natural-is-good.

Phil Hedrick, a conservation geneticist renowned for his knowledge of wolf genetics, offered an alternative reaction to the unlikely possibility that the genetic heritage of Isle Royale wolves is impure: "So what? The wolves from Detroit were wolves and all they could have passed along are wolf genes."

Shaming Isle Royale's wolves because they are unnatural works like other *Argumenta ad Naturam,* by equating impurity with being unnatural and presuming that goodness demands purity. The same shocking syllogism is used to support white supremacy.[18] White supremacy is a moral offense; naïvely falling prey to *Argumenta ad Naturam* is not. But moral offenses are often fertilized by deceptive reasoning. So long as racism persists, we are responsible for struggling against it. Rejecting racism includes recognizing and rethinking the values that underlie its corrosive logic. I am not the first person to point out that undue adoration of "purity" is one of those values.[19]

"Island Complications" conjures more antipathy for restoring wolf predation through its concern that Isle Royale's moose are also unnatural:

> Bill Peterson . . . wrote [in a 1998 article]: "In the early 1950's, Dr. Clay [medical doctor] stopped at a gas station in Mafeking, Manitoba. . . . An elderly man . . . informed him [Dr. Clay] that . . . he had lived in Minnesota when he was young and in about 1907 (perhaps 1905) he had been hired by the state of Michigan to work that winter with a crew live trapping moose near Baudette, MN. They captured either 11 or 13 moose but, in late winter,

he became ill and was unable to accompany the others as they hauled the moose to Two Harbors, MN, where they were loaded onto barges and taken to Isle Royale."

Cochrane, the author of "Island Complications," then invokes his own authority on this matter:

> As someone trained to evaluate oral history and narratives, I think there are elements to this story that give it credibility (namely its specificity of place and activity). There is also the possibility that this story is only partially correct: the entity at that time with the money and interest to make this happen was not the state of Michigan, but rather the Washington Harbor Club, a private club with some of the most well-to-do Duluth businessmen of the day. The club owned various buildings on the southwest end of Isle Royale. . . . Club members also owned railroads that ran from Baudette to Duluth and Two Harbors, and so had the physical means to transport moose by railcar to Two Harbors.
>
> This alternative story . . . provides a more practical explanation . . . than does the prevailing narrative. Further, this explanation does not depend on the exceptional event of a male and female moose swimming miles to a grey mass on the horizon (Isle Royale) that they might not be able to smell in the wave troughs of Lake Superior seas.

This account contrasts with accounts by people who were as knowledgeable as any, wrote at times much closer to the putative event, and offered no hint that people brought moose. Those people seem to take for granted that moose arrived on their own. This is the impression one gets from Allen writing in the 1970s,[20] Mech writing in the 1960s,[21] and Murie writing in the 1930s.[22]

In summary, the evidence for humans establishing Isle Royale's moose population is that some guy (B. Peterson) says that another guy (unnamed, elderly man) told another guy (Clay) in the parking lot of a gas station that it was so. If that's not right, then the evidence is that some people with strong connections to Isle Royale had the "physical means" to do so.

To capture a dozen moose during late winter in the early 20th century in a remote corner of Minnesota and move them to a more remote island is no mean feat. The kind of people interested in performing such feats would also be interested in an audience for their accomplishment and not shy about sharing the good news. Nevertheless, no written record of the event—not in government archives of Michigan or Minnesota, not in newspapers of the day—has been found.

The establishment of a species in new place far from where they had otherwise lived is impressive. When I reflect on those cases, I think, what an exceptional event. Pine martens colonized Isle Royale. Tiny, weak-winged fruit flies colonized the Hawaiian archipelago, some of the most remote islands in the world. It's no less remarkable that humans colonized Hawaii. The list of dumbfounding dispersals is dumbfoundingly long. What's remarkable is how many times these exceptional events have occurred. Finally, moose are capable of swimming to Isle Royale and have been observed often enough swimming in cold waters far from shore.[23]

I cannot document the means by which moose arrived to Isle Royale. No one can. But it's far from clear that humans brought them. Nevertheless, let's stay with the concern of "Island Complications" for just a moment more. How would humans having brought moose to Isle Royale become the linchpin for a reason to allow the wolves of Isle Royale to go extinct? No one has said, so I can only suppose that one would have to argue that the moose of Isle Royale are unnatural by virtue of a rumor that humans brought them. That unnaturalness infects anything requiring the presence of moose. That means wolves are doubly unnatural—they and their food were both brought by humans. Extol the demise of that unnatural monstrosity.

Good Lord.

In 2012, the NPS solicited the general public for their views on what, if anything, should be done about the collapse of wolves on Isle Royale. Of the 1,000 or so people who contributed thoughts, about 12% advocated doing nothing, no matter what.[24] Of those folks, a large share recited, verbatim, a message from Wilderness Watch, an NGO and self-identified guardian of the wilderness: "Nearly all of Isle Royale's 134,000 acres is Wilderness, a fact that should guide any future management decisions, including a possible reintroduction [of wolves]. Wilderness Watch is strongly urging the NPS to refrain from reintroducing wolves, and rather let Nature take her course, even if that means the wolf population might become extirpated."[25]

The Wilderness Watch spokesperson for the Isle Royale case, Kevin Proescholdt, believes wilderness is a landscape "unrestrained, [and] that these areas would be unconfined, unmanipulated, free to take their own course."[26] The *Lansing State Journal* elaborated on Proescholdt's beliefs:[27]

> People tried to talk the writer [of the Wilderness Act]—Howard Zahniser—out of using "untrammeled" in the law, but he did so precisely because it has such a specific meaning, Proescholdt told [an] audience in June. Zahniser, he said, wanted to avoid exactly the kind of situation being debated on Isle Royale today. He understood the human temptation to try to fix things, when all too often we just make things worse, he said. Proescholdt calls it the slippery slope of manipulation. . . . Restraint is difficult, he said. Especially when we're talking about a species as iconic as wolves. But it's what the Wilderness Act requires of us. "To know wilderness is to know profound humility," Proescholdt said, quoting Zahniser. "It's to know one's littleness. Perhaps in the end, this is the distinction of wilderness to man."

The essence of Wilderness Watch's view is *if wolves go extinct, then so be it, because Isle Royale is a wilderness, a place where we should let nature take its own course.*

Some of the grandest *Argumenta ad Naturam* are dressed in the ideas of wilderness. Wilderness is also easily snarled by its three interwoven strands. First, wilderness is a type of land. Of all the types, wilderness land is extreme for being the most natural and the most protected from humans. Wilderness is also a legal designation in the United States, per the Wilderness Act of 1964. Isle Royale is both a national park and a legally designated wilderness. According to the Wilderness Act, for a parcel to be legal wilderness, it must be at least 5,000 acres,[28] the "imprint of man's work [must be] substantially unnoticeable," and it must provide "outstanding opportunities for solitude or a primitive and unconfined type of recreation." That's what the law says. What that language means has been largely left to policy makers who have formed committees, held hearings, and produced stacks of three-ring binders to say what it does and does not mean. Many Americans have never seen any of the more than 750 tracts of legally designated wilderness in the United States. Some humans hold no interest to visit wilderness—in spite of genuinely loving nature. Many do not have the financial means to readily spend the night in a wilderness. Depending on what one thinks wilderness is for, limited attendance may not be a bother and might even be an asset. In any case, few claim significant time in wilderness.

Wilderness is, secondly, a kind of experience. You do not have to be in a wilderness to have a wilderness experience, and you can be in a wilderness and not have the experience. This is a feeling. Sometimes it is a sweaty, hungry, and dirty-with-the-earth kind of feeling. Sometimes it is a sense of immutable timelessness and connection with the cosmos. Other wilderness experiences tend toward physical terror or existential alienation. A unifying feature of wilderness experience is a distinct awareness of being alive. While those sentiments characterize wilderness experience, they do not distinguish it. Wilderness experience is distinguished in being triggered by one's surroundings, though the threshold that triggers a wilderness experience varies. Some need death-taunting feats of athleticism performed many hours or days from medical assistance. Others have it by their second mosquito bite.

Finally, wilderness is an abstraction to which the whole of our relationship with nature is anchored. Wilderness, like other powerful abstractions—justice, freedom, balance of nature—has a history and distinctive cultural morphs. Wilderness is an untamed kaleidoscope of ideas that flickers insight when the barrel is held to the light and turned.

A good place to begin is quite a few years ago when the people lost their trust in Yahweh. He got irritated and made them wander through the wilderness for 40 years. Hundreds of years later, Jesus prepared to become a fisher of men by subjecting himself to the devil's temptations. The venue for this dedication was, of course, 40 days and nights in the wilderness. Many more hundreds of years later, Puritans colonized America with their biblical notions of wilderness, believing that life is a "terrific drama in which God and the devil were joined in struggle toward a divinely appointed resolution. . . . Here in the New World wilderness, amidst wolves, bleak woods, swamps, and cruel Indians, the drama was more fearsome than it had ever been in those lands from which the Puritans had chosen exile. Here the drama was . . . stripped to the essentials."[29]

While there is a distinct Judeo-Christian sense of wilderness, other cultural morphs bear similarity. The selfish Prince Siddhartha was transformed into the Buddha after many dark nights in the wilderness. Vision quests and related ceremonies, practiced by some indigenous cultures, help one enter adulthood or discover an elusive meaning in life. The metamorphosis is often facilitated by occurring in a sacred and natural place, away from humans—in other words, in the wilderness.

These notions of wilderness emphasize struggling and suffering—punitive suffering, transformative suffering, redemptive suffering, various kinds of suffering. But wilderness is not any kind of suffering. It is suffering prompted by deprivation but not any kind of deprivation. It is engendered by stripping oneself from civilization, and the reward for this divestment is the impact on the soul and moral character of some.

Having the bravery to find one's own moral compass is genuine freedom and maturity, according to Søren Kierkegaard, 19th-century philosopher and "father of existentialism." He illustrates this freedom with the Old Testament account of Abraham's near sacrifice of his son, Isaac. On their way to Mount Moriah, where the deed would be done,

Abraham took Isaac on a three-day trek through the wilderness to dispossess himself of human culture. Only by repudiating civilization could Abraham endure the insanity of embracing the commitment to kill his son and then not doing it. Moral freedom and maturity are found in the wilderness, exiled from society.[30]

Wilderness is characterized not by the abundance of nonhuman life but the absence of humanity, culture, and civilization. Wilderness is less a physical place and more a psychological experience that happens to be greatly aided by a landscape where the trappings of humanity are less noticeable.[31] Similarly, Ralph Waldo Emerson and his transcendental followers believed the human spirit is corrupted by society and restored by wilderness.

The three strands of wilderness—land, experience, and abstraction—can converge to become a land valued for its capacity to offer a kind of experience that resonates with some prized abstraction. Follow Kierkegaard and Emerson and we'd manage wilderness lands to provision mental health by fasting from civilization.

The Wilderness Act is not far from this sentiment for its focus on solitudinous re-creation of the human spirit in landscapes where the "imprint of man's work [is] substantially unnoticeable." Attending divine recreation is laudable but sometimes results in amusing irony. For example, Isle Royale is mostly, but not entirely, designated as legal wilderness. About 1% of Isle Royale is excluded from the wilderness designation to allow for some modern comforts—showers, gift shops, frozen pizzas, beer, and other trademarks of humanity.[32] Maps help distinguish wilderness from the civilized areas. Foot trails pass through both areas. The civilized segments of foot trail are accompanied by interpretive signs highlighting geological, zoological, and botanical insights about Isle Royale. On the approach to one of the wilderness boundaries, not far from a sign interpreting remains of a 19th-century mine pit, is another sign announcing the wilderness boundary: "Beyond this point there will be no more interpretive signs . . . you will have to provide your own interpretation."

The visitor is asked to focus on the imminent absence of "man's work" (interpretive signs) and expected to let power boats go unnoticed, while in sight and sound of them from either side of the sign. Furthermore, to preserve the wilderness experience of powerboaters, backpackers are not allowed to set a tent that can be seen from the water. Wheels are generally forbidden in wilderness; kites and frisbees are discouraged. As potent reminders of human culture, they too easily ruin wilderness experience. But pay no attention to GPS units, space-age camp stoves, and 0.2-micron water filters.

I'm fine with those inconsistencies, and the salience of these examples lies elsewhere. First, wilderness is often managed to guide recreators toward a perceptual experience of deprivation in a land untouched by humans. Shortly, we'll see how that guidance can become hurtful. Second, knowing *how* to treat a wilderness area depends on knowing *why* it is valued. Why comes first; how is second. Recreation by the deprivation of humanity is one value. When that value conflicts with another, it is necessary to prioritize the values. For example, the convenience of powerboating around Isle Royale is prioritized over traditional wilderness values. That's not a justification of the prioritization, just a description of it. Soon we'll discuss how prioritization of values influences judgments about Isle Royale wolves. For now, let's look at some of these other wilderness values.

Some say the rugged, independent character of the American people is a gift from wilderness. Bob Marshall, who founded the Wilderness Society in 1935, believed wilderness is important for its "fundamental influence in molding American character."[33] Aldo Leopold believed wilderness is "the very stuff America is made of."[34] They echoed the scholarship of American historian F. J. Turner, who convinced many that subjugation of the wilderness is how European colonists were transformed into true-blooded Americans.[35]

As the American frontier dissolved, some feared that America would lose its national character. Wilderness recreation was a reaction to that fear and a critique of weaker forms of outdoor recreation, like cottaging and car-camping, that were becoming popular during the 1920s. Leopold said as much when he wrote that wilderness recreation is a "means for allowing the more virile and primitive forms of outdoor recreation to survive."[36] These virile activities required outdoor knowledge and skill, like building a fire without a match, constructing a lean-to from fresh-cut saplings and grasses woven into twine, and nourishing one's self by hunting and gathering. Wilderness was solace for a counterculture reacting against the modern life that deprived itself of nature.

For many, wilderness recreation was less a pastime and more an orientation toward life. Kindred souls found each other beneath the banner of woodcraft, "the art of getting along well in the wilderness by utilizing nature's store house."[37] Woodcraft thrived as a counterculture for a few decades and then merged with the mainstream and faded. It found expression, for example, in Boy Scout manuals such as The "How" Book of Scouting (1938), which instructs boys, "Make it yourself. A real red blooded, HE boy would make his crotched supports, trammel bar, tongs, pot hooks, forks, and spoons when and where he needs them."[38]

Wilderness for mental health slumped to wilderness for American nationalism, and now it descends to nauseating machismo. If the Book of Scouting isn't "man" enough, there is Sigurd Olson's "Why Wilderness?," published the same year:[39]

> In some men, the need of unbroken country, primitive conditions and intimate contact with the earth is a deeply rooted cancer gnawing forever at the illusion of contentment with things as they are. . . . These men whereof I speak. I have seen the hunger in their eyes, the torturing hunger for action, distance and solitude, and a chance to live as they will. I know these men and the cravings that is theirs; I know also that in the world today there are only two types of experience which can put their minds at peace, the way of wilderness or the way of war.

For Bob Marshall, the power of wilderness rises from being the "moral equivalent of war."[40] Teddy Roosevelt's wilderness experience seems to have been an irresistible urge to carry a gun to kill as many living creatures as his skill and supply of ammunition would allow.[41]

Conquering and suffering epitomize the common experiences of wilderness on Isle Royale today. After lugging packs over rocky trails, through oppressive humidity and soul-draining rain in the company of black flies, civilization-deprived wayfarers can rest their blistered feet and enjoy a tiny recess from wilderness in one of the 50 or so shelters that are sprinkled across the island. Each is about 8 feet by 16 feet and consists of a planked floor and roof held up by three walls and a screened front.[42]

The insides of these shelters are dimly lit by spruce-filtered sunlight and are saturated with graffiti. In some shelters the only place for a new expression is atop the oldest graffiti, barely legible and dating to the early 1980s. Isle Royale may be the most graffitied wilderness in the world—I mean the expressions on the structures, not the structures themselves.

The park has an artist-in-residence program for the most gifted conveyors of human experience and sign-in books at visitor centers for the rest of us with normal powers of expression. But graffiti is different—it's anonymous, uninhibited, and primal—like the pictographs of Lascaux.

Certainly not every kind of visitor graffities the wall. The most prudish are unlikely participants. However, the abundance of graffiti in relation to the annual number of visitors suggests that graffiti writers are far more common on Isle Royale than in the general public, or that many lose their graffiti virginity on Isle Royale. The graffiti is the expression of hundreds, if not thousands, of common experiences in the wilderness.[43] The common styles of graffiti are all represented here in the wilderness— aphorisms, initials, poems, nonsense, drawings, and simple proclamations of love, like Tom + Mary, circumscribed by a heart.

The subject matter of this graffiti is what interests me. Some subjects are banal, like urban graffiti, and suggest we are the same people in the same state of mind while in the wilderness, as we are at 2 a.m. in a public restroom. Some of the graffiti is transcendental extolment of nature's sublimity, such as a Whitman poem. Some of it is silly. "We tried to sail a tent." "Al lights his farts" and "poop rainbows." Not especially funny, just silly.

My nonscientific survey indicates that the most common category of wilderness graffiti is represented by personal achievement:

I HIKED IT, JD, 7-6-94

I CAME, I SAW, I CONQUERED!—JUNE 2011

WE HIKED 110 MILES IN 7 DAYS. SUCK ON THAT.—KIM

And other low-grade expressions of conquest and physical accomplishment:

3 MILE

MOSKEY BASIN

W. CHICKENBONE

HATCHET LK.

LITTLE TODD

N. LK. DESOR

WASHINGTON CREEK

HUGINNIN COVE TRAIL LOOP-WASH. CK.

BOAT TO RH-3 MILE

TOBIN HARBOR TRAIL-RH-SCOVILLE POT. RH

11 DAYS

70 MILES +

3 HALF GALLONS OF WHISKY

The shelter walls are littered with hundreds of simple lists of days spent in the wilderness, miles hiked, sites reached, numbers of fish caught, and numbers of moose spotted. If I am wrong about the conquering theme, then the most common subject matter of wilderness graffiti is represented by this specimen:

I WAS MANGLED BY THE WOODS

HERE I LIE

A TWISTED PIECE OF FLESH

J. B.

The shelter walls are marked with hundreds of references to cold, rain, mosquitoes, physical toil, and sore feet.

MY TRIP OF A LIFETIME TURNED OUT TO BE THE JOCK-ITCH JOURNEY!

2011

ROSES ARE RED, VIOLETS ARE BLUE, "MOUNTAIN HOUSE" FOODS SUCK

SO COLD, WET AND RAINY

The old abstractions of wilderness—suffering, deprivation, and conquest—live strong in common wilderness experiences of today.

One of the darkest wilderness themes is found in the words of Chief Luther Standing Bear, who explained how his people saw what white people claimed for wilderness:[44]

> We did not think of the great open plains, the beautiful rolling hills, and winding streams with tangled growth as "wild." Only to the white man was nature a "wilderness" and only to him was the land "infested" with "wild" animals and "savage" people. To us it was tame. Earth was bountiful and we were surrounded with the blessings of the Great Mystery. Not until the hairy man from the east came and with brutal frenzy heaped injustices upon us and the families we loved was it "wild" for us.

Yellowstone National Park had been inhabited by Native Americans for millennia before the arrival of white settlers. Within five years of Yellowstone being declared a national park, peace-abiding Native Americans were evicted because they frightened park visitors.

Yosemite was inhabited for seven millennia before white soldiers settled in the valley in the 1850s. At that time, the residents called themselves Ahwahnechee. Their neighbors called them "those who kill," which in their language sounds like *Yos s e'meti*.[45] Yosemite's main supporter, John Muir, evinced streaks of racism, one example of which is relayed here:[46]

> In "Our National Parks," a 1901 essay collection written to promote parks tourism, [Muir] assured readers that, "As to Indians, most of them are dead or civilized into useless innocence." This might have been incisive irony, but in the same paragraph Muir was more concerned with human perfidy toward bears ("Poor fellows, they have been poisoned, trapped, and shot at until they have lost confidence in brother man") than with how Native Americans had been killed and driven from their homes.

The wilderness that European colonists found in southeastern North America was occupied by people of the Mississippian culture, which flourished for 800 years (between about 800 and 1600 CE). The center of that culture appears to have been Cahokia, located near present-day St. Louis, which had grown to 20,000 people by about the year 1200 CE. European colonists took more than 250 years to build a city that large in North America.[47]

Many of the wilderness lands that colonists found must have seemed genuinely void of humans, and disease contributed importantly to that void. Native Americans had never before been exposed to European diseases, especially small pox. These diseases raced ahead of the colonist like a wind-driven prairie fire, obliterating many Native American communities ahead of the colonists themselves. As many as 90% of Native North Americans are believed to have been killed by European diseases by 1650.[48]

Adoring wilderness as landscape untouched by humans is harmful for denying the humanity of Indigenous Peoples. Some environmental philosophers think the stains of hyper-androcentrism and racism make wilderness an unredeemable abstraction. I don't know. I believe in redemption, but that requires sensitivity, ferreting out shameful residue of past ideals, and supersession by an impeccable ideal.

In any case, wilderness ideals evolve inexorably and one of the most recent ideals emerged in the mid-20th century as humans became more numerous and wilderness became less abundant. There were too many wilderness recreationists, and they were beginning to ruin the remaining wildernesses upon which they depended. By the 1960s and '70s, many wilderness managers were actively engaging the problem. Most approaches involved limiting access in one way or another, and no approach seemed acceptable.[49]

Then came the bright idea to educate backpackers to be lighter on the land. Tents would be traded for lean-tos, backpacking stoves would replace campfires, synthetic sleeping bags would substitute for beds of cedar boughs, and dehydrated meals in a foil bag would stand in for hunting and gathering. This was the birth of a new wilderness ideal, known today as "Leave No Trace," which allowed for "long hikes, temporary camps, and . . . observing, appreciating, [and] visiting," but "above all else, the objective of Leave No Trace was leaving the wilderness unchanged."[50] Wilderness recreationists loved this new ethic, which seemed to preclude the need to meter people's access to wilderness.

But devotion to Leave No Trace is troubling for refocusing attention from knowledge about living well from nature's store house to an obsession with gadgets that insulate from nature.[51] Aldo Leopold offers a prescient warning against such a development in *A Sand County Almanac*: "A gadget industry pads the bumps against nature-in-the-raw; woodcraft becomes the art of using gadgets."[52] The obsession with wilderness toys is attended by consumerism: "On this point, a comparison between the 1933 Boy Scout Manual's wonderfully detailed instructions on making a pack from scratch (which included which types of wood to select for the thwarts versus the ribs) gave way to the 1979 Boy Scout Manual's imperative 'buy the best . . . you can afford.'"[53]

According to the Wilderness Act, the first purpose of wilderness is recreation. As such, Leave No Trace is laudable for enabling more wilderness recreation. The irony of Leave No Trace is that wilderness advocates had fought against consumerism and isolation from nature throughout the first half of the 20th century. Now you don't need to know much about nature, you only need to be able to afford the right gear.

I have a friend, Michael Paul Nelson, who may know more about wilderness philosophy than anyone else. He co-edited the two meatiest anthologies on wilderness ever to be compiled, and he once wrote an essay cataloging all the reasons people have offered for valuing wilderness. He described 30 reasons.[54] I have examined several of those reasons in this chapter with the aim of understanding the merits of wilderness as a reason to let wolves go extinct and what the shortcomings of that reasoning might be.

To briefly recap, the impurity of wolves and moose on Isle Royale is not a reason to celebrate their loss. The dominating ideals of wilderness—nature should be allowed to take its course, humans should leave no trace, we should not intervene—are not reasons to allow wolf extinction. Humans already intervened and left more than a trace when we introduced disease in 1980, when three wolves drowned in a mine pit in 2011, and when we warmed the planet, reducing the frequency of ice bridges. Those affairs charted a course to extinction. Restoring wolves would be a remission not an intercession.

The rejoinder to that recap is that only an addict would prescribe additional interference as a cure for an obsession with meddling in the affairs of nature. Hold fast to humility and know that ignorance outstrips good intention. Wilderness advocates and managers are no longer naïve about the impact of past and present humanity on wilderness areas. Many have lowered their aspiration to *minimizing* the influence of humans and fostering experiences that emphasize the untouched elements of a wilderness area.[55]

Thankfully, wilderness is not the only abstraction to which our relationship with nature might be anchored.

10 |

Restoring the Balance

Every gardener knows that rabbits eat vegetables, and too many rabbits result from not enough neighborhood foxes. Experienced gardeners know that a little fox urine applied to the boundaries of the garden encourages rabbits to tend the neighbor's garden. That folk knowledge was not soundly endorsed by science until the end of the 20th century.

The idea really sunk in during the 1980s[1] after two ecologists, James Estes and John Palmisano, showed how the influence of one particular species can cascade throughout an entire food web, affecting many species. They made their remarkable observations in the coastal waters of the northeast Pacific, where sea otters eat sea urchins and urchins eat kelp. That much was common knowledge. What they discovered is that kelp forests persist because of the sea otters. Populations of sea otters were sprinkled up and down the coastal waters of California, British Columbia, and Alaska. Where sea otters lived, Estes and Palmisano noticed that urchins were rare, and rocky reefs were covered in lavish kelp forests extending 50 or 100 feet up toward the surface. Where otters were absent, urchins were 10 to 100 times more abundant, kelp forests were razed, and denuded rock barrens were all that remained.

Kelp forests slow the ocean current, dampen waves, and provide infrastructure for fish to forage and hide. Kelp forests of the Aleutian Islands support 10 times the abundance of fish, like rock greenling, as compared to urchin-dominated rock barrens. In otter-tended kelp forests, glaucous-winged gulls dine daily on fresh fish. When urchins take over, gulls must settle for a menu of intertidal invertebrates. Eagles, whose nests overlook kelp forests, keep a pleasantly mixed diet of fish, marine mammals, and sea birds. Eagles living near the urchin barrens endure a monotonous fare of sea bird day after day.

Otters compete with and, consequently, limit the abundance of diving sea ducks, like eiders and scoters. Otters eat starfish, depressing their abundance and shortening their lives. The latter limits the size to which their bodies can grow. That effect ripples outward and the creatures upon which starfish dine, such as mussels and barnacles, experience less predation and live longer, allowing them to grow three to four times faster in kelp forests so conveniently maintained by otters.[2]

Cataloging all the particular consequences of particular cases—like that involving otters, urchins, and kelp—is a prime directive of science. But a greater value of science lies beyond. That is, showing how all the particular cases represent different instances of a single, underlying phenomenon. Because the underlying phenomenon is often abstract, the parsing of cases sometimes results in incomprehensible

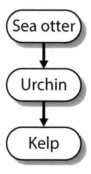

Sea otter food chain

jargon jammed into impenetrable sentences. Other times, however, the result is brilliant. Here's one of the splendid examples.

Begin with a diagram of labeled nodes connected by arrows. Each node represents a species and each arrow represents a relationship between the species.

When an arrow connects predator and prey, the relationship is predation. Herbivory is represented by arrows emanating from herbivores (like urchins). These arrows point to herbivore food, commonly plants. But some herbivores feed on algae, which are not plants. What plants and algae have in common is that both are autotrophs. An autotroph is a creature that creates its own food, and the suffix "troph" is related to "trophic," a word that means "relating to nutrition." Each species in the diagram is said to occupy a certain *trophic level* (decomposer, autotroph, herbivore, predator, scavenger) and each relationship (decomposition, herbivory, predation, parasitism, scavenging) is a *trophic interaction*. Due to its appearance, the sample diagram (and all the words that could be used to describe it) are called a sea otter *food chain*.

Scuba diving through a cold, oceanic maze of kelp is wholly different from hiking beneath the canopy of a boreal forest on a warm day. But the difference is superficial. Food chain diagrams show the two places to be instances of exactly the same thing.

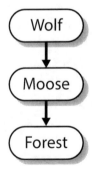

Wolf food chain

Food chains are typically embedded in (and often cannot be extricated from) a *food web*. Food chains and food webs facilitate further abstraction. A *trophic cascade* is a sequence of relationships that "cascade" down the trophic levels, as when otters eat urchins, urchin abundance is depressed and the kelp flourish along with all that depend on the kelp. Trophic cascade is the abstract shorthand for all the events set off by a significant change in otter (predator) abundance.[3]

Trophic cascades require a trigger, some original cause or exogenous force. For the sea otter food web, the exogenous force had been the 19th-century maritime fur industry, which left otter populations few and far between. The otter slaughter ended many decades prior to Estes and Palmisano's observation. The effects of maritime industry persisted for many decades because otters are slow to recolonize places from which they have been extirpated. Recolonization is slow because an otter tends not to travel far from his or her natal waters. Humans were the external perturbation that set motion to the first trophic cascade scientists would ever formally observe. Scientists finally figured out what gardeners had long known about foxes, rabbits, and gardens.

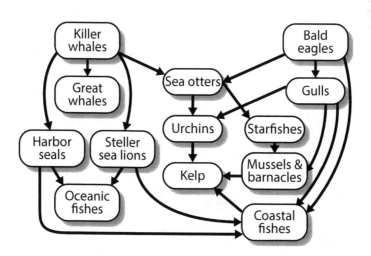

Food web diagram. Adapted from Estes, Peterson, and Steneck 2010

Reflecting on the devastation that otters endured, James Estes later called the circumstance a "fortuitous natural experiment." He meant fortuitous for some scientists, but I'm not sure what essence is communicated in calling the experiment "natural." Nevertheless.

By the mid-1990s, ecologists enflamed themselves in debate about the importance of trophic cascades. The effect of otters on kelp ecosystems was undeniably dramatic, but maybe that was an anomaly. Questions surfaced about the importance of trophic cascades for other kinds of ecosystems. Do they occur in the open ocean? Do they occur in lakes and streams? Forests? Grasslands? How often do they occur in any particular kind of ecosystem—frequently or only occasionally? Are they always as intense as in the case of the sea otters? By the early 1990s, trophic cascades had been identified in high latitude intertidal waters, temperate streams, and freshwater lakes. The known cases were all wet. Some ecologists wondered whether trophic cascades were unimportant in terrestrial ecosystems.[4] If otters affect kelp indirectly by eating urchins, then shouldn't wolves affect the forest indirectly by eating moose?

Rolf, my mentor and longtime partner in Isle Royale research, says he never set out to observe a trophic cascade on land, but he did listen for any story the forest might share. In 1991, he began working with a student, Brian McLaren, who had recently acquired the skill of dendrochronology, which involves measuring and dating the width of rings in the cross section of a tree. The width of each ring indicates the tree's overall growth for a particular year, which depends on how good life had been to the tree that year. Intense herbivory is one of the possible experiences in a tree's life that can reduce overall growth and the width of tree rings.

So, tree rings end up being a good place to listen for tree stories, and balsam fir has a good story to tell because a large share of a moose's diet during the winter is composed of the twigs and needles of balsam fir. McLaren measured the rings of a couple dozen fir trees and found a sustained period of slow tree growth in the early 1990s that was preceded by an increasing moose population of the late 1980s, which was preceded by a declining wolf population of the early 1980s. That period of low wolf abundance had been triggered, you may recall from Chapter 5, by a wolf disease brought to Isle Royale by humans. The tree rings revealed a similar succession during the 1970s—low wolf abundance, followed by high moose abundance, followed by slow forest growth. Those findings are now remembered as the first demonstration of a trophic cascade on land.[5] In retrospect, Rolf hadn't anticipated such a lofty contribution, he "just figured some patterns would show up in the trees."[6]

The dog owners who brought canine parvovirus to Isle Royale in the early 1980s could not have known, and probably have never known, that they initiated an important "experiment" in the history of trophic cascade research and one of the most important observations ever made about the wolves of Isle Royale. That the experiment was unintended and unsanctioned does not diminish its importance. The dog owners' vacation may be properly classed as an exogenous perturbation.

A young, wild-grown balsam fir bears a distinct resemblance to a Charlie Brown Christmas tree—Christmas-tree shaped for sure, but thinner branching than expected for prime holiday decor. When a young fir tree is severely browsed by moose, it gradually transforms, over several years, into . . . well, envision bonsai by moose.

As connoisseurs of moose-produced art, Rolf and I have learned to interpret past and present trends in moose browsing from the shapes of balsam fir. Each year for the past few decades, we have recorded the condition of thousands of fir trees, including total height, height added in the past year, browse damage incurred during the past year, and overall degree of bonsai-ness. From those observations, another important story.[7]

Some balsam firs are full grown with trunks the diameter of a record album and tip-tops rising 30 to 40 feet over the ground, swaying in the wind with other trees that have grown into the canopy—cedar, aspen, spruce, and a half dozen other species on Isle Royale. These canopy trees are protected from moose herbivory by their height. Each year the balsam firs rain seeds onto the forest floor, and the seeds sprout into little seedlings. For the first couple of years, fir seedlings are protected from moose herbivory by their diminutive size. Once they rise six inches or so off the forest floor, moose begin their relentless cropping. Each year, just a nip here or a nibble there. Year after year. They slowly become bonsai trees—never growing more than about three feet tall. Some of these short, abused balsam firs are 30, 40, even 50 years old. Eventually, they give up and die, never growing into the canopy, never tall enough to produce seeds.

Fir trees do not begin producing seeds until they are about 12 feet tall—coincidentally the height at which their tops begin to grow beyond the reach of gnawing moose teeth. Much of Isle Royale's forest is inhabited by balsam firs of two sizes: big canopy trees and saplings less than about three feet tall—nothing in between.[8]

A few years before I started working on Isle Royale, in 1988, Rolf fixed small, numbered, aluminum disks to every fir tree that had grown into the canopy along a 10-mile stretch of trail. He labeled 479 trees. Each year since, we have walked that trail to record which are still alive and which have died. By 2019, only 21 were still alive. The last canopy trees—the ones that produce seeds of the next generation—are dying. They are dying of old age, as balsam fir trees only live for about 100 or 125 years. The canopy trees we had been observing escaped into the canopy during the early 20th century, about the time that moose first arrived to Isle Royale. Since that time, no balsam fir has escaped into the forest canopy.[9] That's why fir trees across so much of Isle Royale come in only two sizes—mature canopy trees and stunted bonsai trees.

When all the canopy trees die, the seed source will be terminated. No more new seedlings. The bonsai fir will gradually die without being replaced. Moose will have undercut the capital that has been producing between a third and half of their winter diet.

In some cases, the opening left by a deceased fir tree will be replaced by a spruce or just grass, turning the forest into a moose-spruce savannah.[10] Other openings will be replaced by deciduous trees, such as aspen or birch. Those tree species produce fine food for the summer but not winter,[11] and at no time during the year can moose support their magnificent mortal flesh on spruce or grass. During the first decade of the 21st century, there was growing reason to think that moose could change the forest in a way that would preclude supporting enough moose to support a wolf population. Whatever the outcome, it would unfold slowly, over several decades. And, by the time the outcome became a certain or obvious, it would be too late to do anything about it.

Then, one bright winter morning in 2012, I skied past evidence favoring an alternative possibility. Rolf and I were out to find and necropsy a moose that the wolves had killed and that we had discovered during a flight several days earlier. We began the day's trek skiing a few miles down Washington Harbor. When the map indicated that we were close to the site, we left the harbor and pressed into the forest. Just a few meters from shore I noticed a balsam fir that rose above the rest. Never before had I seen in these forests a balsam fir that was five and a half feet tall. Neither had Rolf. We looked around, and there were others. We stood in a patch of short, but vivacious, balsam firs that were making a run for it, racing for the canopy at a tree's pace. In the years since we skied through that patch, we've noticed many more fir trees—hundreds, in fact. By 2018, we were finding fir trees that were six or seven feet tall and pushing a few inches closer each year to that 12-foot threshold beyond which they begin to escape moose.

Here's what led to the emergence of these ebullient firs. In the late 1990s, the moose population had been overabundant and then crashed when confronted with a severe winter. Right after the moose population crashed, and by coincidence, the Old Gray Guy arrived. His genetic infusion reinvigorated the wolf population, which subsequently kept moose abundance low for about a decade, 2004–2012. (See the time series of wolf and moose abundances on page 146 in Chapter 5). Those years saw the fewest moose for the longest period, likely, since the early 20th century when moose first arrived to Isle Royale. The fir trees noticed. For the first time in a century, fir saplings slowly awoke from their stupor and grew toward the canopy. Before the saplings could become trees, the genetic benefits of the Old Gray Guy were expiring. The wolf population stumbled, and moose abundance began rising again. Each year from 2012, moose ate more than the year before.

The forest balanced on a knife's edge. Fir saplings were growing fast but so too was the moose population. If enough fir trees grow tall enough before moose become too abundant, then the population of fir trees will be rejuvenated. If not, if moose abundance increases faster than the balsam firs, then the fir trees will be cut back one tree at a time by starving moose, grabbing the main stems of 10-foot trees with their mouths. The twist of a massive head, snap, and another 10-footer becomes a six-footer. The snapping doesn't begin until after moose exhaust all the bites that are otherwise within reach. Tree snapping began during the winter of 2017–18. The kismet of balsam fir is prognostic for many plant species in the forest.

In 2012, the wolf population sank lower, the forest's future became hazier, and the Park Service frittered. I implored the Park Service to take the circumstance seriously, and I failed for several years. When they did begin to act, it was to assure themselves and the public that extinction is a natural occurrence on islands. Then they spent a few years rediscovering what I'd explained back in 2011 when I anticipated the collapse before it occurred. By May 2018, the Park Service appeared poised to decide to restore wolf predation on Isle Royale. Implementing the restoration of wolves would likely take a couple of years or more. No one knew how long it would take for a fledgling wolf population to reestablish a predatory influence on the moose population. The island would, in the end, have been without a functional wolf population for approximately a decade. The long-term consequences are presently unclear.

This account of wolves, moose, and the forest may well be understated. At least that's the impression I get from reading what other

Balsam fir saplings broken by moose. J. A. Vucetich

good biologists have said of similar situations. In Yellowstone National Park during the early 20th century, wolves were exterminated and elk abundance soared. The Park Service scientist, George M. Wright, observed the impact firsthand, prompting him to write the following:[12]

> The one outstanding thing was the frightful conditions of the range. Throughout this section one cannot find a juniper, or Douglas fir that has not been browsed to the reaching limit. Many trees are dead from this. There is no reproduction. Willows are browsed and battered. The sagebrush

has been hammered down. . . . The soil has been packed by countless game trails and is badly cut up. Truly this range looks worse than anything I have seen on the Kaibab.

Moose had lived on Isle Royale for 50 years before the mid-20th-century arrival of wolves. In 1930, Adolph Murie, also a Park Service biologist, described the moose's brunt this way: "[There are some areas] in which the top of every tree is broken off, and there is little else to eat except bark. No poplar reproduction was noted. The winter moose food is practically gone from the island."[13]

An important element of the National Park Service's history can be told as a history of damage caused by overabundant large herbivores resulting from the extermination of predators. Wind Cave, Great Smoky Mountains, Indiana Dunes, Catoctin Mountain, and Rocky Mountain National Park, to name just a few.

With that context about trophic cascades and those details about balsam fir, the reason to restore wolves on Isle Royale is simple. *The health of ecosystems inhabited by large herbivores depends on the cascading trophic effects of predation.* This is such a great reason in so many ways. It applies not only to Isle Royale, where the herbivores are moose and the predators are wolves, but also to any large carnivore in need of restoration or conservation anywhere in the world. The broad reach of this rationale is important because carnivores across the planet are in need and there is value in one potent rationale that covers all cases.

This thinking has added oomph because it's rooted to science and endorsed by conservation biologists around the world. Public decisions are supposed to be based on facts and science, and scientific reason is supposed to be more robust than value-laden reason—especially when the values are dear to some and despised by others. After all, we're entitled to our own values but not our own facts. This "trophic cascade argument" is the go-to reason: simple, scientific, and applicable to carnivores everywhere.

Wait a minute.

The trophic cascade argument *is* well exercised by carnivore advocates. But, if this thinking is so compelling, then why are carnivores around the world so beleaguered and worse off with each passing year? The sorry state of carnivores justifies at least a little interrogation of the prime reason offered for their conservation. Pinpointing its limits and shortcomings cannot hurt, and it might help.

The cross-examination can begin by acknowledging that carnivores are important in the specific sense that they regularly have widespread effects on the ecosystems they inhabit. That is a scientific fact.[14] But "important" should not be elevated to "good," not so quickly and uncritically. What, if anything, is good about trophic cascades or the carnivores that trigger them? Might the world be just as fine with fewer carnivores and trophic cascades? If not, why not?

Careful. Turn the corner and we're liable to find the gleaming, mesmerizing eyes of an *Argumentum ad Naturam*: X is natural, what is natural is good, therefore X is good—where X is short for "trophic cascades and the carnivores that trigger them." We need a richer reason for thinking X is good.

We might slip past that siren by shifting the question from *why is it good?* to *what is it good for?* Like this: the material well-being of many human communities depends on healthy forests and grasslands.[15] Those ecosystems can be degraded with overbrowsing and overgrazing by large herbivores—moose, deer, elk, gazelles, and so forth. The risk of detrimental overconsumption is greatly reduced when the abundance of large herbivores is limited by large carnivores—wolves, lions, lynx, wolverine, bears, and so on. What are carnivores good for? They're good for humans.

Nevertheless, human communities often find it difficult to live near populations of large carnivores. Carnivores kill domestic livestock. That's not good for humans. Carnivores kill wild prey. Some humans think that's not good because they'd prefer to reserve for themselves all those hunting opportunities—sometimes for recreation and sometimes for subsistence. Some species of large carnivores, on some occasions, threaten and take human lives. That's not good for humans.

While these conflicts are real, they are also frequently and grossly exaggerated. Regardless, the genuine and perceived threats fuel persecution against carnivores,[16] leading humans to kill carnivores at unsustainably high rates through illegal poaching, legal culling, and recreational hunting. The result of all this killing is that two-thirds of the world's carnivore species are threatened with extinction, and most places on the planet have lost most of their native carnivores. That situation is bad and getting worse.

Americans drove to extinction most of the large carnivores that inhabited most of eastern North America. Britons assailed their large carnivores—wolves, lynx, and brown bears—to extinction centuries ago. Africans and Asians pummeled lions and tigers into living in just a pittance of the places where they once lived. It's hard to make a case that the material well-being of those humans is worse as a result of those extirpations. When something (think, carnivore) is valued only for what it might do for me, its utility may go unrecognized, be outweighed by the cost of keeping it around, or be replaced by a substitute.

This is not a denial of the utility of carnivores, just acknowledgment of the risk in valuing something only for its utility. The evidence seems painfully clear. The utility of carnivores is insufficient to motivate adequate human coexistence with carnivores.

The shortcomings with utility extend well beyond carnivores.[17] Many experts on the conservation of biodiversity think that most biodiversity is of insufficient utility to motivate humans to conserve it.[18]

Forget utility and the economic calculus implied by invoking utility. Humans do lots of things that are not economically sensible. For example, many find wonder and beauty in wolves, carnivores, and the entire cornucopia of biodiversity—regardless of economic value. This should be more than enough reason to impose coexistence with carnivores on those who'd rather not coexist with them. Fine reasoning, except we currently share the planet with enough humans who do not enjoy that enthusiasm. If the conservation of carnivores and other inspiring, albeit less useful, biodiversity is reduced to a popularity contest, then I'm afraid those who survive the popularity contest will be largely relegated to zoos. This is no reason to stop professing the wonder and beauty of nonhuman life. To the contrary, it is occasion to dig deeper for the articulable reasons that burgeon inspiration.

The concern for Isle Royale wolves is intertwined with concern for carnivores wherever they live and for the whole of biodiversity. Humans are pummeling the planet with a biodiversity catastrophe that extends well beyond carnivores, and the reasons used to judge the fate of wolves on Isle Royale intermingle with reasons for averting the biodiversity catastrophe. Before probing this nest of reasons, there is value in quickly reviewing the biodiversity catastrophe. Doing so highlights the blurring of concerns.

A really long time ago, the first vertebrate species evolved into existence from some creature that did not have a backbone. This first vertebrate looked something like a hagfish. If you are unfamiliar, think eel but less attractive. Aside from the looks department, this first vertebrate species was successful, made a good living (probably as a scavenger), and eventually evolved to become two species. Then two species became four, and four species become eight.

But no species persists forever. (Extinction, as we've been told, is a natural process.) Vertebrate species go extinct a million years or two or three after evolving into existence.[19] So, extinctions have always occurred at some *rate*, but those losses have been more than offset by the higher rate at which new species evolve into existence. Over time a few ancestral species of pale, scavenging vertebrates that slithered across the ocean floor, diversified into 40,000 species of fish, amphibians, reptiles, birds, and mammals, including primates, great apes, and one great ape in particular. Us.

What's important here is how long it took for all that vertebrate biodiversity to accrue—about 500 million years. During that time, on average and very roughly, one new species was added about every 400 years, and one existing species went extinct about every 450 years.[20] The rate of extinction was only a bit less than the rate of speciation. That's how it worked until humans got arrogant about their use of the planet. What humans have done, according to best estimates, is increase the rate of extinction by about 1,000 times.[21]

The mathematics of *rates* are sometimes easier to intuit when they govern something that we care more deeply about. Suppose you have a savings account where money is added each quarter according to an investment rate of return that is, say, 10%. And suppose money is removed from that account at a rate of 7% by a loan company. Because the rate of return exceeds the rate charged by the loan company, you'll slowly accrue money over the course of years. That is, until the loan company increases by 1,000 times the rate it charges. Then you'll be filing for bankruptcy before the next quarter.

Because humans have increased the rate of extinction by 1,000, an estimated 20% of the 40,000 species of mammals, birds, reptiles, amphibians, and fish that inhabit the earth are at significant risk of extinction over the next century. We've treated invertebrates and plants similarly.

That is the ongoing biodiversity catastrophe.

This present-day mass extinction is not the planet's first. Mass extinctions have struck the planet on five previous occasions. They were caused by asteroids, a drop in the concentration of atmospheric oxygen, and the onset of an especially cold ice age. Each time the planet recovered with the evolution of new species. But evolution is a slow process, and if past mass extinctions are a guide, then a reasonable guess on recovery time for the current biodiversity crisis is in excess of 3 million years and, perhaps, up to 10 million years.[22] The last time the earth recovered from a mass extinction, good things followed. Sixty-five million years ago, extinction hit several hundred, perhaps 1,000, species of dinosaur.[23] The vacancy made space for the evolution of all the mammals that currently grace the earth—including and eventually a conspicuously social species of naked ape. Something good is certain to follow the obliteration of all those mammals.[24]

That perspective is pertinent but dizzying and possibly demoralizing. Another perspective lessens the risk of paralysis by gesturing toward a strategic, achievable goal for lessening the catastrophe. Long

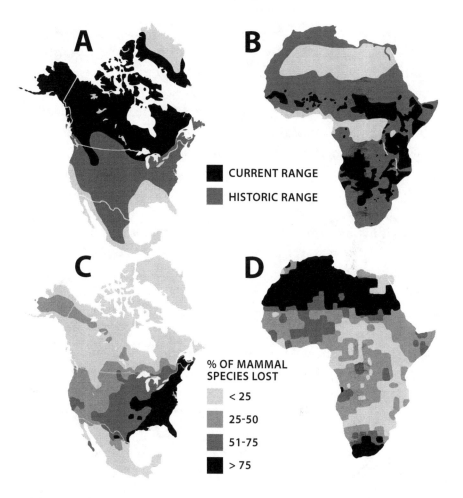

Prior to total extinction, imperiled species are often subject to contraction of geographic ranges, as seen in wolves in North America (A) and lions in Africa (B). The cumulative effect of many species' ranges contracting is that a large portion of species have already been driven to (local) extinction from most places on the planet. The magnitude of the loss is indicated, for example, by the percent of mammal species that have been lost from particular regions of North America (C) and Africa (D). Adapted from Ceballos and Ehrlich 2002, Bauer et al. 2016, and Bruskotter et al. 2014 before any imperiled species goes completely extinct, its demise is preceded by local extinctions, which cause a species' geographic range to contract. Today, the *average* mammal species has been driven to extinction over more than two-thirds of its historic range.[25]

The cumulative effect of these local extinctions is staggering. For example, at least 25% of the mammal species have been driven to regional extinction from large portions of the United States and Africa. More than 75% of mammal species have been lost from regions east of the Appalachian Mountains (USA) and the Saharan portion of Africa. The march toward global extinction is preceded by widespread local extinctions, which have already resulted in shocking losses of biodiversity for most regions of the planet.

The causes of population loss are mainly pollution (air, water, and land), overexploitation (a.k.a., too much killing, often to provide humans with food), the introduction of certain nonnative species (as when, for example, humans inadvertently enable lion fish to colonize new coral reefs and outcompete native species),[26] and humans' propensity to change the land by converting, for example, forests and grasslands into agricultural fields, cities, neighborhoods, and parking lots. Those causes—pollution, overexploitation, nonnative species, and land conversion—are driven by the basic needs of billions of humans, the greed of some of those people, and the exigent urge for people to parent to more than one child.[27]

With that rap sheet, some think humans are no more than a cancerous blight to the earth. While the thought is a genuine temptation, it is best to resist.

Casting the biodiversity catastrophe in terms of contracting geographic ranges is staggering, but it also hints at where to begin pressing against the catastrophe. The strategic goal, if I might suggest one, is to alleviate threats to populations at least to the point of allowing most species to begin reinhabiting most of the places where they lived prior to humans having been so profoundly and collectively gluttonous.[28] That strategic goal is already codified in American law, though no American government has been earnest enough to implement it.[29]

If that goal is too ambitious, then another should be proposed. In any case, an overarching goal is essential for guiding adequate progress toward sustainable coexistence. Overarching goals also require supporting reasons—good, solid reasons that persuade those less concerned with the catastrophe and withstand the objection of detractors.[30]

Progress in resolving enormous problems is sometimes aided by mucking around smaller manifestations of the problem. That's what we're doing by focusing on the foreground and background simultaneously. Understanding the reasons for what to do about the wolves of Isle Royale is perched in the foreground, and the biodiversity catastrophe lords over the background. To be patent, failing to solve a simpler problem correctly for the right reason bodes poorly for the bigger, thornier problems.

One reason that we considered for restoring Isle Royale wolves had prima facie potential to cover the full depth of our focus: we should restore and conserve carnivores and other biodiversity because doing so would serve the needs and desires of humans. That reason fell short. We had also worked a more basic reason: *the health of ecosystems inhabited by large herbivores depends on the cascading trophic effects of predation.* It, too, seems to have potential, especially for its connection to science. But, this oft-cited reason hasn't brought about a lessening of the catastrophe.

Maybe there is an overlooked weakness swirling in that reason's core motivation—the health of an ecosystem? A stupefied advocate may become exasperated. Who could possibly object to healthy ecosystems?

Ah yes, we may have touched a soft spot. Very tenderly now, what exactly is ecosystem health?

Turns out, there are nearly as many answers to that question as people who have written on the topic. Conveniently, most answers sit comfortably in one of two camps.[31] In the first camp, we hear that ecosystems are healthy to the extent they provide humans with what they want (in perpetuity).[32] That idea is, for many people, the essence of sustainability. In the second camp, ecosystems are said to be healthy to the extent that humans have not affected them. Wilderness Watch[33] maintains this camp.

With the first answer, there is no doubt. Human well-being is important and depends on ecosystems producing things that promote human well-being. But to understand ecosystem health *only* in terms of human interests is too narrow, too human-centered. To think humans are the only ones on the planet who matter is a mental disorder that academic ethicists have labeled anthropocentrism. Humans cannot be the only ones who matter, can they? There has to be more. Ecosystem health must be a richer concept, right?

On the second answer, again, no doubt. Humans are plainly and unquestionably capable of bruising and destroying ecosystems. But not every consequence of every human action involving nature should be tallied as harmful. There is a need to discern benevolence and benignity from malevolence and destructiveness. To think, so simply, that ecosystems are healthy to the extent they are unaffected by humans is misanthropic.[34] By that thought, humans are little more than a cancerous blight to all the nonhumans. There has to be more to it. Ecosystem health must be a richer concept, right?

We have two conceptions of ecosystem health, each with its own nub of insight, but neither is right on its own.[35] Maybe we can find some judicious middle ground between these two conceptions.

Unfortunately, I'm afraid that neither hybridization nor compromise will work. The one notion of ecosystem health is rooted to anthropocentrism and the other is rooted to misanthropy. Anthropocentrism and misanthropy are something like dengue fever and yellow fever. One is either infected or not. There are mild cases and severe cases, and it's better to have a mild case. But our ambition should be complete rejection of misanthropy and anthropocentrism, not a joint case of both. Maybe a mild case of both is better than a severe case of either, but this is not the ground on which to build aspirations.

Setting aside the wisdom of infectious disease, I had been content in thinking that ecosystem health is a concept like justice, which is neither precise nor definitive. It's meaning has been and will be debated for millennia. Nevertheless, we have systems of justice, including the administration of civil and penal codes that fend away many injustices. We also seem to understand justice well enough to, at least, lessen the grossest deficiencies in systems of justice, on some notable occasions and over long arcs of time.

Likewise, we don't know precisely what ecosystem health is, but whatever it is, it certainly encompasses the inclusion of predators in ecosystems inhabited by large herbivores. So, ecosystem health might be like Justice Stewart's understanding of obscenity: we'll know it when we see it. (To paraphrase what the justice of the US Supreme Court said in a 1964 legal opinion on a case about free speech.) The quip betrays a serious legal philosophy, by which we may not know the precise line that separates obscenity from decency, but that doesn't preclude identifying instances of obscenity that are far past the line.

I am no longer satisfied with those attempts to rescue ecosystem health from anthropocentrism and misanthropy. Besides, a good relationship with nature might even be understood without referring to ecosystem health. To this end, we may be inspired by a belief of the Ojibwe, Native North Americans whose homeland happens to include Isle Royale. They believe the wolf is their brother—not metaphorically but literally. Shortly after creating heaven and earth, Gichi Manito (the Great Spirit) created Anishinaabe (the original human) and Ma'iingan (the wolf). They were created as brothers and tasked with traveling throughout all creation together to honor the plants and animals by giving them names. Anishinaabe and Ma'iingan were close and realized they were siblings with all the animals and plants. When the two had completed their task, Gichi Manito informed them that they would go separate ways. The Great Spirit foretold that both would be hated, hunted, and misunderstood; ultimately, they would share the same fate.[36]

The essence of this Ojibwe understanding is fully reconcilable with Western thinking. Where the Ojibwe understand us to be literal siblings with all the denizens of nature, the scientific understanding is that all organisms—animals and even plants and microbes—have a common ancestor.[37]

Common ancestry is the scientific way of saying we are all siblings. Perhaps, "cousins" is a better word. Taking no liberties with the English language, common ancestry is acknowledgment that all of the earth's organisms, including humans, are members of the same family—not metaphorically but literally.[38]

Kinship is a scientific truth that is also accompanied by a simple, but far-reaching, obligation. Family members are entitled[39] to be treated fairly and with concern for their interests. To see how, start on solid and familiar ground: We know that all humans are entitled to fair treatment. But why? "Because we're human" is an inadequate answer and begs the same question, why? What is it about humans that imbues us with entitlement to fair treatment? "Because God made us this way" can be a fine answer, but what exactly is it about how we were made that imbues us with this entitlement?

Scholars have been thinking about the answer for a long time. One of the most important answers to the question is that each human has interests. Take one of the simplest interests, to avoid pain. If you want to be treated with fair concern for your interest (avoiding pain), then as a matter of basic consistency wouldn't you also be obligated to treat others with that interest in the same way that you wish to be treated? This, of course, is what Judeo-Christian people call the Golden Rule. Most cultures have some variant of the Golden Rule. Academic ethicists refer to this as the principle of ethical consistency.[40] Notice, nothing in the logic says fair treatment ends with humans. Rather, fair treatment applies to anyone—human or otherwise—with interests. This more than covers the obligation to creatures like wolves and moose—though the logic extends far beyond mammals.[41]

This obligation has its detractors. Some say it means no human can ever act against the interest of another creature. Fiddle-faddle. I'm obligated to treat humans fairly but that does not mean I am never justified to act against the (genuine or perceived) interest of a human.

The obligation simply means I cannot hurt another creature (human or nonhuman) without good reason.

What counts as a good reason is sometimes easy to judge and sometimes not. But all the cases are well judged according to a marvelous set of four virtues. These virtues are most easily apprehended by a heuristic. Suppose you and I both want to eat the last piece of chocolate cake. To honor the virtue of *equality*, we could split the piece of cake. But you might say that you have been an especially good person today and by the virtue of *equity*, you should have the last piece.[42] Though, I might not have eaten at all today, invoke the virtue of *need*, and ask to have the entire piece. Finally, you might remind me that it's your birthday. By the virtue of *entitlement*, the cake should be yours.

The banality of this concocted dispute over cake is an asset. It implies that we exercise these virtues frequently—probably several times a day—and many of us are pretty good at knowing which virtue best fits which situations in life. Here's the wild part. I don't know of any reason those virtues should apply only to humans. They would seem to apply to the whole family—any creature with an interest. This is shaping up as a powerful obligation. If I understand it properly, I don't think it's possible to opt out of this obligation without rejecting the Golden Rule.[43]

It only seems that we've drifted far from the initial concern. In fact, virtues and kinship may be key for understanding how to treat the wolves of Isle Royale. Because the wolves of Isle Royale deserve fair treatment, and because knowing what counts as fair treatment often requires taking account of the context surrounding the particular situation, we should remind ourselves of some context that swirls round the relationship between humans and wolves. Three points seem relevant.

First, shortly after European colonists arrived to North America, they began to systematically slaughter wolves with the intent of complete obliteration. A bounty on wolves was the ninth law ever passed by the fledgling government of Michigan during its first year of state-hood in 1837. Of all the responsibilities that an infant government might take on, government-subsidized extermination of wolves was a top priority for Michiganders. Wolf bounties existed in Michigan until 1965.

Second, in recent years wolves have experienced impressive expansion in the northern Rockies and western Great Lakes. Today they occupy more territory than at any point in the past century and a half. This is one of the greatest successes in carnivore conservation. Still, within the coterminous United States, wolves inhabit less than about 15% of their former range, though they could readily live in many more places.[44] Each year, hundreds of wolves are killed in the United States (and elsewhere). The persecution is (and was) ostensibly motivated by competition between humans and wolves for livestock and wild prey and, preposterously, to protect human safety. The underlying motivations, then and probably more so now, seem more basic—hatred, callous disregard for life, and culpably delusional fear. That hatred (largely enabled by state and federal government) is what prevents wolves from further restoration.[45]

Third, and most specifically, the wolves of Isle Royale suffered severe inbreeding as an indirect effect of anthropogenic (human-caused) climate change. While climate change began as an unanticipated consequence of industrialization, its profoundly harmful nature has been known for decades. Yet we continue releasing too much greenhouse gas in a rush of systematized and collectivized avarice.

With that context, what precisely is our obligation to the wolves of Isle Royale? What counts as fair treatment of those wolves on that island?

Humans had been allowing the last two wolves on Isle Royale to live their lives on Isle Royale with dignity and respect.[46] Obligation fulfilled, right? Maybe not.

Many wolves living at other times were not treated fairly and should have been. But they are gone, and there's no way to go back to them. Many wolves living today in many places are denied fair treatment that they deserve. But that's not Isle Royale. The last two Isle Royale wolves were treated as fairly as we could have treated them.[47] When the last two wolves of Isle Royale die, then there would be no wolves on Isle Royale to whom dignified treatment is owed. There were harms (disease and climate change), but any obligation to those wolves decomposes with their corpses because corpses don't have interests.

The Ojibwe-inspired kinship beliefs for treating all living things with respect—that's solid. And when we decide to live by it, the biodiversity catastrophe will lessen greatly. But this line of thought doesn't seem to help us understand what to do about the wolves of Isle Royale.

By now, you may recognize that I am rifling through the arsenal of reasons for caring about nature. I am reviewing the major themes of environmental ethics. Doing so is important because, as we are about to see, the loss of wolf predation on Isle Royale is critically similar to many instances of biodiversity loss. If we fail to find a clear, solid reason to protect wolves on Isle Royale, then we should prepare to lose copious biodiversity to ambiguity and indecision.

But we're not done yet. We still have a few more ideas in the storehouse.

The next idea is tethered to Frederic Clements. He is the fellow from Chapter 4 who counted plants beneath the hot, Nebraskan sun until he saw that ecosystems are so inextricably interconnected as to be a superorganism, "a live, coherent thing."[48] Seventy years later, in the 1970s, Lynn Margulis and James Lovelock observed something similar, but on a much grander scale. The microbiologist and chemist observed that the sun's output of heat has increased by 25% to 30% since life first appeared on earth. Yet, the earth's temperature did not increase but instead fluctuated within a range suitable for life, and the stable fluctuations were maintained by biological processes. The concentration of salt in the ocean and oxygen in the atmosphere has also remained favorable for life and been maintained by biological processes.

Your body sustains the favorable conditions it needs—it maintains temperature and blood glucose, creates the impetus to drink when dehydrated, and fights infections that threaten vital balances. No one organ is capable of such homeostasis. This stability emerges, almost magically, from *interactions* among the body's systems—nervous, circulatory, immune, and digestive, to name a few.

Similarly, life on earth seems to maintain favorable environmental conditions. This environmental stability is not the property of organisms. Rather, it emerges from *interactions* among organisms. This balance of nature seems to rise ex nihilo because interactions seem ethereal compared to the corpora of organisms. Margulis and Lovelock named this emergent, environmental homeostasis after the personification of planet Earth. They called it the Gaia hypothesis.[49]

Ah-ha! The entire planet is a superorganism with interests of her own. If so, then Gaia herself is a family member.[50] She deserves, like any other member of our family, to be treated fairly.

Fine, except for the wrinkle that most life scientists would probably object. They would insist on reserving the word "organism" to refer to things that reproduce and pass genetic material on to offspring. Neither ecosystems nor planets do that. This may seem an exceedingly fine point, but it's important to them, and they have their reasons.[51]

The hang-up is, I think, genuinely semantic. Perhaps living things come in different varieties, organisms are but one variety of living thing, and ecosystems and Gaia are a different kind of living thing. The hallmark of these last kinds of living thing is not reproduction or genetic material but rather a rich, internal interconnectedness from which homeostasis emanates.

If this all checks out, then ecosystems and Gaia are not merely things for humans to use. They are instead members of the family. And these members might even have interests and perspectives. This possibility is evocatively conveyed by another scientist in an essay describing, of all things, his regret for having killed a wolf:

We reached the old wolf in time to watch a fierce green fire dying in her eyes. I realized then, and have known ever since, that there was something new to me in those eyes—something known only to her and to the mountain. I was young then, and full of trigger-itch; I thought that because fewer wolves meant more deer, that no wolves would mean hunters' paradise. But after seeing the green fire die, I sensed that neither the wolf nor the mountain agreed with such a view.[52]

This oft-quoted passage is from an essay, "Thinking Like a Mountain," by Aldo Leopold, and the expression cannot be dismissed as artistic license. Leopold inspires devoted scholarship on ecocentrism, which is a kind of environmental ethic that posits that ecosystems are entitled to direct moral consideration, not because they might be valuable to humans but because they are intrinsically valuable, in their own right, like any other member of our family.[53]

If mountains have a perspective that can be gleaned by thinking like a mountain, then I imagine islands do too. Furthermore, ecosystems have—what is at least loosely analogous to—an anatomy. This anatomy is represented by food web diagrams like that depicted toward the beginning of this chapter. When an ecosystem loses a species belonging to this anatomy, its health is diminished, not unlike the detriment you would experience if you were to lose some part of your anatomy.

This is sounding very much like the idea we already considered and dismissed: *the health of ecosystems inhabited by large herbivores depends on predators.* Except, this approach seems to slide past the two troubling hang-ups. First, it invokes a sense of ecosystem health without rousing anthropocentrism or misanthropy. Second, it seems to deke *Argumenta ad Naturam.* Both snags were slipped by recognizing ecosystems as living things in their own right, and hence members of our family and entitled to dignified treatment.

There is value in momentarily setting aside this idea that "ecosystems are living things" and returning to the Ojibwe-inspired ideas, where there is more insight to distill. Recall that the essence of those ideas is that we should treat all living creatures fairly.[54] And recall that a useful framework for thinking about fairness is the set of virtues: equality, equity, need, and entitlement. Those are the virtues that guide social justice, also known as "distributive justice" because it concerns the allocation of stuff among the users of stuff—stuff like money, possessions, land on which to live, food, and water.[55]

Distributive justice stands beside the other basic kinds of justice, punitive justice and restorative justice. Aristotle launched the discussion of restorative justice for Westerners in his magnum opus of ethics, the *Nicomachean Ethics*: "For in the case also in which one has received and the other has inflicted a wound . . . the suffering and the action has been unequally distributed: but the judge tries to equalize things by means of the penalty, taking away from the gain of the assailant." After much elaboration over the centuries, restorative justice has become a core basis for resolving civil disputes among humans. We might also find that restorative justice has something useful for understanding the restoration of balance to nature. Let's see.

Restorative justice is easiest to apprehend when one individual has harmed another, the two individuals are still alive, and the ill-gotten gain is restored to the victim at the expense of the transgressor. Restorative justice is complicated by two common circumstances. First, how can justice be restored when the involved parties are not individuals but entire groups, with one group having a substantial and adverse impact on individuals of another group? The groups may be characterized by race, gender, nationality, or religion. And, I suppose, species?

The second complication for restorative justice is what, if anything, does restorative justice require after the transgressor(s) and victim(s) die but their descendants live on? Many present-day injustices (between races, genders, nationalities, religions, and species) are perpetuated by present-day actions, but they were catapulted by the actions of those now long dead.

The thorniness of those complicating circumstances is reflected, for example, in Yahweh's apparently inconsistent views on the matter. In

Deuteronomy, he says, "Fathers shall not be put to death because of their children, nor shall children be put to death because of their fathers. Each one shall be put to death for his own sin." In the Book of Numbers, we are told, "The Lord is slow to anger and abounding in steadfast love, forgiving iniquity and transgression, but he will by no means clear the guilty, visiting the iniquity of the fathers on the children, to the third and the fourth generation."

Since the Old Testament, more thought has been given to these more complicated cases. The best-available thinking seems to be, in a nutshell, if the transgressors' descendants have undue advantage as a result of the transgressors' actions and at critical cost to the victims' descendants, then restorative justice calls for the transgressors' descendants to atone with the victims' descendants.[56] The meanings of "undue" and "critical" are subject to deliberation, and the form of atonement is open to discussion, but the overall obligation is not.[57]

Insomuch as the principles of restorative justice apply to the relationship between wolves and humans, we have an obligation to right the past wrongs perpetrated against wolves by humans. That is, persecution that led to the unjust extermination of wolves over significant portions of their geographic range. Restoring a healthy population of wolves to Isle Royale, where they can live free of ongoing persecution, is a meager gesture in the direction of righting those past wrongs.

We will soon be as far as I can bring us. As we arrive, let's scan the entire scene one last time. Wolf predation in a protected area was lost to inadvertent and indirect effects of anthropogenic climate change. Restoring that predation was feasible, and the Park Service eventually decided to do so. But was it the right thing to do? That question is important for several reasons. The Park Service seemed to struggle with the question. The question is also emblematic of many instances where the biodiversity of a protected area is, or is about to be, harmed by climate change and other large-scale anthropogenic assaults to Gaia.[58] Finally, answers to the question imply much about what our judgments will be for protecting and restoring biodiversity that is not especially valuable to humans—that is, much if not most biodiversity. Frankly, the questions stemming from the Isle Royale case are more important than the Isle Royale case itself. In exploring the questions, we considered several principles, each with wide jurisdiction:

- Treat the members of our family fairly, that is, by judicious application of the virtues associated with equality, equity, need, and entitlement.
- Fair treatment includes righting past wrongs, even when doing so doesn't advance the material well-being of the transgressor.
- We have a large family. It certainly includes all vertebrates and quite likely more than just vertebrates. Our family may include not only the creatures that inhabit ecosystems but also the ecosystems themselves, including the biggest ecosystem of all, Gaia.
- Gaia is increasingly crowded, and her homeostasis increasingly stressed, as indicated by the biodiversity crisis. This crowding and stressing are the result of too many people having taken more than their fair share and too many people having given birth to more than one child.

The fairest way for a human to live on such a planet is by this maxim: *do not infringe on the well-being of others any more than is necessary to live a healthy, meaningful, and flourishing life.*[59] Many common views for what is required to flourish are almost certainly wrong, and getting it right will require serious collective soul-searching.[60]

Those four principles and the maxim are also ensconced with this positive aphorism: *live, let live, and right past wrongs.*

I think these principles are reason enough to have restored wolf predation to Isle Royale. But the reasoning includes at least one soft spot with broad consequence.

If wolves should be treated fairly because they are members of the family, shouldn't moose also be treated fairly? If so, why is it okay to enable packs of wolves to tear apart many dozens of moose every year?

While the tearing apart of moose, carefully metered, might be good for the moose *population*, it's not obviously in the interest of the *individual* moose being torn apart. The suffering of a moose is real, and her interest to live another day is genuine. But it's not obvious that ecosystems have genuine interests. The "suffering" of an ecosystem that loses predation is more abstract, less real. It's fine to be concerned with intact ecosystems, but not at the expense of an individual's suffering.

This line of thinking taps a puissant and neglected concern in humans' relationship to nature. The concern pertains to conflicting obligations, and we can nudge toward the concern with a simple example.

Humans routinely have multiple obligations, such as to family and the workplace. Meeting both obligations is often easy, and meeting one (say, work) regularly contributes to fulfilling the other (in this case, family). Occasionally, however, the obligations seem to conflict, as when illness in one family member obligates another family member to stay home from work on a day when there is an especially important meeting. In most cases, it's easy to prioritize conflicting obligations, or even find a creative way to meet both obligations, though not always. What ensues is a moral dilemma.

With that simple example in mind, recall that we have obligations to treat individual humans and nonhumans fairly and with concern for their interests. We also have obligations to treat ecosystems appropriately. If we have obligations to a moose and her interest to live another day, obligations to right past wrongs to wolves, and obligations to maintain and restore intact ecosystems, then might it be reasonable to prioritize obligations to the moose over the other obligations?

My intuition tells me that's the wrong prioritization. However, putting individuals over ecosystems seems to be supported by some sharp and compassionate thinkers who have worked hard to give sincere reasons for thinking that to be the right prioritization.[61] Sociological surveys also suggest that views among the general public are quite varied on such matters.[62]

The legitimate difficulty of such prioritizing is laid bare by another example. For the past decade, the Canadian province of British

Columbia has been killing a couple hundred wolves each year to prevent the extinction of woodland caribou in southern British Columbia. Persistence of these caribou (who are preyed upon by the wolves) would likely require killing wolves in perpetuity. The wolves in this part of the world are abundant, so concerns about the well-being of the wolf population are minimal. Which obligation should be prioritized? The interest of a wolf to live another day and not be poisoned or gunned down from a helicopter by government agents? Or the interest to prevent the caribou population from going extinct?

The conservation of biodiversity is inundated with cases that seem to pivot on conflicting obligations to biodiversity and individual organisms of which that biodiversity is composed. Many cases arise because one of the greatest anthropogenic threats to conservation is that some species, in some places, have become overabundant to the point of being detrimental to other species.[63] The most common method for dealing with this situation is to kill the individuals belonging to the overabundant species. Examples include killing barred owls to conserve spotted owls, killing brown-headed cow birds to conserve warblers, killing harbor seals to conserve salmon, killing ravens to conserve greater sage grouse and Mojave desert tortoises, killing red foxes or rats because they kill rare birds, and killing mink where they imperil water voles.[64] That's a small sampling from among hundreds of cases.

The sorting of conflicting obligations is sometimes aided by knowing the context that surrounds a case. To see how, let's give another look at the wolves and caribou of British Columbia. Wolf abundance in the region is relatively high because the wolves are supported by an abundance of moose in the area. Caribou are an incidental source of food to wolves, but every caribou lost to wolf predation is a big deal. The circumstance begs the question, why are moose so abundant? Well, that's because moose do best not deep within the forest (like caribou do) but on the forest edge, as where a forest abuts a clearing. Furthermore, some humans have been eager to promote oil and gas exploration. To satisfy that urge, they have cut miles and miles of strips through the forest to conduct seismological analyses. Those strips make outstanding moose habitat. In one breath, humans log the forest, which leads to an abundance of moose, which results in too many wolves, and then humans kill wolves because the wolves kill too many caribou. The salient point is that wolves are not culpable for killing caribou, but humans are culpable for the debacle and its perpetuation.[65]

When humans kill in the name of conservation, typically they are culpable for the circumstances that motivate the killing. Culpability presents itself to the Isle Royale case, as when a person thinks, *Humans are culpable for the widespread poor treatment of wolves and the loss of wolf predation on Isle Royale because it was the result of anthropogenic climate change. Humans should atone for those culpabilities. Furthermore, wolves are not culpable for killing moose, nor are humans culpable for enabling wolves to do so. For those reasons, restoring wolf predation should be prioritized over concerns about the suffering to result from wolf predation.*

Notice the assertion that forms the last part of that reason, about human inculpability for enabling wolves to harm moose. That assertion does not deny what happens (humans enabling the suffering of moose via wolf predation). Rather, it denies that humans are blameworthy for doing so.

My intention is modest—not to analyze the last part of that reasoning—but merely to highlight two examples. One example about wolves in Isle Royale and another about wolves in British Columbia, both of which suggest that right relationships with nature depend not only on knowing when and how humans cause harm to nature but also on knowing which harms should be followed by what kind of atonement.

Culpability plays exactly this powerful role among human relationships, why not between humans and nature? Culpability's reach extends far beyond who caused what harm to whom. It aims to identify when an agent of harm is blameworthy for that harm. That's important because the assignment of blame greatly influences right responses to the harm.

For example, the harm exacted by every homicide is identical—one human has been killed by another. But culpability and the attending justice varies according to the kind of homicide—first-degree murder, second-degree murder, manslaughter, and so on.

Even 10-year-old children invoke culpability when they plead to their parents, "I didn't mean to." That petition is not a denial of who got hurt or who did the hurting. It's a plea for exoneration.

We invoke culpability across the diversity of human relationships. I'll go no farther down this path, except to restate that right relationships with nature, especially from this point forward, likely depend on some, as of yet, unarticulated principles involving culpability. It's complicated and not well explored.

The future of human relationships with nature will also be increasingly defined by moral dilemmas that require adjudicating which of two conflicting obligations ought to be prioritized. For Isle Royale, the conflict is between the well-being of a wolf population and the well-being of individual moose. For British Columbia, the conflict is between the well-being of a caribou population and the well-being of individual wolves. In both cases, the conflicting obligations are to non-humans. Cases involving conflicting obligations *between* humans and nonhumans are also, sadly, on the increase. To offer a sense, here are a few examples:

- Humans have reduced the abundance of wild elephants by about 90% over the past century, and most of the remaining elephants on the planet are enslaved for cheap entertainment. One of the planet's last populations of wild elephants depends on seasonal migration routes over the Bangladesh-Myanmar border. By 2018, that border also became the de facto home for nearly a million refugees—the Rohingya people, who have faced decades of systematic discrimination and violence in Myanmar. It is virtually impossible to do good by those refugees without harming the elephants and vice versa. The urgent injustice is beyond words.
- At a much larger scale, African wildlife is on the ropes—lions, rhinos, gorillas, cheetahs, African wild dogs, pangolins, and

addaxes to name a few. Human destitution in Africa is un-bounded anguish. In many places, human well-being competes directly with the well-being of wildlife. In a few decades, the number of humans in Africa will quadruple. My heart shudders. Throughout it all, some wealthy humans (from Africa and across the planet) have worked hard to make profits from the tragedy.
- At the broadest scope, the United Nations has carefully crafted goals for alleviating poverty and protecting biodiversity. They have also secured at least lip service from most nations to work toward those goals. Amazing. Except there is good reason to think it impossible to meet both sets of goals.[66]

Our future relationship with the human and nonhuman members of our family will depend, sadly but inevitably, on how we prioritize conflicting obligations. We have choices about how to respond to such dilemmas and choices that influence the chances of creating more dilemmas.

Intuition is critical for discovering how to make such choices, and thankfully we are awash in intuition. Collectively, our intuitions point in virtually every direction. We also need hard-nosed, compassion-soaked reasoning. My fear is for the short supply of that kind of reasoning. It's scarce for an innocent reason: it's really hard and it's not a lot of fun.

Some of you began reeling when I considered prioritizing the well-being of an individual moose over the well-being of Isle Royale's eco-system. Those agitated eye twitches are triggered, at least in part, by a belief that individual moose are better off, on average and over the long term, in an ecosystem where they are subjected to wolf predation.[67]

The vertigo may have worsened when I considered prioritizing the well-being of some individual wolves in British Columbia, where wolves are abundant, over the well-being of an entire population of caribou, whose extinction would be effectively irreversible. Surely, the concerns surrounding conservation—the maintenance and restoration of healthy ecosystems and populations—override concerns for the well-being of individual animals.

The rightness of that thinking is poignantly captured in *Star Trek II: The Wrath of Kahn* with the death of Mr. Spock, when he sacrifices his life to save the entire ship's crew. As Spock is dying, he says, "It is logical. The needs of the many outweigh . . ." Captain Kirk continues for him, "the needs of the few." Spock finishes, "Or the one."

The many outweigh the few. The moral calculus of Spock's death seems simple. But that calculus is greatly complicated if, for example, Kirk had decided against Spock's will that Spock should die to save the crew. And it would be further complicated if we then learned that Kirk had been responsible for creating the predicament. Maybe Spock should be killed in these alternate storylines. If so, one would have some very difficult explaining to do. But then again, maybe Spock shouldn't be killed, even if it means the death of the crew. That judgment would also involve some very difficult explaining.

Humans have a long-running tension between individualism and collectivism. As when some Americans revolt over stay-at-home orders during a pandemic. Different cultures manage those conflicts differently, and members of one culture are often either unaware that such differences could exist or are unable to comprehend how any such difference could be sensible. For example, in individualistic cultures, marriage is seen to be a union between two individuals. In collectivists cultures, marriage is a union between families. Cultures also change over time with respect to the relative valuing of individualism and collectivism.[68]

This tension between individualism and collectivism exists not only in relationships among humans but also when humans relate to nonhumans—as when humans decide to kill animals in the name of conservation. *The future of human relationships with nature depends on resolving moral dilemmas, and key to sorting these dilemmas is wise handling of culpability and tensions between individualism and collectivism.*[69]

These last several pages are a tornado of unresolved ideas. I could press those ideas further, but I have not, at least so far, seen a wholly satisfying resolution to share.

Furthermore, I never really expected (or promised) to sketch any grand resolution. I've done my best to share my experience on Isle Royale and to explain what I think are essential principles for resolving good relationships with nature, especially the four bulleted points and their attending maxim. I may have pinpointed some of the ideas most in need of resolution, especially the influence of culpability and the individual-collective spectrum. But my deepest hope is to have shown the value and difficulty of reasoning.

Human nature drags us, often without our consent or awareness, to uncritically honor the reasons held by those with whom we most closely identify and to prejudicially dismiss the reasons of those who seem different. Breaking that grip requires special effort.

I am confident that my reasonings fall short in various places, and that some of those shortfalls will be flagged by my friends and antagonists (some of whom are the same people). When they do, they will prove my point. Reasoning is hard.

Furthermore, while reasoned criticism is essential, the opportunity for growth and honor rests not with criticism but with the response. A cordial response is necessary but far from sufficient. Capitulation is sometimes key, but not always. Growth and honor rise and fall in empathizing with critics until they are understood, perhaps even better than the critics understands their own view.

This response is as virtuous as it is practical, insomuch as resolving pernicious problems depends on seeing the paleness in both an adored reason and a cherished critique. The required wisdom often lies elsewhere altogether. In any case, reasoning is—like it or not—a team sport, and we don't get to pick our team mates. They were assigned.

My faith in reason is not a pipe dream to sort it all, once and for all, with a flower-waving think-in. We face at least two syndromes. The first is inequality in affluence and political power that corrupts governments around the world, even the most democratic governments. That inequality syndrome hog-ties the pursuit of fair policies. I have little idea how to solve that problem.

The second syndrome features moral dilemmas, for which the first challenge is to accurately identify a genuine dilemma. That is, to accurately identify when we are and are not playing a zero-sum game, as in cases where we can meet some particular human interest or some need of biodiversity but not both.[70] I don't think we're very good at telling the difference.[71] The second challenge in this dilemmatic syndrome is to adjudicate the genuinely zero-sum games. I don't think we're very good at that, even for relatively modest cases, such as whether to restore wolves to Isle Royale.

It's this dilemmatic syndrome to which I think reason has much to contribute. Reason even has a role to play in confronting the inequality syndrome. The people with more than their fair share of things and power are motivated by their reasons. It's their reasoning that's weak not the power they wield, and the best antidote to weak reasoning is strong reasoning. Faith in reason is not a Panglossian delusion. It's history. Reason has played instrumental roles in various movements to advance fairness in opposition to those wielding brute power in a sea of complacency. Mary Wollstonecraft, Mahatma Gandhi, Ruth Bader Ginsburg, and Desmond Tutu have all been effective, in part, because of their exceptional facilities with, and faith in, reason.

Some think it's too late for reasoning. The biodiversity crisis ratchets on. The planet's physical systems, on which all biodiversity depends, stomp toward tipping points of no return. We must act. We must act now.

I, too, feel the pressure of that temptation, but I disagree. First, thinking and acting are not mutually exclusive. Second, the margins for error are shriveling and the trade-offs that ensnare our choices are vertiginously steep. Getting it right will require more thinking, not less.

I have no idea how it will turn out. I suspect worse than expected in some ways and better than expected in other ways. Leastwise, that's been the pattern throughout human history. But my hope doesn't commune with daydreams about how it might turn out. That's the kind of hope that taunts despair.[72]

My hope, even when the evidence suggests otherwise, rests with the ability of any human, myself included, to discover and endeavor for a right purpose in life. Because that requires reasoning, my blind faith is in hard-nosed, compassion-soaked reasoning. Embracing this hope may be essential for favoring fair coexistence with biodiversity. At least, clinging to the pursuit of right purpose and reason is the only way for anyone to preserve their humanity.

We can end the way we began, by asking, *Why wolves?* The answer hasn't changed. Because they remind us to think.

Coda

Having addressed the most important issues concerning the restoration of balance, we can conclude with what some of you may have been wondering for the past few chapters. What's happened on Isle Royale since Chapter 7, when we left the wolf population with just two wolves in early 2018?

Stepping back just a few years, here's how it played out. I anticipated the wolf collapse in 2009, just before it began. The National Park Service seemed reluctant to engage the concern. I spent a couple of years trying to call attention to the problem, despite having been given the unambiguous impression that I'd have hell to pay if I did so. In 2013, I continued advocating for public discourse and also began to advocate for genetic rescue, which would have involved bringing one or two wolves from the mainland to Isle Royale. That light touch would likely have rejuvenated the genetic health of the wolf population and done so before the moose population had grown to such high abundance.

Eventually, the NPS solicited comments from the public for views on whether the wolf population should be aided. At the end of that solicitation, in early 2014, the NPS issued a press release stating their decision that "the park will not bring wolves from outside to Isle Royale."[1] The document also stated that "for the past two years, park managers have discussed island and wolf management with wildlife managers and geneticists from across the US and Canada and have received input during public meetings and from Native American bands of the area," and that the "decision on a way forward is supported by our review of the best available science, law and policy."

I asked to see a copy of the analysis based on that best-available science, law, and policy. I was told, no. Eventually, results of the public comments were made available with assistance from the US Freedom of Information Act. An analysis of those comments indicates that more than 80% of the commenters believe that it is a good idea to restore wolf predation to Isle Royale.[2]

In 2016, the NPS felt obligated by federal law to develop an environmental impact statement (EIS), which is a formal decision-making process. NPS released a draft EIS in early 2017. The draft EIS left some careful readers with the impression that the NPS was leaning against restoring wolf predation on Isle Royale. Somewhere between the draft and final EIS, the NPS seems to have changed its leaning. In June 2018, the NPS made its final decision to take definitive action to restore wolf predation on Isle Royale.

I believe the NPS made the right decision, but if you ask me why the NPS decided as it did, I'm not sure that I can tell you. The reasoning offered in the EIS seems opaque.

Nevertheless, the NPS began implementing its decision in September 2018. I was excluded from significant elements of the restoration process, in favor of a researcher from another university. (I suppose that was the hell I was promised if I spoke out about the Isle Royale case.) Consequently, much of what I know about the NPS restoration effort is what I have read in NPS press releases and the media.

In a nutshell, this is what's happened. Between September 2018 and September 2019, the NPS captured 20 wolves that it deemed fit for relocation. They came from the "arrowhead" region of Minnesota; from Michigan—a little south of Hancock, where I live; from a site near Wawa, Ontario; and from Michipicoten Island, which is part of Ontario and nestled in the northeast corner of Lake Superior. Of those 20 wolves, eight were dead by February 2020. That death rate was described by one wolf expert as "surprising."[3] The causes of death include being killed by wolves after being released on the island, stress from having been captured and relocated, and unknown causes.

Wolf M183 (the last male from the previous population) was killed by some of the relocated wolves in October 2019.[4] Wolf F193 (the last female) was likely alive when M183 was killed, but she was never seen after his death. She likely died or was killed shortly afterward.

For the newly translocated wolves, those first months were a pandemonium of disrupted families, unfamiliar landscapes without the territorial boundaries that provide so much order, and strange neighbors willing to kill out of their own senses of desperation. By February 2020, we perceived the formation of four nascent social groups, each with just two or three wolves. One group established a territory at Isle Royale's east end. Another staked a claim on the island's west end. The two remaining groups, under duress from their neighbors, were squeezed into thin strips of real estate along the shoreline.

In spite of the chaos, the newly arrived wolves fed themselves well. We observed those four social groups kill two dozen moose during the winter field season in January and February of 2020.

Then COVID struck, stole our ability to be in the field, and blinded us to what we would otherwise have seen during the winter of 2020–21. A year later COVID lightened its grip, and we were back in the field.

While we were gone, much happened. East Pack had grown to about a dozen wolves. West Pack did the same.[5] Clearly, a great deal of reproduction occurred during the springs of 2020 and 2021.

East Pack and West Pack maintained their dominance into February 2022, by which time at least one of the shoreline groups had slipped into oblivion and at least one new group formed. In a sign that the population was taking root, one of the new social groups was composed, not of translocated wolves, but rather of their offspring who had matured into wolves of their own. That spring, pups were born to three social groups, and possibly to the fourth group.

By the following winter, the two less secure groups pressed East and West Packs into smaller homes and into tussling over the no-man's-land that formed between those packs. I expect only one of those packs to survive. Meanwhile, East Pack's membership held steady at about a dozen, but the wolves of West Pack stumbled down to just five.

Four years following the translocations, the wolves were still sorting their social lives. But the population had certainly reestablished itself in terms of total abundance. In February 2023, the population included about thirty wolves—more wolves than the island typically experiences during any given year.

By this time, most of the translocated wolves had died. The population was now largely comprised of their descendants, unaware of how they came to be where they were, unaware that so many cared about their fates, or why.

To assess the wolf population's influence on Isle Royale's ecosystem, we studied the predation rate, which is a statistic that indicates the proportion of moose killed each year by the wolf population. When predation rate is in the neighborhood of 9–11%, the moose population tends to hold steady by producing enough new calves to offset losses to predation. When predation rate is greater, moose abundance tends to decline. When lower, moose abundance tends to increase. Not by coincidence, predation rate is about 10%, on average, over the long term.

With that context, it's meaningful to know that from 2013 to 2019, the annual predation rate was estimated at a measly 2%. A year after the translocations, predation rate swelled to 5%—then to 9% in 2022 and 10% in 2023. The increase in predation rate was not due to wolves filling their bellies with more and more meat. Each wolf was—on average—eating about the same throughout this increase in predation rate. Rather, predation rate was increasing because wolf abundance was increasing and moose abundance was decreasing. Most importantly, those predation rate statistics tell us that it took about three years for wolf predation to be just on the cusp of powerful enough to reduce moose abundance.

Moose abundance peaked at about 2,000 moose just as the translocations were occurring. Over the next four years, the population fell by 50% to about 950 moose in 2023. The first two years of decline were almost entirely attributable to starvation, and the more recent years of decline were a concoction of predation and malnutrition.

Starvation and malnutrition were rampant because the vegetation had been thoroughly pummeled by the time of the translocations. As moose abundance declined, browsing pressure lightened. But forests recover slower than a wolf population can. It will take time for the plants to reawaken. When and how the forest recovers are some of the bigger unknowns—unknowns that percolate up the food chain to uncertainties about future wolf and moose dynamics.

The other great unknown is the genetic health of the new wolf population. After only one generation, they show worrying levels of inbreeding. If this early trend continues, then within the next five years or so, the wolf population will reach levels of inbreeding that typically cause geneticists to sound alarms. I fully expected inbreeding to occur—but not this fast.

The rapid inbreeding occurred because too much reproduction was concentrated among the wolves that had been translocated from the same location. In particular, five of the six wolves to reproduce and pass genes into the new population's next generation were from Michipicoten Island. In contrast, none of the four wolves from mainland Michigan contributed to the new population's gene pool. They all died before reproducing. None of the three wolves from mainland Ontario lived long enough to reproduce. Of the five wolves from Minnesota, one left the island on an ice bridge and three others died before reproducing. Only one of the five Minnesotan wolves contributed genes to the new population.[6]

It would have been better for a couple of wolves from each of the four locations to have contributed to the gene pool. The challenge is that human translocators only get to decide who is translocated. Beyond that, the wolves decide who mates with whom. The precise social dynamics among the wolves that led to so much inbreeding are not known with great certainty, but a plausible account begins by noting that the eight wolves from Michipicoten Island were an alpha female, her mate, and six of their offspring. The wolves from other locations did not have the benefit of being moved as a large family group.[7]

Members of the Michipicoten family had a higher survival rate than did wolves from any other location, plausibly because they had each other to rely on during their first months on Isle Royale. They likely were more comfortable banding together for protection against being killed by other wolves and to share the burden of killing moose.

In March of 2020, a brother and sister from the Michipicoten family made the fateful decision to mate with each other. For whatever reason, they thought that to be their best option. That brother and sister have given birth to more wolves in the new population than any other pair of wolves. They are the wolves of East Pack.*

The mother of that brother-sister pair is the alpha female who was translocated from Michipicoten. That makes her the grandmother of every wolf known to have been born so far in the new population. Soon, she is likely to be the grandmother and great-grandmother of all the wolves born on Isle Royale. All of the wolves born on Isle Royale since the translocations are also at least as related to one another as first cousins. This sounds much like the story from twenty-five years prior of the Old Gray Guy and his descendants (Chapter 6).

If the translocations had been designed so that none of the translocated wolves were related to each other, none would have had the upper hand of coming with family, and a mating between full siblings would have been impossible. Those circumstances would have

---

*To further contextualize the near-total dominance of Michipicoten wolves, note that the alpha male of West Pack since 2020 is the brother of East Pack's alpha pair.

significantly elevated the chances of more even contributions to the gene pool and slowed rates of inbreeding. While that is how I would have designed the translocation, no one knows for sure how that would have turned out either.

Because the new wolves of Isle Royale are the old wolves of Michipicoten Island (with a sprinkling of genes from one Minnesotan wolf), your curiosity may be piqued at the prospect of reflecting on the lives these wolves had on Michipicoten. The first point of interest is that the alpha pair brought to Isle Royale from Michipicoten were not born on Michipicoten. Rather, they walked to Michipicoten on an ice bridge in 2014. When they arrived, they found no other wolves. They were the first wolves to live on Michipicoten in a very long time. But they did find an abundance of caribou—somewhere between 700 and 1,000 caribou on an island that is about a third the size of Isle Royale.

These caribou have their own interesting story. In 2014, they were virtually all that remained of the many, many thousands of caribou that once roamed the vast forests north of Lake Superior. Caribou had been brought to Michipicoten in the early 1980s, because it was about the only place they could persist. This population's mainland range had been ruined by excessive logging, and efforts to restore those forests continue to be too little and risk being too late.

In any case, caribou can under certain ecological circumstances be especially vulnerable to wolf predation. One of those circumstances is being confined to a small island with wolves. Biologists knew what would happen the moment that wolves arrived to Michipicoten. But I'm not sure anyone expected the speed of what came next. Within four years of wolves' arrival, caribou abundance went from about 800 to less than 20. The government of Ontario acted quickly and airlifted the 15 caribou that they could capture and moved them to the only other islands in Lake Superior where they could possibly live.[8]

Without the caribou on Michipicoten, those wolves were fated to starve. Expecting the government of Ontario to move those wolves to the mainland would have been farcical, given that its primary relationship with wolves is to administer an annual hunt that kills many hundreds of wolves for reasons that critics describe as cruel and contrary to well-established principles of science.

The Michipicoten wolves were left to starve at the very same time that the NPS was looking for wolves to bring to Isle Royale. Private funds were raised, and the move was made, though at least one wolf was left behind. So, the wolves of Michipicoten were brought to Isle Royale, but not because they were part of a well-developed set of genetic considerations. Rather, they were brought to Isle Royale to save them from starvation.

Those pitiful wolves and caribou, moved around like pieces in a board game, somehow remain the only way we seem able to hang on to valuable elements of biodiversity.

In closing, we should ask: What can the Isle Royale translocation experience teach us? While the answers will evolve, some insight may already be emerging.

With respect to scientific knowledge about predation, I think the Isle Royale experience has been largely confirmatory. Scientists have known for some time now that wolves limit prey according to their predation rate. The predation rate dynamics of Isle Royale's new wolf population have unfolded pretty much as expected. We should carry such knowledge with confidence to consider the value of restoring other carnivores.

With respect to scientific knowledge of herbivory, we have known for a long time that large mammalian herbivores living without predators can adversely impact ecosystems. What we know too little about is the extent to which ecosystems can recover after predation is restored, or the extent to which those places are forever changed. Isle Royale will contribute insight on those questions as the years unfold.

The translocation experience also teaches about the importance of considering the genetics of translocated individuals. Genetic considerations are often, though not always, given due attention when translocating animals for conservation.[9] In any case, there is no simple recipe for how to best take account of genetics. I expect that Isle Royale and other translocation experiences will contribute to the development of guidelines or principles on how to best account for the genetic aspects of translocation.

The high rate of inbreeding already observed in the new wolf population indicates that now is the time for leaders to think in earnest about whether humans should continue to conserve predation on Isle Royale. Nature will likely force answers sooner than some might previously have expected. The essential questions to begin answering include: Was the interest to conserve predation limited to that single set of translocations? If so, why? If not, then to what extent should the National Park Service be proactive or reactive in conserving predation as inbreeding inexorably cinches its grip on the wolf population?

A deeply reactive approach might involve doing nothing until the wolf population is on the precipice of failure in a way that is obvious even to a skeptic. That approach would likely unfold as the first set of translocations did—that is, occurring about a decade after predation was effectively lost and requiring considerable effort to start an entirely new population.

By contrast, a proactive approach would require actions that are less effortful and less invasive, but probably more frequent. A proactive approach could involve translocating a single wolf who then becomes a reproductive member of the population anytime the population's inbreeding level reaches a point that causes geneticists to worry.[10] This proactive approach is also a reasonable match to processes by which populations maintain genetic health when they do not need assistance from humans.

Other considerations would further indicate the pros and cons of being proactive or reactive. In any case, the essential point is, now is the time to expeditiously develop those considerations, because the time to make well-considered decisions is very likely to soon pass.

The most important lesson from the restoration experience on Isle Royale may begin by recalling that the biodiversity crisis is—at its core—the monumental, unprecedented loss of species' geographic range (Chapter 10). Even the faintest mitigation of that crisis requires restoration at a monumental, unprecedented scale.

Most of these restorations will be difficult because they will require some group of humans to sacrifice some interest of theirs to make way for the restoration. This is so, because the biodiversity crisis has always been fueled by humans pursuing their interests. Anytime a group of humans (with sufficient political power) is asked to sacrifice something dear, conflict tends to follow. So, we can expect most future decisions about restoration of any consequence to be rife with conflict.

The decision to restore wolf predation on Isle Royale was burdened and slowed by the attention it gave to conflicting views about the purpose of wilderness (Chapters 8 and 9). At the same time, it might also be said that restoration on Isle Royale was planned too quickly because not enough consideration was given to genetics. The concern is that the Isle Royale case was easy—like a kindergarten exercise—compared to the myriad restorations required for mitigating the biodiversity crisis. If we are to even slow the crisis, then we'll have to become much wiser *and* much quicker at adjudicating conflicts that stand in the way of restoring biodiversity that humans are responsible for having destroyed.

John A. Vucetich
September 27, 2023

Chapter 1. Why Wolves?

1. Extensive research demonstrates and explains the nature of grief and separation anxiety in a variety of nonhuman animals, including dogs (e.g., Panksepp 2004, esp. chapter 14, which is entitled "Loneliness and the Social Bond: The Brain Sources of Sorrow and Grief").

2. Allen 1979.

3. Wolves on Isle Royale have lived in packs with as few as one other wolf and as many as 17 other wolves. Typical pack sizes vary among wolf populations. For example, wolves in north-central Canada that feed on bison (bison are a third larger than moose) often live in packs of 10 to 12 wolves. As a wolf population expands into new regions, like the wolves in the western Great Lakes region during the 1990s and early 2000s, the older offspring have a greater tendency to disperse from the pack because the promise of establishing their own family is greater. The packs in these populations are commonly just three or four wolves. Those packs in the western Great Lakes region also feed primarily on white-tailed deer, which are considerably smaller than moose.

4. Wolves typically give birth to between six and eight pups in late April. Commonly, only three or four survive to see their first winter. Pups are born in an altricial condition; that is, relatively little development occurs in the womb. At birth wolves are about the size of a large potato, blind and deaf. They are only about 1% the mass of their mother (compared to human babies, which are 4% the size of their mothers). Mortality rates are high under the best of circumstances during the first year of life.

5. Some believe it wrongfully anthropomorphic to recognize individual personalities in wild animals. Experts in animal behavior demonstrate otherwise. Scientific evidence suggests that even hermit crabs possess personality. For a recent review, see Carere and Maestripieri (2013).

6. For more on the importance of metaphor in understanding biological phenomena, see Keller (2002, esp. p. 146).

7. Evolutionary ecologists have a long track record of studying families in a broad context that includes human and nonhuman animals. For example, Emlen 1995; Royle, Smiseth, and Kölliker 2012.

8. *Merriam-Webster's Collegiate Dictionary*, 11th ed. (2020), s.v. "person."

9. This refusal is also the subject of formal scholarship (e.g., De Waal 1999). For an accessible treatment of these ideas, see De Waal (1997).

10. Exaggerating differences between humans and nonhumans routinely results in wrongful neglect. However, believing in imagined similarities between humans and nonhumans risks caring for another in a way that is unwittingly harmful. Lennie Small, from John Steinbeck's *Of Mice and Men*, for example, believed that little rabbits would appreciate big strong hugs as much as he did.

11. Humans, for example, discovered the neural basis for empathy from a macaque who was empathizing with a human. For anecdotal examples of nonhuman animals empathizing with other species, see Cronin (2014). For a broader treatment of empathy in nonhuman animals, see De Waal (2010).

12. The rumen is the first chamber of a moose's four-chambered stomach, where food is partially digested by a diverse community of microbes. After microbes do their job, the food wad is regurgitated and moose chew their meal a second time. This second round of chewing is referred to as *rumination*. After ruminating, the food is swallowed again and sent to subsequent chambers of the stomach. This entire digestive process (also known as rumination) may be nature's most elaborate digestive process, and it occurs in cows, deer, sheep, giraffes, and about 200 other species of large herbivore.

13. Tyson 2012.

14. Rolf Peterson is second in the succession of scientists to have led research on the wolves of Isle Royale. Officially retired since 2006, he still contributes greatly the project. I'll tell you much more about Rolf in Chapter 4.

15. While the waters of Lake Superior are cold, that water is often warmer than the air during the winter. When colder air blows over warmer waters during the winter, the common result is lake-effect snow. More precisely, moisture evaporates from the lake into the layer of air that is closest to the lake. The evaporating moisture is often visible. That lower layer of air is also warmed by the lake. The warm, moist layer of air begins to rise, because warm air is less dense than cooler air. As the warm air rises, it cools again. In that cooling process, the moisture is squeezed out and falls as snow.

16. That satisfaction is born from a basic research objective to obtain an accurate estimate of per capita kill rate, which is a statistic that measures, in a sense, the rate at which moose flow into wolves. More precisely, this statistic is the number of moose killed, divided by the number of wolves making the kills, divided by the number of days over which the kills were made. That statistic is a foundation for understanding how and why the abundance of wolves and moose fluctuate over the years. Accurate estimates of kill rate depend on detecting every kill.

Chapter 2. Thoughts of a Moose

1. Lexico.com, s.v. "parasite." Accessed November 2, 2020. https://www.lexico .com/definition/parasite.

2. Set two American football fields side-by-side and you would have an area that is approximately a hectare in size.

3. Inbreeding occurs when related individuals mate and pass their related genes to their offspring. As such, inbreeding is an intergenerational process. Because short-lived species have short generation times, the effect of inbreeding can reveal itself in shorter periods of time.

4. There are many species of bot fly, and one, *Cuterbra fontinella*, is partial to deer mice.

5. Munger and Karasov 1994.

6. Metabolism increases by about 9% for a mouse that is hosting a bot fly larva, which is comparable to the metabolic cost of pregnancy (Munger and Karasov 1994).

7. For example, Tabatabaie et al. 2011.

8. In a study of mouse survival, of those mice uninfected with bot flies, 80% were dead within 15 weeks of having first been captured. Among mice that had been infected with one bot fly, 20 weeks passed before 80% had died. Among mice with two bot flies, 45 weeks passed before 80% of the mice had died (Burns, Goodwin, and Ostfeld 2005).

9. Bradshaw, Casey, and Brown 2012, 47.

10. Another reason for thinking so is that deer with the fewest ticks groom the most, but moose with the most ticks groom the most (see Samuel 2004). While this idea is consistent with the hypothesis, I am not so sure it is inconsistent with fall grooming by deer being the result of stimulus-driven grooming.

11. See Han et al. (2013) and Johns Hopkins Medicine (2013) for an accessible explanation. To appreciate the importance of having more or less tactile neurons, know that you have more pressure sensing neurons on your fingertips than your back. As a result, you can distinguish two closely positioned pin pricks on your fingertips but not your back.

12. O'Connell-Rodwell 2007.

13. Humans immigrated to North America (though perhaps not for the first time) on the same bridge.

14. Sanguivory is also called hematophagy. The proboscis of a vampire moth (*Calyptra spp.*) is a minor evolutionary modification of the design used by closely related moths who feed on juice by puncturing the skin of fruits. The vampire finch (*Geospiza difficilis septentrionalis*) lives on the Galápagos Islands and pecks the feet of blue-footed boobies and then laps the blood. Vampire snails (*Cancellaria cooperii*) live off the coast of California and suck the blood of electric rays.

15. When you give a pint—about 10% of your entire supply—it takes about eight weeks to replace the lost red blood cells. That recovery would be longer if you were poorly nourished, like a moose in early spring.

16. This view of life is an attractive description of a *food web diagram*, like that shown on page 289 (Chapter 10). Food web diagrams are not simple, physical facts of the matter, like lead is denser than aluminum. Food web

diagrams are abstract templates onto which ecologists set and compare their empirical observations.

In sharing this view of life—with its connections among wolves, moose, ticks, and the forest—ecologists use strategic location of qualifiers like *can*, *are likely*, and *contribute* (as opposed to fully determine). The empirical work of ecology includes quantifying the *contributions* (weak or strong) and converting the *can's* and *are likely's* into probabilities (which may range from near nil to almost unity).

While this view of life is genuinely ecological, it may be more metaphysical than empirical.

17. That statistic corresponds, for a typical-sized pack of, say, six wolves, to 4.5 moose per month for the entire pack, or about one moose every seven days. These statistics also pertain to kill rates during the winter months. The best evidence suggests that kill rates during the summer are only half. For technical details, see Vucetich et al. (2011).

18. The age structure of a population refers to the relative abundance of different aged individuals. For example, on Isle Royale, the proportion of a moose population that is old (older than nine years old) may be as low as 10% and as high as 40%. The difference is important because old moose are easier for wolves to kill, and the predation of an old moose is less consequential to population dynamics because those moose are soon to die anyway. For details, see Hoy, MacNulty, et al. (2020).

19. Vucetich and Peterson 2004.

20. We estimated a population-wide kill rate by observing 24 of the 29 wolves that lived on Isle Royale in 2000. The other five wolves were observed too infrequently for us to be confident about the frequency of their kills.

21. Young et al. 2011.

22. Mining and packaging the earth for human consumption is an industry in places like west Africa and an undocumented industry in very remote places like southern Georgia, USA (see Schmidt 1984).

23. The truth is slightly more complex. Nevertheless, most plants need so little sodium as to be negligible. See for example, Pardo and Quintero (2002) and Maathuis (2014).

24. Risenhoover and Peterson 1986.

25. Plato theorized that our ideas of the world are more real, perfect, and timeless than physical manifestations that we associate with those ideas or "Forms."

Chapter 3. Beginnings

1. Mech 1966, 24; Peterson 1977, 1.

2. This quote and the previous three are from the preface of Allen (1979).

3. In a typical year we collect about 10 to 20 scats per wolf, which provides reasonable assurance of collecting at least one scat from each wolf. For technical details, see Marucco et al. (2012).

4. These quoted passages are from an interview with George Desort conducted in 2008. Transcript on file at the Michigan Technological University Archives and Copper Country Historical Collections, Houghton, MI.

5. Mech (1966) cites 14 papers on the ecology or behavior of wolves.

6. Mech 1966, 57.

7. Mech 1966, 45.

8. Mech 1966, 50.

9. Mech 1966, 126.

10. Mech 1966, 151.

11. Wolves' tendency to kill weakened moose depends on the relative abundance of weakened moose. In some years, weakened moose are more common, for example, due to a severe winter. In those cases, wolves tend to kill more frequently, and a greater portion of their kills are weakened moose. In other years, weakened moose are less common. For example, in some years calves and old moose (which are easier to kill) are less common than moose in their prime (which are tougher to kill). During these years, wolves tend to kill less, and weakened moose make a up a smaller portion of those killed.

12. Mech 1966, 171.

13. Allen and Mech 1963, 219.

Chapter 4. Balance of Nature

1. Liordos et al. 2017.

2. This myth was collected by a 19th-century anthropologist and retold by Raley (1998).

3. The quoted text is from Gregory Nagy's translation of the *Homeric Hymn to Demeter* (brackets in the original), see https://uh.edu/~cldue/texts/demeter.html.

4. Cicero, *De Natura Deorum (On the Nature of the Gods)*.

5. Plotinus, *Ennead* III, 2:15.

6. Allen 1979, xviii.

7. Derham 1714, as quoted by Simberloff 2014.

8. Derham 1714, as quoted by Egerton 1973, 333.

9. Mayr 1982, 318.

10. Barrow 2009, 39.

11. By the early 19th century, geologists were also beginning to understand and accept stratigraphy, whereby different layers of rock represent different periods in the earth's history. They also knew the fossils embedded in one layer represented species different from those found in other layers. Dinosaurs for example, first appear in rocks formed during the Triassic period, but never before, and last appear in rocks that formed during the Cretaceous period, but never afterward. These ideas added to evidence for the extinction of older species and evolution of newer species.

12. Mayr 1982, 349. Louis Agassiz, another highly reputed geologist, is among those who believed extinctions might have been the result of God occasionally destroying the earth.

13. Darwin 1872, 57.

14. Wallace (1853) 2010, 400–401.

15. Wallace (1858) 2003. The similarity between Wallace's and Darwin's texts was pointed out by Carroll (2009).

16. Darwin 1987, 175. This quote was also brought to my attention by Carroll (2009).

17. McKinney 1966, cited in Simberloff 2014.

18. Fox 2011, 67.

19. D'Ancona 1954.

20. Marceline Desbordes-Valmore, 19th-century poet.

21. The method described here is how one would create a graph of Volterra's equations if they were translated from differential equations (continuous time) into a difference equation (discrete time). If we were to stick with Volterra's differential equations, then we'd need to invoke a little calculus.

22. Volterra 1926, 560.

23. Kadlec and Wallace 2008, 486.

24. For a technical, but accessible, treatment of the interplay between ecological theory and observation as it pertains to predation theory, see Gatto (2009).

25. Huffaker 1958.

26. Utida 1957.

27. Elton 1930.

28. Nicholson 1933, 133.

29. Nicholson 1933, 134.

30. The governor analogy is from Nicholson (1937).

31. As the rotation of the engine's shaft increases, centrifugal force increases. That force is converted into a motion that controls a valve, reducing the amount of steam entering the engine, thereby limiting its speed.

32. Nicholson 1933, 136.

33. Nicholson 1933, 133.

34. Pound and Clements 1900.

35. Clement and Pound developed their quadrat method when they were 20 and 24 years of age, respectively. More than a century later, in a world that makes routine use of GPS, satellite imagery, and infrared spectrophotometry, the quadrat remains a cornerstone of plant ecology. Within a few years of counting plants in the hot Nebraskan sun, Pound apparently succumbed to tedium and changed course. A few years after that, he became dean of the University of Nebraska's Law School and eventually became dean of the Harvard Law School.

36. As Clements's belief was summarized in Worster 1993, 159.

37. Forbes 1883, 33.

38. McIntosh 1998, 426.

39. The complete quote is found on page 25 of the second edition of *Fundamentals of Ecology* by Eugene Odum, printed in 1967. The first edition, printed in 1959 and coauthored with Howard Odum, includes only the text after the ellipsis.

40. No effort is required to see basic differences between an organism's homeostasis and an ecosystem's homeostasis. More insight, however, is required to see the similarities. The comparison reminds one of an analogous comparison made by Darwin in *The Descent of Man, and Selection in Relation to Sex* (1871, 105): "Nevertheless the difference in mind between man and the higher animals, great as it is, certainly is one of degree and not of kind."

41. For different perspectives on balance of nature, see Simberloff (2014) and Botkin (1990).

42. Allen 1979, 343.

43. On this point, the metaphysics of ecology touch the metaphysics of Buddhism by placing so much emphasis on relationships, as opposed to the relators enmeshed within relationships (Priest 2014).

44. Allen 1979, 28.

45. Allen 1979, 90.

46. Darwin's evolutionary theory—whereby every species, including humans, is the result of the same mindless and purposeless process—is supposed to have humbled humans and to have made us less anthropocentric. Perhaps it has. In any case, I think we might have more distance to travel in this direction.

47. More on this topic in Chapter 10.

48. Allen 1979, 343.

49. Allen 1979, xviii.

50. I had the privilege of contributing an editorial to the *New York Times* on the wolves and moose of Isle Royale (Vucetich, Nelson, and Peterson 2013). The editor retitled it (without our approval or notice) as "Predator and Prey, a Delicate Dance." I do not recall what I had entitled the editorial, but I am certain it did not contain any reference—direct or remote—to a delicate dance. I was able to say what I wanted in that editorial without invoking notions of balance or delicacy.

51. Pierce et al. 1987. Survey results like that have been documented on dozens of occasions. For a review see Hawcroft and Milfont (2010).

Chapter 5. Exogenous Forces

1. The technique involves counting rings in the teeth, not unlike the manner by which a tree can be aged. Michael Wolfe aged more than 350 dead moose whose teeth had been collected (with skulls and mandibles) over the previous decade. Sixty years later, we still use the same technique.

2. Allen happened to be present on February 25, the day of that discovery. After 1962, he was present for about a month of each winter field season.

3. Allen's primary interest, reflected in *Wolves of Minong* (Allen 1979), was to understand how fluctuations in the abundance of wolves and moose could be explained by mechanisms such as birth rates and death rates, and to understand how those processes could be explained by predation and properties of the moose population, such as its age structure and nutritional condition.

4. Allen believed it was possible that one black wolf died, and he believed that it was possible that one of the black wolves was present but appeared lighter in color than previous years (see Johnson, Wolfe, and Allen 1968, 15).

5. A beta wolf is dominant to all wolves in the pack except the alpha.

6. I have tried to convey these dynamics of the 1960s and early '70s as Allen might have. The notion that wolves *occasionally* come or go across an ice bridge was taken for granted; it would have been worth careful documentation but not occasion for surprise. The presumed isolation of Isle Royale wolves rose to prominence with a genetic analysis that would come in the late 1980s, which (misleadingly) suggested that the wolf population had been isolated. More on that later in this chapter and the next.

Finally, black wolves (whose actual color varies from darker-than-typical gray to plain-old black) are uncommon and had not been seen before on Isle Royale. Allen's exclamation on February 25 was genuine and, I believe, caused by detecting a specific instance of immigration because of the immigrants' color. Allen's more enduring interest in this particular pack seems largely limited to being able to distinguish them from the gray-colored wolves, which, for example, is a great aid in deducing the total number of wolves in the population (e.g., Johnson, Wolfe, and Allen 1968).

7. According to information available at the time, the estimated abundance of moose increased from 600 to 848 between 1960 and 1962, declined for two years, and then rose to 1,015 by 1968. Allen was wary of the reliability of those early estimates. For details, see figure 1 of Johnson, Wolfe, and Allen (1968). Those estimates have been replaced by an alternative means of deriving retrospective estimates of abundance. For details, see Hoy, MacNulty, et al. (2020).

8. My relationship with Rolf Peterson began in 1989, when I was 18 years old, and we have shared some of the most precious experiences of our lives. So, it'd be difficult for me to refer to him as "Peterson." I'll call him Rolf. By contrast, I never met Allen, who was deep in retirement by the time I came to Isle Royale. And I knew Mech as "Mech" for a number of years through his scientific writing before developing a personal relationship.

9. The quote continues "or it might be said that we returned to a continuation of the intensive wolf investigations that were carried out by L David Mech from 1958 to 1961" (Peterson et al. 1971, 1).

10. Those patterns were more thoroughly tested in Montgomery et al. (2013). That paper analyzed the locations of death for more than 700 moose that died between 1959 and 2008.

11. An entire section entitled "Snow Studies" opens with "snow is an extremely important factor . . . affecting . . . certain facets of the wolf-moose relationship, such as calf vulnerability and kill distribution."

12. Keith and Bloomer 1993.

13. Predation is experienced from two very different perspectives—the prey's and the predator's. At the population level, the predator's perspective is

quantified by the per capita kill rate, which describes the rate at which each predator acquires food. The prey's perspective is quantified by predation rate, which is the proportion of the prey population killed each year by predation. Predation rate and kill rate are typically not related to one another. That is, a good year for wolves (high kill rate) may or may not be a bad year for moose (high predation rate). For a technical account, see Vucetich et al. (2011). For an accessible summary see Vucetich and Peterson (2012, 9).

14. The notable ecologist Sir Robert May led ecologists into a period of focused and formal attention on the influence that abiotic processes could have on biotic systems. One of his earlier pieces (May 1973) is explicitly motivated by observations from Isle Royale.

15. The "carving" metaphor is attributable to Pearl (2009, 348). The metaphor is also valuable for understanding how to distinguish between exogenous and endogenous forces.

16. For a technical perspective on this approach to science, see Vucetich, Nelson, and Bruskotter (2020).

17. Evidence to this effect is found, for example, in Peterson (2008, 7).

18. And we will return to the close, personal experiences of wolves in Chapters 6 and 7.

19. Peterson and Stephens 1980.

20. Arnett served in the US presidential administration of Ronald Reagan, which was not especially friendly to environmental concerns. Before serving as assistant secretary, Arnett led a team of geologists in the first drilling for oil and gas on Alaska's Kenai Peninsula. His obituary recalls that he had been "affiliated with many conservation organizations including California Rifle and Pistol Association, California Waterfowl Association, Congressional Sportsmen's Caucus Foundation, Ducks Unlimited, Game Conservation International, International Order of St. Hubertus, Mzuri Safari Foundation, National Wild Turkey Foundation, Ruff Grouse Society, United Conservation Alliance, Wildlife Legislative Fund of America, World Wilderness Congress and the Virginia Rifle and Revolver Association" (*Bakersfield Californian*, July 7, 2019). Arnett equated conservation, more or less, with hunting. Many with that mindset (right up to the present day) oppose predators. For a formal treatment on the relationship between hunting and conservation, see Treves and Martin (2011) and Bruskotter, Vucetich, and Nelson (2017).

21. A young field assistant who would go on to lead research on the wolves of Yellowstone.

22. As a sign of how much has changed in 40 years, the National Rifle Association also provided funding for the wolf-moose project from 1975 to 1979. Today there is a tendency for gun advocacy to be associated with antipathy toward predators.

23. Peterson and Stephens 1981.

24. Peterson, Page, and Stephens 1982.

25. Peterson 1984, 1.

26. It was reported as 11 wolves in Peterson (1988) and corrected to 12 in Peterson (1989).

27. Peterson 1988, 1.

28. Aerial observations are limited to the daylight hours of the seven weeks that make up winter study, which represent 5.5% of the hours in a year. Aerial observations are further restricted by inclement weather, which brings the total time in the air to about 100 hours per year.

29. Peterson and Stephens 1981, 11.

30. Oelfke et al. 2000.

31. For an accessible overview of how viruses can spillover from one species to another, see Quammen (2012).

32. Details about the Duluth dog were brought to light 20 years later, after Dr. Larry Anderson—the vet who cared for the Duluth dog and former board member of the International Wolf Center—shared his recollections with Rolf, prompting a reexamination of Dr. Anderson's medical records from 1981. For details, see 2001-02 Annual Report (Peterson, Rolf, and Vucetich 2002).

33. Some research suggests that CPV can reduce the 16% to 58% growth rate of a well-fed wolf population down to 4% (Mech et al. 2008). Other recent research indicates that CPV is deadlier among wolves with impoverished genetic diversity (Hedrick, Lee, and Buchanan 2003).

34. Last reverberations, that is, until 2009, when CPV would strike a second time.

35. The population-level consequences of inbreeding are typically understood by comparing sets of population that are identical in all ways except that some are inbred and others are not.

36. The acknowledgment of possible immigration and emigration was made in several Annual Reports prior to the findings of Wayne et al. (1991).

37. While seemingly plausible and parsimonious, complete isolation turned out to be the wrong conclusion. But we'll save those details for the next chapter.

38. This sentiment is the substance of chapter 8 of Peterson (1995b).

39. Peterson 1991, 9.

40. Peterson 1995b, 171.

41. Even a meager understanding of a population's ecology requires estimates of annual population growth rate, which requires estimates of abundance every year. On several occasions so far during the 21st century, the park indicated that it was considering reducing the frequency of winter study to alternate years.

42. Dave Mech was among this group of scientists.

43. Fritts and Carbyn 1995; Fuller, Mech, and Cochrane 2003.

44. US National Park Service 1999.

Chapter 6. The Old Gray Guy

1. Predation rate is the proportion of moose in the population that die from predation. It is a cause-specific mortality rate for moose, and it quantifies the

impact of wolf predation on a moose population. Predation rate dropped from 11% to 6.5% during the late 1980s and early 1990s (Peterson et al. 2014).

2. A moose necropsy involves collecting a piece of bone marrow, weighing it, drying it at about 65°C for a few days to drive off all the water, and then weighing it again. From the difference between the two weights, we estimate the fat content of the marrow, which is the last fat reserves that a moose (or any mammal) draws on when confronting starvation.

3. Hoy, Peterson, and Vucetich 2018.

4. Peterson et al. 2010.

5. Peterson 1995a.

6. Balsam fir is the name of a species of evergreen tree. It is an especially important source of food for moose, but only during the winter.

7. Concerns about inbreeding in the moose population are limited, in part, because the moose population is larger.

8. Peterson 1995a.

9. See graph in Chapter 5.

10. Administration and fundraising have also grown to be stifling parts of the job.

11. Peterson et al. 1998.

12. The rate at which prey are lost from the population is also indicated by the predation rate, which is equal to the kill rate times the number of predators, divided by the number of prey.

13. They do not get equal shares. Nevertheless, the average (or per capita) rate is still a useful notion.

14. Kill rate was not systematically observed until about 1970.

15. At the time I did this work, there was controversy about whether kill rates were more appropriately predicted by prey density or the ratio of predator-to-prey. For details, see Abrams and Ginzburg (2000).

16. Vucetich, Peterson, and Schaefer 2002; Sand et al. 2012.

17. The conversion is based on having calculated an index value for 13 carcasses and then measuring the weight of those carcass remains. The correlation between the index values and weights was 0.93. For details, see Vucetich, Vucetich, and Peterson (2012).

18. They almost always (90% of the time) consume at least 73% of the edible portions of a carcass.

19. Gunners 2012.

20. The pattern might also seem consistent with the first explanation, but there is a complicating consideration that leads to this decision-rule (see Vucetich, Vucetich, and Peterson 2012).

21. Gjerris and Gaiani 2013.

22. It is plausible that wolves live in groups because larger groups are better able to defend territories that are large enough to house all the moose that a pack needs. It is also plausible that wolves live in groups that are larger than is efficient because the parents (alpha wolves) make sacrifices for their offspring,

which can be explained by kin selection. For details on how these explanations fall short, see Vucetich, Peterson, and Waite (2004).

23. Vucetich, Peterson, and Waite 2004.

24. Based on past patterns of wolf mortality on Isle Royale, a normal number of deaths would have been about seven.

25. Overheated moose experience elevated heart rates and respiration rates. They adjust their behavior, for example, by seeking cooler microhabitats and by foraging less. When the overheating is chronic, occurring week after week, the nutritional condition of a moose declines. These responses are similar to that of any heat-stressed endotherm (including humans). The difference is that moose experience heat stress during the summer when temperatures exceed about 55°F. During the winter, when their fur is more insulative, they get hot when it's warmer than about 25°F. For technical details, see Schwartz and Renecker (1998).

26. See page 37 of Chapter 2; see also Hueffer, O'Hara, and Follmann (2011).

27. A couple of years later, in 2003, we recovered the carcass of a wolf that had been killed by Middle Pack when he had trespassed deep into their territory. We recovered the carcass within a day or two of the wolf's death. He is memorable because the toe pads on both his front feet were fused—essentially like a duck. We took it to be a harmless genetic anomaly. Later, a colleague told us that fused toe pads were common in the small, inbred population of wolves in Italy.

28. The vertebrae of a mammal are divided into several kinds: cervical (neck) vertebrae, thoracic vertebrae (which support the ribs), lumbar vertebrae of the lower back, and the sacral vertebrae, which connect to the pelvis. Due to their slightly different functions, the kinds of vertebrae have acquired—over the hundreds of millions of years since we diverged from our fish ancestors— slightly different shapes.

29. Wolves recolonized southern Scandinavia in the early 1980s, when a pair of wolves immigrated from the Finnish-Russian population. Today population abundance is controlled through hunting, leaving the population at about 300, which is small enough to raise concerns about inbreeding (see Laikre et al. 2013).

30. Räikkönen et al. 2009.

31. To highlight a distinction, inbreeding is the mating among close relatives, and inbreeding depression is the adverse consequences of inbreeding. These adverse consequences can range from malformations to reduced lifespan.

32. Before this discovery, some natural resource professionals had become convinced that the wolves of Isle Royale demonstrated that being severely inbred is no reason to be concerned with inbreeding depression. See note 42 from Chapter 5.

33. Marucco et al. 2012.

34. But take a look at the graph in Chapter 5 and note that the Old Gray Guy's tenure is from 1998 to 2006, during which time wolf abundance made a rough rebound. More important, kill rates were high throughout that period, which seems to have driven the moose population down.

35. For more on other instances of genetic rescue, see Frankham (2015) and Whiteley et al. (2015).

36. Geffen et al. 2011.

37. See page 134 of Chapter 3.

38. See Anderson et al. 2009.

39. The complete case is made in Hedrick et al. (2014).

40. This understanding is aligned with Kuhn's (1962) account for the development of scientific knowledge.

Chapter 7. The Unraveling

1. Her death was attributed to "uterine inertia," meaning the uterus stopped contracting during labor. We are unaware of any other reported cases anywhere involving the death of a wolf while giving birth. All eight pups died.

2. Snow fleas are tiny invertebrates just a few millimeters in length. In spite of their name, they are not fleas, or arthropods or even insects. They are a different class of creature all together called *Collembola*, or more commonly springtails. A cubic meter of soil may be inhabited by thousands of springtails. Some species have an ability to stay active in cold temperatures without freezing. On a warmer winter day, hundreds of snow fleas sometimes gather in the tracks we leave behind in the snow.

3. Tracking is greatly aided by bright sun, which creates shadows that darken the depressions of a track, which then stands out in sharp contrast to the bright surface of the snow (see page 79, Chapter 3).

4. In a typical winter field season, we collect skeletal remains from one to two dozen wolf-killed moose. Most of the bones have at least some flesh, and some have considerable flesh. The flesh is removed by boiling the bones in a 35-gallon garbage can over a propane burner. The effort entails removing the partially boiled flesh with a knife and making sure that no bone is ruined by over-boiling.

5. When biologists speak in their vernacular to describe intercourse between wolves, they say the wolves are "tied." The male mounts the female from behind. After doing so, he dismounts, but remains connected. The wolves' rear-ends abut together and the anatomy is such that they are not easily separated in those moments, which can last for more than 15 minutes. The tied pair does not stand still throughout this time. They shuffle around on the snow-covered ice, leaving a distinctive track, recorded with eight paws.

6. The winter field season runs from mid-January to early March. The "summer" field season runs from late April (as soon as the harbors are ice free) until late summer. The winter field season is focused on counting wolves and moose, estimating kill rate, performing necropsies, collecting urine and pellets of moose (to analyze their nutritional condition), and making behavioral observations. The summer field season is more focused on finding and necropsying moose carcasses and measuring vegetation to better understand the relationship between the moose and the forest (see Chapter 10). Pilot and plane

are central to the winter operation. Summer has more to do with walking and boating.

7. At this point in the attack, Pip was no longer visibly present. Recall the Trio is composed of Pip, his brother, and his brother's new love interest. Also keep in mind that Pip's brother and Isabelle are of the opposite sex and probably half-brother and sister. Of the Trio, it is the female who would have had venom for Isabelle. Pip has the least interest in contributing to the demise of Isabelle.

8. Normally, one of the first things wolves do when feeding on a freshly killed moose is remove the rumen, an organ about three times the size of a basketball, full of fermented, partially digested vegetation. Wolves have no interest in eating it or spilling its contents over the rest of their meal.

9. These wolves were father and daughter as well as half siblings because they had the same mother.

Chapter 8. Sense of Place

1. Evidence suggests that an immigrant arrived approximately once a decade or so. See Hendrick et al. (2014) for technical details.

2. The NPS knew of the question's importance. They knew the issue presented itself in other cases around the nation (see Chapter 10). They also knew that they were unprepared to confront the question.

3. Vucetich, Peterson, and Nelson 2013b; see also Vucetich 2016a.

4. The NPS press release is dated April 9, 2014, and is included with the archived collection of wolf-moose documents at the Michigan Technological University Archives and Copper Country Historical Collections, Houghton, MI.

5. After the press release was issued, the NPS would not share documentation of the evaluation upon which they claimed to rest that decision. Also, while the press release makes reference to having "discussed" the problem for two years, it seemed that NPS spent at least a year working hard to avoid addressing the issue.

6. For technical details, see Gore et al. (2011).

7. In particular the presentation stated, "So, you can see in these trend lines, you can see that beavers are actually been declining for a significant number of years. The population of beaver . . . looks to be . . . not doing as well. Moose and wolves, when you look at the long term average, seem to be rather stable. But we're gonna talk about how that may not be the case, and that's probably why a lot of you are here right now." The quoted material is from notes taken of a transcript from one of the NPS presentations recorded in Houghton, MI (July 27, 2015). Never does the presentation discuss the concern that the wolf population was at considerable risk of extinction due to inbreeding. The notes made from the transcript and NPS educational posters may be found in the wolf-moose collection of the Michigan Technological University Archives and Copper Country Historical Collections, Houghton, MI.

8. That statement represents a dominant theme in the draft Environmental Impact Statement (US National Park Service 2016).

9. All the quotes in this paragraph are from Mech (2013).

10. The population had ceased to perform its ecological function in the technical sense that predation rate had dropped to less than about 3% by 2011. That rate of predation is negligible. For technical details, see figure 3 of Peterson et al. (2014) and figure 4 of Vucetich et al. (2011).

11. Gostomski 2013. The quotation within this passage is Gostomski quoting Baldwin (2011).

12. Vucetich, Peterson, and Nelson (2013a) is a formal response to Gostomski (2013).

13. Cochrane 2013, 321.

14. For a treatment of this issue in its general form, see Vucetich, Nelson, and Batavia (2015).

15. Bryson 1990.

16. Muir (1918) 2020.

17. James 1910.

18. Twain 1866.

19. Coan 1882.

20. Marshall 2013.

21. Rocks are classified according to three basic means of formation. Sedimentary rocks represent one such classification and are formed when layers of sediment—often deposited by wind or water—are compressed and hardened into rock.

22. Fellman 2014.

23. That erosion is now memorialized by reddish sedimentary rocks that can be seen along the south shore of Isle Royale, from Malone Island to just past Cumberland Point.

24. The only features on the earth larger than these glaciers were the continents themselves. The glaciers' height, in excess of a mile, exceeded the prominence of mountains in Colorado. They were massive enough to depress the earth's crust into the mantle. Today, the earth's crust in the Lake Superior region still rebounds by several millimeters each year as it recovers from the preponderance of ice.

25. Isle Royale and her sister ridges on Keweenaw Peninsula, which probes Lake Superior from the south.

26. Poirier and Taylor 2007.

27. With the Treaty of La Pointe, the Ojibwe people ceded all of western Lake Superior, the western portion of the Upper Peninsula of Michigan, and a northern portion of Wisconsin.

28. Donovan 2012.

29. In addition to the cottagers, another 250 people on Isle Royale were living much tougher lives as commercial fishing families. They removed hundreds of tons of trout from the waters of Isle Royale.

30. Today, approximately 80% of Americans live in cities.

31. The two operations took place near Washington Harbor and Long Point.

32. Today we have 59 national parks and more than 350 other monuments, sites, and other various units.

33. Roosevelt had also created five national parks, bringing the nation's total to nine. But the parks were suffering from not having an agency to manage them.

34. Baldwin 2011, 27.

35. Leopold 1921.

36. Baldwin 2011, 39.

37. Public Broadcasting Service 2009.

38. Murie 1930.

39. Murie 1935, 39.

40. Murie 1935, 43.

41. Wright 1932.

42. Erickson 1981, cited in Wagner 2006.

43. Wagner and Sax 1995.

44. Baldwin 2011, 14.

45. Baldwin 2011, 23.

46. Stoll 1921, as cited in Baldwin 2011, 14.

47. Murie 1935.

48. Murie 1934, 42.

49. Quoted in Meine 2009.

50. Backes 1999, 122.

51. Steinhart 1995.

52. Unpublished correspondence, wolf-moose collection of the Michigan Technological University Archives and Copper Country Historical Collections, Houghton, MI.

53. Cochrane 2013, 321.

Chapter 9. All Natural

1. Cochrane 2013, 313.

2. For more on logical fallacies, see Withey (2016).

3. Consumer Reports National Research Center 2016.

4. Forester 1778, ii.

5. Forester memorialized his experiences in *Observations Made during a Voyage round the World* (1778). The history of island biogeography that I convey follows Quammen (1996).

6. Darlington 1957.

7. MacArthur and Wilson 1963, 386.

8. MacArthur and Wilson 1967.

9. Preston 1962b, 427, as cited by Quammen 1996.

10. Isle Royale may well be the last such forest. For context, in Yellowstone wolves and their prey (elk and bison) are among of the best protected in the world. Yet human killing has a significant impact on all three populations. Wolves and elk are killed when they leave the park; bison are culled as part of a

controversial government program motivated in large part by concerns from owners of cattle that graze just outside Yellowstone. The concern is about bison infecting cattle with brucellosis. I cannot think of a place where predation and herbivory are as little affected by human killing or logging as on Isle Royale.

11. Allen is referring to what was known by 1979 with respect to wolf social behavior, compared to what was known about wolves' social behavior in the 1950s.

12. Allen 1979.

13. Cochrane 2013.

14. The quoted material in this paragraph is from an NPS poster entitled "Species Introductions/Population Dynamics" (Wolves of Isle Royale, Michigan Technological University Archives and Copper Country Historical Collections, Houghton, MI).

15. For decades it had been presumed that Queenie was shot and killed. That presumption was questioned in 2013 when an NPS memo written in 1952 was made available. The memo was written by chief Ranger Henderson and addressed to the Superintendent of Isle Royale. Its purpose was to provide a final report on the fate of the Detroit zoo wolves, and it gives the overall impression that Jim was the only unaccounted for wolf. However, some readers of the memo emphasize a single, grammatically ambiguous phrase, "shot at." That phrase was transformed by the NPS into "shot at (but not hit)." That is the totality of evidence used to support the NPS's recent contention that Isle Royale wolves descended from Queenie.

By contrast, Dave Mech who had worked with those involved with the wolf-release fiasco and who is meticulous in documenting historical accounts of wolves, writes in the mid-1960s, very clearly and without qualification, that both female wolves (Lady and Queenie) were "disposed of" (Mech 1966). Park Ranger Bob Linn was present at the fiasco, and according to Allen (writing in the 1970s), "As Bob Linn recalls it, Queenie was shot later that same day" (Allen 1979).

The letters and educational materials referred to in this passage may be found at Michigan Technological University Archives and Copper Country Historical Collections, Houghton, MI.

16. By 2011, 60% of the population's gene pool descended from the Old Gray Guy. Prior to that—in the 1980s and 1990s—the population was descended from one female and two males (Adams et al. 2011). Those ancestral wolves came from the western Great Lakes region. It is also known that Big Jim's mother was a wolf from Saskatchewan, Canada, and his father was from Upper Michigan (Allen 1979). Wolves from Saskatchewan tend to have a genetic signature that is distinct from wolves of the western Great Lakes region, and it is not found in the wolves of Isle Royale.

Furthermore, it is plausible for a wolf to have contributed genes to a population that are later lost due to genetic drift or natural selection. For example, the Old Gray Guy's genomic sweep led to loss of genetic contributions from other wolves, and those loses occurred within just 10 years.

17. I realize that an advocate for purity might not accept absence of evidence for the purity of a population. Some eugenics programs have, in the past, reversed the burden of proof, requiring a demonstration of purity of heritage.

18. When white supremacists test their DNA, they routinely discover that their heritage does not pass their own standard of purity. Impurity among white supremacists is common because their standard for purity is conceptually impossible, given what is known about the human tendency to procreate with unrelated individuals and the inheritance of DNA. For technical details, see Yudell et al. (2016). For an accessible account, see Howard (2016).

19. Berthold (2010), for example, states: "As inheritors of this racist culture, we are all lovers of purity, and we are all responsible for rethinking this value." See also Shotwell (2016) and Falkof (2016).

20. Allen (1979) writes:

> [Moose] do not venture far out onto open ice. In studies since 1958 we probably have never seen a moose a mile from shore, and half a mile would be the usual limit for such forays. It is much more likely that the initial breeders colonizing the island swam across the channel. The distance would have been as little as 13 miles from the nearest Canadian islands (Spar or Victoria), but it is likely to have been considerably farther. . . .
>
> Pertinent to this is an observation described to me in 1961 by James M. Godbold, who was then director of photography for *National Geographic*. In May 1956 he employed a bush pilot in Grand Marias, Minnesota, to fly him around Isle Royale in a Seabee. Ten miles out from shore they saw four moose swimming toward the island. The pilot said he had seen this more than once in the month of May.
>
> The temperature of Lake Superior is around 54 degrees F. This level of exposure is quickly lethal to human being, but the moose is a large animal . . . and conserves heat efficiently, as we will have occasion to point out in other connections. It is at home in the water, and the swim from Canada or Minnesota is well within its capabilities.

21. Mech (1966) writes: "Since moose are excellent swimmers and have been seen swimming in Lake Superior several miles from shore (Hickie, n.d.), it appears more likely that they reached Isle Royale by swimming from Canada. Indeed, P. M. Baudino of Calumet, Mich., told me that in the early 1930's in late June he observed a bull moose about half-way between Amygdaloid Island (part of Isle Royale) and Sibley Peninsula, swimming toward Canada."

My colleague, Rolf Peterson, has also interviewed two people who describe firsthand accounts of moose swimming in Lake Superior far from shore.

22. Murie (1934) never mentions the possibility that people brought moose, but he does write: "In former years moose apparently crossed over from the mainland at intervals. . . . According to persons long familiar with Isle Royale, the last influx of moose occurred during the winter of 1912–1913."

23. See notes 20, 21, and 22, above.

24. See Nelson et al. (2016) for methodological details. See Gore et al. (2011) for a similar analysis.

25. That quoted text appears in Wilderness Watch (2014). Awareness that those opposed to restoring wolf predation tended to recite Wilderness Watch's position arose in the development of Nelson et al. (2016) (Michael Paul Nelson, personal communication).

26. Knott-Ahern 2013.

27. Knott-Ahern 2013.

28. 4.5 miles on a side if the parcel is square.

29. US Historian, Frederick Jackson Turner, quoted in Vickery 1994, 24.

30. Vucetich and Nelson 2008.

31. Vucetich and Nelson 2008.

32. In 1967, the Park Service proposed that 119,612 acres of Isle Royale (~90% of the island) be declared federally designated wilderness. That proposal was followed by concerns that much of the unprotected areas would be developed. After considering various proposals, Congress declared 131,880 acres of Isle Royale as wilderness in 1976.

33. Turner 2002, 467.

34. Solomon 2014.

35. While limitations of F. J. Turner's "Frontier Hypothesis" are now better understood (e.g., Popper, Lang, and Popper 2000), the salient point is that architects of wilderness believed the myth.

36. Leopold 1925, 401.

37. Turner 2002, 464.

38. Boy Scouts of America 1938, as cited in Turner 2002, 466.

39. Olson (1938) 1998.

40. Marshall (1930) 1998.

41. Milman 2017.

42. Most, but not all, of the shelters are in the nonwilderness exclusion zones of Isle Royale. The shelters were constructed prior to Isle Royale being designated as a federal wilderness. While wilderness designation has often led to the removal of structures (bridges, fire towers, etc.), these shelters were allowed to remain. I suspect not because they belong in wilderness but because they are such a marvelous convenience. For broader context about how human infrastructure ends up in wilderness, see Nickas and Proescholdt (2005).

43. The ubiquity of Isle Royale's wilderness graffiti may make the inside of a shelter feel like an open invitation, a community space for public expression. However, I feel duty-bound to point out that leaving graffiti in a shelter is vandalizing federal property and officially punishable.

44. Standing Bear (1933) 1998.

45. Anderson 2004.

46. Purdy 2015.

47. For a broader treatment of wilderness and racism see the collection of essays in Part 2 of Nelson and Callicott (2008).

48. Krech 2000; Mann 2005.

49. Proposed solutions included reservation systems, quotas, a two-tiered system of land management involving strictly protected wilderness and less protected "backcountry" areas. According to Turner (2002, 471): "The biologist Garrett Hardin suggested that wilderness access be limited to 'physically vigorous people.' Paul Petzoldt, the founder of the National Outdoor Leadership School, proposed a meritocracy, giving priority to those best educated in wilderness skills."

50. Turner 2002.

51. Turner 2002.

52. Leopold (1949) 2020, 156.

53. Turner 2002.

54. The essay is Nelson (1998), and the two anthologies are Callicott and Nelson (1998) and Nelson and Callicott (2008).

55. For a broader sense of this concern, see Nickas and Proescholdt (2005).

Chapter 10. Restoring the Balance

1. These indirect effects of predators had long been appreciated through informal observations, such as those relayed in Darwin's *On the Origin of Species* and by Charles Elton (1927), but rigorous scientific evidence for the idea was not widely appreciated until toward the end of the 20th century.

2. This passage is a summary of Estes, Peterson, and Steneck (2010).

3. The term "trophic cascade" was coined by Robert Paine in 1980.

4. Scientists would seem not to have taken gardeners' observations seriously—probably because those observations were not made by scientists.

5. McLaren and Peterson 1994.

6. Pers. Comm., July 12, 2016.

7. Elements of this story are drawn from Brandner, Peterson, and Risenhoover (1990) and Vucetich and Peterson (2014).

8. This size distribution of balsam fir characterizes the western forests of Isle Royale. Balsam fir is altogether rare in the middle region of Isle Royale due to a forest fire in the 1930s. Balsam fir is present in all size classes in the eastern region. The reason for these differences is unclear.

9. There is reason to believe that aspen (and other trees species) have been able to escape into the forest canopy only during periods of low moose abundance. The dynamics of these other tree species are, however, more difficult to document. An important exception is spruce, which moose do not eat.

10. See Rotter and Rebertus 2015.

11. The twigs of deciduous trees (e.g., birch and aspen) and shrubs (e.g., red-osier dogwood and beaked hazel) are more nutritious, on a per gram basis, than balsam fir. However, deciduous twigs are small compared to the needles and twigs in a single bite from a balsam fir. Consequently, during the winter, the net energy gain from balsam fir is likely greater than from deciduous forage.

12. Wright 1932.

13. Murie 1930.

14. For full disclosure, some ecologists are concerned that other ecologists exaggerate the influence of predators. A good example of the concern is Allen et al. (2017). For a rebuttal to that concern, see Bruskotter et al. (2017).

15. This passage is a summary of Vucetich and Macdonald (2017).

16. Important examples of genuine threats include human-eating lions in portions of Africa and human-eating tigers in portions of south Asia. In the United States, however, the impacts of wolves on livestock and hunting are very small, and wolves do not pose a threat to human safety (see Vucetich 2016b).

17. Vucetich, Bruskotter, and Nelson 2015.

18. I led a sociological survey, the results of which are soon to be published, that asked a sample of approximately 450 conservation professionals from around the word a number of questions, including this question: "Of all Earth's biodiversity, what portion is likely to be protected by economic valuation?" About two-thirds of the respondents said, "Less than half." My collaborators on this survey are J. T. Bruskotter, L. Van Eeden, and E. Macdonald.

19. May, Lawton, and Stork 1995.

20. Alroy 2009.

21. Pimm et al. 2014; see also Mace 1998.

22. Chen and Benton 2012; Davis, Faurby, and Svenning 2018.

23. Le Loeuff 2012.

24. Brocklehurst et al. 2015.

25. Ceballos and Ehrlich 2002; Ceballos, Ehrlich, and Dirzo 2017; Wolf and Ripple 2017.

26. See Simberloff (2013) for a broad and accessible review of invasive species.

27. Murtaugh and Schlax 2009; Ganivet 2020.

28. Structuring the overarching goal of conservation in this way is valuable because it unifies two important concepts: population viability and ecosystem health. This unification is also aligned with the notion that ecosystems are healthy when inhabited by fully functioning populations of species that had been native to a particular ecosystem.

29. This paragraph is aligned with the US Endangered Species Act of 1973 (see Vucetich, Nelson, and Phillips 2006; Vucetich and Nelson 2014).

30. This section is largely excerpted from Vucetich and Nelson (2018).

31. Vucetich and Nelson 2010.

32. This understanding of ecosystem health is also tied to the notion that sustainability and conservation are about the provisioning of ecosystem services (Braat and de Groot 2012).

33. See Chapter 9.

34. Misanthropy is to disparage, neglect, or have undue disregard for the intrinsic value of humans or human interests. Because the limits of misanthropy may be genuinely complicated (e.g., Gibson 2017; Cooper 2018), to

unwittingly fall prey to a misanthropic thought does not necessarily make one a misanthropic person.

35. Some consider ecosystem health to be no more than a poor metaphor because health is a coherent concept only when describing an organism. I think health is more versatile. It can, for example, describe a relationship between humans. "What is a healthy ecosystem?" is a way of asking a perfectly coherent question: "What counts as a right relationship with nature?"

36. Various tellings of this myth exist. For one, see Benton-Banai (2010). I know the story from Jimmie Mitchell, member of the Little River Band of Ottawa Indians.

37. You know of course that third cousins have a common ancestor, that is, a great-great-grandparent. Likewise, evolutionary science says that any two creatures, no matter how distantly related, share a common ancestor that lived a long time ago. For example, a human and a hawk have a common ancestor that would have looked something like a lizard and lived 300 million to 350 million years ago.

38. The idea that humans are, in an ethically relevant sense, siblings with nature also has roots in medieval Christianity with *Canticle of the Sun* by St. Francis of Assisi in the 13th century and with the *Little Flowers of St. Francis*, recorded in the 14th century.

39. As political theorists use the terms, an entitlement is similar to a right (such as the right to vote), though there are important differences. In particular, rights imply a social contract but entitlements may not. For example, the right for women to vote in the United States was an entitlement before it was a right (created in 1920). The existence of the entitlement motivated the creation of the right. It is somewhat inconvenient that this technical distinction between rights and entitlements does not match more common senses of those words, where, for example, entitlement is sometimes associated with the negative connotation of behaving (inappropriately) entitled.

40. Gensler 2013.

41. Vucetich et al. 2018.

42. This taxonomy of virtues represents formal theory from the scholarly field of social justice (Hülle, Liebig, and May 2018). Equity calls for the allocation of benefits and burdens according to individuals' contributions and efforts. Equity is distinct from equality, which calls for everyone being allocated the same share. Need calls for allocations according to individuals' needs. Entitlement calls for allocations according to ascriptive characteristics, such as social origin, or characteristics acquired from the past, such as an inheritance. Fairness is the wise application of those four virtues.

43. One can choose to not fulfill the obligation, of course, but that doesn't absolve one from the obligation.

44. See Bruskotter et al. 2014. Also, genetic analyses suggest that the western coterminous United States had been inhabited by 300,000 to 500,000 wolves prior to persecution by humans (Leonard, Vilà, and Wayne 2005). Today the

western United States is inhabited by 1,500 to 2,000 wolves—fewer than 1% of the number prior to persecution.

45. Vucetich et al. 2017.

46. Wolves on Isle Royale have had three of the requirements for a dignified life: food, space, and freedom from human persecution. Knowing all that is required for wild animals to have a dignified life is a topic of some discussion (e.g., Nussbaum 2009).

47. Except, perhaps, if we had mitigated inbreeding depression in a timely manner through genetic rescue. But let that idea rest for a moment (see Coda).

48. For more on Clements' metaphysical beliefs about nature, see Worster (1994).

49. Ruse 2013.

50. This simple idea comes with technical details and caveats in philosophy. See Shrader-Frechette (1996).

51. Trifonov 2011. See Gabbatiss (2017) for a popular treatment.

52. Leopold (1949) 2020, 121.

53. The Gaia hypothesis is also aligned with another kind of environmental ethic, Deep Ecology (James 2000). While Deep Ecology differs from ecocentrism, both pay special attention to holistic understandings of nature (Keller and Golley 2000; Nelson 2010). For an accessible introduction to both ethics, see Keller (2010).

54. Callicott and Nelson 2004.

55. Hülle, Liebig, and May et al. 2018 and references therein. See Sandel (2009) for a popular treatment.

56. See Hill 2002; Cohen 2009; Khatchadourian 2006; Collste 2010.

57. Restorative justice is, in part, about sorting conflicts among the values of distributive justice, especially when and how to prioritize entitlements (particularly, inheritance, broadly construed) over needs, equity, and equality. Furthermore, restorative justice should not be reduced to a form of collective punishment.

58. Haiken 2019; Dayer et al. 2019.

59. Vucetich et al. 2018.

60. While flourishing involves an inescapably normative element, there is an impressive amount of objective, scientific knowledge about what makes people flourish (Diener 2009).

61. Nassbaum 2009; McMahan 2010, 2016; see also Keulartz 2016.

62. Here I refer to what are likely varied views about the general prioritization of concerns relating to conservation and animal welfare (Bruskotter, Vucetich, and Nelson 2017).

63. Simberloff 2013.

64. Proulx et al. 2017; Cornwall 2014; Yurk and Trites 2000; Rothstein 2004; Zuckerman 2014; Kristan and Boarman 2003.

65. It's no small detail to have portrayed "humans" in this example as one undifferentiated group, when they might more appropriately be portrayed as

those who favor and those who oppose the debacle. Taking account of that difference does not, however, lead to a clearer moral calculus.

66. Pradhan et al. 2017; Bowen et al. 2017.

67. At this point, some argue, "If the goal is to minimize suffering, then forego wolves and let humans hunt moose, because humans can kill a moose more humanely than wolves." That idea is very unlikely to work on Isle Royale for logistical reasons and is probably not a good idea in its general form. For emphasis, this response is not a claim against hunting in general, it is caution against thinking that hunting is an equivalent replacement for predation.

68. Triandis 2018.

69. For what it's worth, my intuition is that it is right to prioritize collectivism over individualism in the Isle Royale case and to prioritize individualism over collectivism in the British Columbia case. The moderating consideration is culpability.

70. Pradhan et al. 2017; Bowen et al. 2017.

71. Vucetich, Nelson, and Bruskotter 2017.

72. For more on hope and despair in the context of the biodiversity crisis, see Nelson and Vucetich (2009).

Coda

1. National Park Service press release dated April 9, 2014, available through the Wolves of Isle Royale Collection, Michigan Technological University Archives and Copper Country Historical Collections, Houghton, MI.

2. Nelson et al. 2016.

3. Mlot 2019.

4. I apologize for referring to the wolves only by their assigned numbers. By the time I began writing this Coda, I had been instructed by the NPS to not use more humane names for the animals studied on Isle Royale.

5. For decades, my colleagues and I have given the names "East Pack" and "West Pack" to the packs living on the eastern or western portions of Isle Royale, even though the wolves belonging to those packs change over time. This East Pack and West Pack consist entirely of reintroduced wolves or their descendants.

6. It is possible that M16 (a black, male wolf from Wawa) produced pups with F11 (female from Michipicoten). If so, it is likely that those pups died young, that is, before reproducing. It is also possible that M19 (Michigan) mated with a female born to East Pack in 2022, but he died the same month that she would have given birth. I don't think those pups survived. In any case, if there was a more even contribution of genetic material from the translocated wolves, it will be revealed by forthcoming genetic analysis. Finally, one of the five Minnesotan wolves died while being held by the NPS before arriving to Isle Royale.

7. For context, two of the three wolves from Wawa, Ontario were brother and sister.

8. This telling of Lake Superior caribou is stunningly brief and leaves much out (for a succinct, but more complete telling of their story, see McLaren 2022).

9. For example, much consideration is given to the genetics of translocation for Mexican wolves and red wolves, perhaps not so much for African lions (see Bertola et al. 2022, 22–39).

10. This proactive approach is not quite so simple as translocating a wolf. Rather, one also needs to determine that the translocated wolf contributed genes to the population by reproducing. This determination can be made by noninvasive genetic monitoring that is already done on an ongoing basis. If the translocated wolf were to die before reproducing, then another would need to be translocated. The level of inbreeding that typically causes worry among geneticists is $F = 0.2$. If recent trends persist, that level of inbreeding will be reached in about 5 years (the year 2028).

Abrams, Peter A., and Lev R. Ginzburg. 2000. "The nature of predation: Prey dependent, ratio dependent or neither?" *Trends in Ecology & Evolution* 15, no. 8: 337–341.

Adams, Jennifer R., Leah M. Vucetich, Philip W. Hedrick, Rolf O. Peterson, and John A. Vucetich. 2011. "Genomic sweep and potential genetic rescue during limiting environmental conditions in an isolated wolf population." *Proceedings of the Royal Society B: Biological Sciences* 278, no. 1723: 3336–3344.

Allen, Benjamin L., Lee R. Allen, Henrik Andrén, Guy Ballard, Luigi Boitani, Richard M. Engeman, Peter J. S. Fleming, et al. 2017. "Can we save large carnivores without losing large carnivore science?" *Food Webs* 12: 64–75.

Allen, Durward L. 1979. *Wolves of Minong: Their Vital Role in a Wild Community.* Boston: Houghton Mifflin.

Allen, Durward L., and L. David Mech. 1963. "Wolves versus moose on Isle Royale." *National Geographic* 123(2): 200–219.

Alroy, John. 2009. "Speciation and extinction in the fossil record of North American mammals." In *Speciation and Patterns of Diversity*, edited by Roger Butlin, Jon Bridle, and Dolph Schluter, 301–322. Cambridge: Cambridge University Press.

Anderson, Daniel E. 2004. "Origin of the word Yosemite." Updated July 2011. http://www.yosemite.ca.us/library/origin_of_word_yosemite.html.

Anderson, Tovi M., Sophie I. Candille, Marco Musiani, Claudia Greco, Daniel R. Stahler, Douglas W. Smith, Badri Padhukasahasram, et al. 2009. "Molecular and evolutionary history of melanism in North American gray wolves." *Science* 323, no. 5919: 1339–1343.

Backes, David. 1999. *A Wilderness Within: The Life of Sigurd F. Olson.* Minneapolis: University of Minnesota Press.

Baldwin, Amalia Tholen. 2011. *Becoming Wilderness: Nature, History, and the Making of Isle Royale National Park.* Houghton, MI: Isle Royale and Keweenaw Parks Association.

Barrow, Mark V., Jr. 2009. *Nature's Ghosts: Confronting Extinction from the Age of Jefferson to the Age of Ecology.* Chicago: University of Chicago Press.

Bauer, Hans, Craig Packer, Paul Funston, Philipp Henschel, and Kristin Nowell. 2016. *Panthera leo. The IUCN Red List of Threatened Species.* Gland: IUCN. https://doi.org/10.2305/IUCN.UK.2016-3.RLTS.T15951A107265605.en.

Benton-Banai, Edward. 2010. *The Mishomis Book: The Voice of the Ojibway*. Minneapolis: University of Minnesota Press.

Berthold, Dana. 2010 "Tidy whiteness: A genealogy of race, purity, and hygiene." *Ethics & the Environment* 15, no. 1: 1–26.

Bertola, Laura D., et al. 2022. "Genetic Guidelines for Translocations: Maintaining Intraspecific Diversity in the Lion." *Evol Appl*. 15: 22–39.

Botkin, Daniel. B. 1990. *Discordant Harmonies: A New Ecology for the Twenty-First Century*. Oxford: Oxford University Press.

Bowen, Kathryn J., Nicholas A. Cradock-Henry, Florian Koch, James Patterson, Tiina Häyhä, Jess Vogt, and Fabiana Barbi. 2017. "Implementing the 'Sustainable Development Goals': Towards addressing three key governance challenges—collective action, trade-offs, and accountability." *Current Opinion in Environmental Sustainability* 26: 90–96.

Boy Scouts of America. 1938. *The How Book of Scouting*. New York: Boy Scouts of America.

Braat, Leon C., and Rudolf de Groot. 2012. "The ecosystem services agenda: Bridging the worlds of natural science and economics, conservation and development, and public and private policy." *Ecosystem Services* 1, no. 1: 4–15.

Bradshaw, John W. S., Rachel A. Casey, and Sarah L. Brown. 2012. *The Behaviour of the Domestic Cat*. Boston: Cabi.

Brandner, Thomas A., Rolf O. Peterson, and Ken L. Risenhoover. 1990. "Balsam fir on Isle Royale: Effects of moose herbivory and population density." *Ecology* 71, no. 1: 155–164.

Braun, Martin. 1983. "Why the percentage of sharks caught in the Mediterranean Sea rose dramatically during World War I." In *Differential Equation Models*, edited by Martin Braun, Courtney S. Coleman, and Donald A. Drew, 221–228. New York: Springer.

Brocklehurst, Neil, Marcello Ruta, Johannes Müller, and Jörg Fröbisch. 2015. "Elevated extinction rates as a trigger for diversification rate shifts: Early amniotes as a case study." *Scientific Reports* 5, no. 1: 1–10.

Bruskotter, Jeremy T., John A. Vucetich, Sherry Enzler, Adrian Treves, and Michael P. Nelson. 2014. "Removing protections for wolves and the future of the US Endangered Species Act (1973)." *Conservation Letters* 7, no. 4: 401–407.

Bruskotter, Jeremy T., John A. Vucetich, and Michael P. Nelson. 2017. "Animal rights and wildlife conservation: Conflicting or compatible." *The Wildlife Professional* July/August: 40–43.

Bruskotter, Jeremy T., John A. Vucetich, Douglas W. Smith, Michael Paul Nelson, Gabriel R. Karns, and Rolf O. Peterson. 2017. "The role of science in understanding (and saving) large carnivores: A response to Allen and colleagues." *Food Webs* 13: 46–48.

Bryson, Bill. 1990. *The Lost Continent: Travels in Small-Town America*. New York: William Morrow Paperbacks.

Burns, Catherine E., Brett J. Goodwin, and Richard S. Ostfeld. 2005. "A prescription for longer life? Bot fly parasitism of the white-footed mouse." *Ecology* 86, no. 3: 753–761.

Callicott, J. Baird, and Michael P. Nelson, eds. 1998. *The Great New Wilderness Debate*. Athens: University of Georgia Press.

Callicott, J. Baird, and Michael P. Nelson. 2004. *American Indian Environmental Ethics: An Ojibwa Case Study*. Upper Saddle River, NJ: Prentice Hall.

Carere, Claudio, and Dario Maestripieri, eds. 2013. *Animal Personalities: Behavior, Physiology, and Evolution*. Chicago: University of Chicago Press.

Carroll, Sean. 2009. "Great Minds Think Alike." NOVA website, September 29, 2009. www.pbs.org/wgbh/nova/evolution/great-minds-think-alike.html.

Ceballos, Gerardo, and Paul R. Ehrlich. 2002. "Mammal population losses and the extinction crisis." *Science* 296, no. 5569: 904–907.

Ceballos, Gerardo, Paul R. Ehrlich, and Rodolfo Dirzo. 2017. "Biological annihilation via the ongoing sixth mass extinction signaled by vertebrate population losses and declines." *Proceedings of the National Academy of Sciences* 114, no. 30: E6089–E6096.

Chen, Zhong-Qiang, and Michael J. Benton. 2012. "The timing and pattern of biotic recovery following the end-Permian mass extinction." *Nature Geoscience* 5, no. 6: 375–383.

Coan, Titus. 1882. *Life in Hawaii: An Autobiographic Sketch of Mission Life and Labors, 1835–1881*. New York: A. D. F. Randolph & Co.

Cochrane, Tim. 2013. "Island complications: Should we retain wolves on Isle Royale?" *The George Wright Forum* 30, no. 3: 313–325.

Cohen, Andrew I. 2009. "Compensation for historic injustices: Completing the Boxill and Sher argument." *Philosophy & Public Affairs* 37, no. 1: 81–102.

Collste, Göran. 2010. "'. . . restoring the dignity of the victims.' Is global rectificatory justice feasible?" *Ethics & Global Politics* 3, no. 2: 85–99.

Consumer Reports National Research Center. 2016. *Food Labels Survey: 2016 Nationally-Representative Phone Survey*. April 6, 2016.

Cooper, David E. 2018. *Animals and Misanthropy*. New York: Routledge.

Cornwall, Warren. 2014. "There will be blood." *Conservation*, October 24, 2014. University of Washington. http://conservationmagazine.org/2014/10/killing-for-conservation.

Cronin, Melissa. 2014. "Twelve remarkable interspecies relationships that prove adoption isn't just for humans." *The Dodo*, April 24, 2014. http://www.thedodo.com/12-remarkable-interspecies-rel-523336558.html.

Curtis, John T. 1956. "The modification of mid-latitude grasslands and forests by man." In *Man's Role in Changing the Face of the Earth*, edited by William L. Thomas Jr., 721–736. Chicago: University of Chicago Press.

D'Ancona, Umberto. 1954. *The Struggle for Existence*. Leiden, Netherlands: E. J. Brill.

Darlington, Philip J. 1957. *Zoogeography: The Geographical Distribution of Animals*. New York: John Wiley & Sons.

Darwin, Charles. 1871. *The Descent of Man, and Selection in Relation to Sex.* London: John Murray.

Darwin, Charles. 1872. *The Origin of Species by Means of Natural Selection, or, the Preservation of Favoured Races in the Struggle for Life.* 6th edition. London: John Murray

Darwin, Charles. 1987. *Charles Darwin's Natural Selection: Being the Second Part of his Big Species Book Written from 1856 to 1858.* Edited by Robert C. Stauffer. Cambridge: Cambridge University Press.

Davis, Matt, Søren Faurby, and Jens-Christian Svenning. 2018. "Mammal diversity will take millions of years to recover from the current biodiversity crisis." *Proceedings of the National Academy of Sciences* 115, no. 44: 11262–11267.

Dayer, Ashley A., Kent H. Redford, Karl J. Campbell, Christopher R. Dickman, Rebecca S. Epanchin-Niell, Edwin D. Grosholz, David E. Hallac, Elaine F. Leslie, Leslie A. Richardson, and Mark W. Schwartz. 2019. "The unaddressed threat of invasive animals in U.S. National Parks." *Biological Invasions* 22: 177–188.

Derham, William. 1714. *Physico-Theology: Or, a Demonstration of the Being and Attributes of God, from his Work of Creation.* London: W. Innys.

De Waal, Frans. 1997. "Are we in anthropodenial?" *Discover*, July 1997.

De Waal, Frans. 1999. "Anthropomorphism and anthropodenial: Consistency in our thinking about humans and other animals." *Philosophical Topics* 27, no. 1: 255–280.

De Waal, Frans. 2010. *The Age of Empathy: Nature's Lessons for a Kinder Society.* New York: Broadway Books.

Diener, Edward. 2009. *The Science of Well-Being: The Collected Works of Ed Diener.* Vol. 37. New York: Springer.

Donovan, Jennifer. 2012. "A mystery solved: 3 wolves drowned in old mine shaft at Isle Royale National Park." *Michigan Tech News*, June 14, 2012. https://www.mtu.edu/news/stories/2012/june/mystery-solved-wolves-drowned-old-mine-shaft-isle-royale-national-park.html.

Egerton, Frank N. 1973. "Changing concepts of the balance of nature." *Quarterly Review of Biology* 48, no. 2: 322–350.

Elton, Charles S. 1927. *Animal Ecology.* London: Sidgwick and Jackson.

Elton, Charles S. 1930. *Animal Ecology and Evolution.* Oxford: Clarendon.

Emlen, Stephen T. 1995. "An evolutionary theory of the family." *Proceedings of the National Academy of Sciences* 92, no. 18: 8092–8099.

Erickson, Glenn L. 1981. "The northern Yellowstone elk herd—a conflict of policies." *Proceedings of the Annual Western Association of Fish and Wildlife Agencies* 61, 92–108.

Estes, James A., Charles H. Peterson, and Robert S. Steneck. 2010. "Some effects of apex predators in higher-latitude coastal oceans." In *Trophic Cascades Predators, Prey, and the Changing Dynamics of Nature*, edited by John Terborgh and James A. Estes, 37–53. Washington, DC: Island Press.

Falkof, Nicky. 2016. "The myth of white purity and narratives that fed racism in South Africa." *The Conversation*, May 15, 2016. https://theconversation.com/the-myth-of-white-purity-and-narratives-that-fed-racism-in-south-africa-59330.

Fellman, Megan. 2014. "Mysterious midcontinent rift is a geological hybrid." *Northwestern Now*, October 16, 2014. https://news.northwestern.edu/stories/2014/10/mysterious-midcontinent-rift-is-a-geological-hybrid.

Forbes, Stephen A. 1883. "The food relations of the Carabidae and Coccindellidae." *Bulletin of the Illinois State Laboratory of Natural History* 1: 33–64.

Forester, John Reinold [Johann Reinhold]. 1778. *Observations Made during a Voyage Round the World, on Physical Geography, Natural History, and Ethic Philosophy*. London: G. Robinson.

Fox, William P. 2011. *Mathematical Modeling with Maple*. Boston: Cengage Learning.

Frankham, Richard. 2015. "Genetic rescue of small inbred populations: Meta-analysis reveals large and consistent benefits of gene flow." *Molecular Ecology* 24, no. 11: 2610–2618.

Fritts, Steven H., and Ludwig N. Carbyn. 1995. "Population viability, nature reserves, and the outlook for gray wolf conservation in North America." *Restoration Ecology* 3, no. 1: 26–38.

Fuller, Todd K., L. David Mech, and Jean Fitts Cochrane. 2003. "Wolf population dynamics." In *Wolves: Behavior, Ecology and Conservation*, edited by L. David Mech and Luigi Boitani, 161–191. Chicago: University of Chicago Press.

Gabbatiss, Josh. 2017. "There are over 100 definitions of 'life' and all are wrong." *The Big Questions*, January 2, 2017. British Broadcasting Corporation. http://www.bbc.com/earth/story/20170101-there-are-over-100-definitions-for-life-and-all-are-wrong.

Ganivet, Elias. 2020. "Growth in human population and consumption both need to be addressed to reach an ecologically sustainable future." *Environment, Development and Sustainability* 22, no. 6: 4979–4998.

Gatto, Marino. 2009. "On Volterra and D'Ancona's footsteps: The temporal and spatial complexity of ecological interactions and networks." *Italian Journal of Zoology* 76, no. 1: 3–15.

Gause, Georgii F. 1934. *The Struggle for Existence*. Baltimore, MD: Williams and Wilkins.

Geffen, Eli, Michael Kam, Reuven Hefner, Pall Hersteinsson, Anders Angerbjörn, Love Dalen, Eva Fuglei, et al. 2011. "Kin encounter rate and inbreeding avoidance in canids." *Molecular Ecology* 20, no. 24: 5348–5358.

Gensler, Harry J. 2013. *Ethics and the Golden Rule*. New York: Routledge.

Gibson, Andrew. 2017. *Misanthropy: The Critique of Humanity*. London: Bloomsbury Academic.

Gjerris, Mickey, and Silvia Gaiani. 2013. "Household food waste in Nordic countries: Estimations and ethical implications." *Etikk i Praksis—Nordic Journal of Applied Ethics* 1 (2013): 6–23.

Gore, Meredith L., Michael P. Nelson, John A. Vucetich, Amy M. Smith, and Melissa A. Clark. 2011. "Exploring the ethical basis for conservation policy: The case of inbred wolves on Isle Royale, USA." *Conservation Letters* 4, no. 5: 394–401.

Gostomski, Ted. 2013. "Are Isle Royale wolves too big to fail? A response to Vucetich et al." *The George Wright Forum* 30, no. 1:96–100.

Gunners, Dana. 2012. "Wasted: How America is losing up to 40 percent of its food from farm to fork to landfill." *Natural Resources Defense Council*. Accessed April 10, 2020. https://www.nrdc.org/sites/default/files/wasted-food-IP.pdf.

Haiken, Melanie. 2019. "How climate change could destroy our national parks." *Sierra*, November 27, 2018. https://www.sierraclub.org/sierra/how-climate -change-could-destroy-our-national-parks.

Han, Liang, Chao Ma, Qin Liu, Hao-Jui Weng, Yiyuan Cui, Zongxiang Tang, Yushin Kim, et al. 2013. "A subpopulation of nociceptors specifically linked to itch." *Nature Neuroscience* 16, no. 2: 174.

Hawcroft, Lucy J., and Taciano L. Milfont. 2010. "The use (and abuse) of the new environmental paradigm scale over the last 30 years: A meta-analysis." *Journal of Environmental Psychology* 30, no. 2: 143–158.

Hedrick, Philip W., Rhonda N. Lee, and Colleen Buchanan. 2003. "Canine parvovirus enteritis, canine distemper, and major histocompatibility complex genetic variation in Mexican wolves." *Journal of Wildlife Diseases* 39, no. 4 (2003): 909–913.

Hedrick, Philip W., Rolf O. Peterson, Leah M. Vucetich, Jennifer R. Adams, and John A. Vucetich. 2014. "Genetic rescue in Isle Royale wolves: Genetic analysis and the collapse of the population." *Conservation Genetics* 15, no. 5: 1111–1121.

Hill, Renée A. 2002. "Compensatory justice: Over time and between groups." *Journal of Political Philosophy* 10, no. 4: 392–415.

Howard, Jacqueline. 2016. "What scientists mean when they say 'race' is not genetic." *Huffington Post*, February 9, 2016. https://www.huffpost.com/entry /race-is-not-biological_n_56b8db83e4b04f9b57da89ed.

Hoy, Sarah R., Daniel R. MacNulty, Douglas W. Smith, Daniel R. Stahler, Xavier Lambin, Rolf O. Peterson, Joel S. Ruprecht, and John A. Vucetich. 2020. "Fluctuations in age structure and their variable influence on population growth." *Functional Ecology* 34: 203–216.

Hoy, Sarah R., Rolf O. Peterson, and John A. Vucetich. 2018. "Climate warming is associated with smaller body size and shorter lifespans in moose near their southern range limit." *Global Change Biology* 24, no. 6: 2488–2497.

Hoy, Sarah R., Rolf O. Peterson, and John A. Vucetich. 2020. *Ecological Studies of Wolves on Isle Royale: Annual Report 2019-20*. Houghton, Michigan: Michigan Technological University.

Hueffer, Karsten, Todd M. O'Hara, and Erich H. Follmann. 2011. "Adaptation of mammalian host-pathogen interactions in a changing arctic environment." *Acta Veterinaria Scandinavica* 53, no. 1: 17.

Huffaker, Carl Barton. 1958. "Experimental studies on predation: Dispersion factors and predator-prey oscillations." *Hilgardia* 27, no. 14: 343–383.

Hülle, Sebastian, Stefan Liebig, and Meike Janina May. 2018. "Measuring attitudes toward distributive justice: The basic social justice orientations scale." *Social Indicators Research* 136, no. 2: 663–692.

James, George Wharton. 1910. *The Grand Canyon of Arizona: How I See It*. Boston: Little, Brown, & Co.

James, Simon P. 2000. "'Thing-centered' holism in Buddhism, Heidegger, and Deep Ecology." *Environmental Ethics* 22, no. 4: 359–375.

Johns Hopkins Medicine. 2013. "Itchiness explained: Specific set of nerve cells signal itch but not pain, researchers find." *ScienceDaily*, January 2, 2013. https://www.sciencedaily.com/releases/2013/01/130102104548.htm.

Johnson, Wendel J., Michael L. Wolfe, and Durward L. Allen. 1968. *Community Relationships and Population Dynamics of Terrestrial Mammals of Isle Royale, Lake Superior, Second Annual Report, Covering the Tenth Year in the Isle Royale Studies, 1967–68*. Lafayette, ID: Purdue University.

Kadlec, Robert H., and Scott Wallace. 2008. *Treatment Wetlands*. Boca Raton, Florida: CRC Press.

Keith, Lloyd B., and Sara E. M. Bloomer. 1993. "Differential mortality of sympatric snowshoe hares and cottontail rabbits in central Wisconsin." *Canadian Journal of Zoology* 71, no. 8: 1694–1697.

Keller, David R. 2010. *Environmental Ethics: The Big Questions*. West Sussex, UK: Wiley-Blackwell.

Keller, David R., and Frank B. Golley, eds. 2000. *The Philosophy of Ecology: From Science to Synthesis*. Athens: University of Georgia Press.

Keller, Evelyn Fox. 2002. *Making Sense of Life*. Cambridge, MA: Harvard University Press.

Keulartz, Jozef. 2016. "Should the lion eat straw like the ox? Animal ethics and the predation problem." *Journal of Agricultural and Environmental Ethics* 29, no. 5: 813–834.

Khatchadourian, Haig. 2006. "Compensation and reparation as forms of compensatory justice." *Metaphilosophy* 37, no. 3-4: 429–448.

Knott-Ahern, Louise. 2013. "Silence of the wolves." *Lansing State Journal*, October 20, 2013. http://archive.lansingstatejournal.com/interactives/isleroyalewolves.html.

Krech, Shepard. 2000. *The Ecological Indian: Myth and History*. New York: W. W. Norton.

Kristan, William B., III, and William I. Boarman. 2003. "Spatial pattern of risk of common raven predation on desert tortoises." *Ecology* 84, no. 9: 2432–2443.

Kuhn, Thomas. 1962. *The Structure of Scientific Revolutions*. Chicago: University of Chicago Press.

Laikre, Linda, Mija Jansson, Fred W. Allendorf, Sven Jakobsson, and Nils Ryman. 2013. "Hunting effects on favourable conservation status of highly inbred Swedish wolves." *Conservation Biology* 27, no. 2: 248–253.

Le Loeuff, Jean. 2012. "Paleobiogeography and biodiversity of Late Maastrichtian dinosaurs: How many dinosaur species went extinct at the Cretaceous-Tertiary boundary?" *Bulletin de la Société Géologique de France* 183, no. 6: 547–559.

Leonard, Jennifer A., Carles Vilà, and Robert K. Wayne. 2005. "Legacy lost: Genetic variability and population size of extirpated US grey wolves (*Canis lupus*)." *Molecular Ecology* 14, no. 1: 9–17.

Leopold, Aldo. 1921. "The wilderness and its place in forest recreational policy." *Journal of Forestry* 19, no. 7: 718–721.

Leopold, Aldo. 1925. "Wilderness as a form of land use." *The Journal of Land & Public Utility Economics* 1, no. 4: 398–404.

Leopold, Aldo. (1949) 2020. *A Sand County Almanac: And Sketches Here and There.* New York: Oxford University.

Liordos, Vasilios, Vasileios J. Kontsiotis, Magdalini Anastasiadou, and Efstathios Karavasias. 2017. "Effects of attitudes and demography on public support for endangered species conservation." *Science of the Total Environment* 595: 25–34.

Maathuis, Frans J. M. 2014. "Sodium in plants: Perception, signalling, and regulation of sodium fluxes." *Journal of Experimental Botany* 65, no. 3: 849–858.

MacArthur, Robert H., and Edward O. Wilson. 1963. "An equilibrium theory of insular zoogeography." *Evolution* 17, no. 4: 373–387.

MacArthur, Robert H., and Edward O. Wilson. 1967. *The Theory of Island Biogeography.* Princeton, NJ: Princeton University Press.

Mace, Georgina. 1998. "Getting the measure of extinction." *People & the Planet* 7, no. 4: 9.

Mann, Charles C. 2005. *1491: New Revelations of the Americas Before Columbus.* New York: Alfred and Knopf.

Marshall, Jessica. 2013. "Geology: North America's broken heart." *Nature News* 504, no. 7478: 24.

Marshall, Robert. (1930) 1998. "The problem of the wilderness." Reprinted in *The Great New Wilderness Debate*, edited by J. Baird Callicott and Michael P. Nelson, 1998, 85–96. Athens: University of Georgia Press.

Marucco, Francesca, Leah M. Vucetich, Rolf O. Peterson, Jennifer R. Adams, and John A. Vucetich. 2012. "Evaluating the efficacy of non-invasive genetic methods and estimating wolf survival during a ten-year period." *Conservation Genetics* 13 no. 6: 1611–1622.

May, Robert M. 1973. "Time-delay versus stability in population models with two and three trophic levels." *Ecology* 54 no. 2: 315–325.

May, Robert M., John H. Lawton, and Nigel E. Stork. 1995. "Assessing extinction rates." In *Extinction Rates*, edited by John H. Lawton and Robert M. May. Oxford: Oxford University Press.

Mayr, Ernst. 1982. *The Growth of Biological Thought: Diversity, Evolution, and Inheritance.* Cambridge, MA: Harvard University Press.

McIntosh, Robert P. 1998. "The myth of community as organism." *Perspectives in Biology and Medicine* 41, no. 3: 426–438.

McKinney, H. Lewis. 1966. "Alfred Russel Wallace and the discovery of natural selection." *Journal of the History of Medicine and Allied Sciences* 21, no. 4: 333–357.

McLaren, Brian E. 2022. "Lake Superior Caribou." *Lake Superior Magazine*, April 7, 2022. https://www.lakesuperior.com/the-lake/natural-world/lake -superior-caribou/.

McLaren, Brian E., and Rolf O. Peterson. 1994. "Wolves, moose, and tree rings on Isle Royale." *Science* 266, no. 5190: 1555–1558.

McMahan, Jeff. 2010. "The meat eaters." *New York Times*, September 19, 2010.

McMahan, Jeff. 2016. "The moral problem of predation." In *Philosophy Comes to Dinner: Arguments on the Ethics of Eating*, edited by Andrew Chignell, Terence Cuneo, and Matthew C. Halteman, 268–294. New York: Routledge.

Mech, L. David. 1966. *The Wolves of Isle Royale*. National Parks Fauna Series no. 7. Washington, DC: National Park Service.

Mech, L. David. 2013. "The case for watchful waiting with Isle Royale's wolf population." In *The George Wright Forum* 30, no. 3: 326–332.

Mech, L. David, Sagar M. Goyal, William J. Paul, and Wesley E. Newton. 2008. "Demographic effects of canine parvovirus on a free-ranging wolf population over 30 years." *Journal of Wildlife Diseases* 44, no. 4: 824–836.

Meine, C. 2009. "Early wolf research and conservation in the Great Lakes region." In *Recovery of Gray Wolves in the Great Lakes Region of the United States: An Endangered Species Success Story*, edited by Adrian P. Wydeven, Timothy R. Van Deelen, and Edward J. Heske, 1–13. New York: Springer.

Milman, Oliver. 2017. "'There is no sport in that': Trophy hunters and the masters of the universe." *The Guardian*, July 27, 2017. https://www .theguardian.com/environment/2017/jul/27/theres-no-sport-in-that-trophy -hunters-and-the-masters-of-the-universe.

Mlott, Christine. 2016. "In reversal, U.S. Park Service aims to move new wolves to Isle Royale." *Science*, December 16, 2016. http://www.sciencemag.org/news /2016/12/reversal-us-park-service-aims-move-new-wolves-isle-royale.

Mlott, Christine. 2019. "Relocated island wolves outlasting mainland wolves in new Isle Royale home." *Science*, December 20, 2019. https://www.sciencemag .org/news/2019/12/relocated-island-wolves-outlasting-mainland-wolves -new-isle-royale-home.

Montgomery, Robert A., John A. Vucetich, Rolf O. Peterson, Gary J. Roloff, and Kelly F. Millenbah. 2013. "The influence of winter severity, predation and senescence on moose habitat use." *Journal of Animal Ecology* 82, no. 2: 301–309.

Moore, Kathleen Dean, and Michael P. Nelson, eds. 2011. *Moral Ground: Ethical Action for a Planet in Peril*. San Antonio, Texas: Trinity University Press.

Muir, John. (1918) 2020. "The Grand Canyon of the Colorado." In *Steep Trails* by John Muir. Eugene, OR: Doublebit Press.

Munger, James C., and William H. Karasov. 1994. "Costs of bot fly infection in white-footed mice: Energy and mass flow." *Canadian Journal of Zoology* 72, no. 1: 166–173.

Murie, Adolph. 1930. "Field Notes, Summer 1930, Isle Royale, Michigan." Unpublished, archived at the Isle Royale Wolf-Moose Collection, Michigan Technological University Archives and Copper Country Historical Collections, Houghton.

Murie, Adolph. 1934. *The Moose of Isle Royale*. University of Michigan, Museum of Zoology, Miscellaneous Publications no. 25. Ann Arbor, MI: University of Michigan.

Murie, Adolph. 1935. "Report on the qualifications and development of Isle Royale as a National Park, June 13, 1935." Box ISRO of the NPS History Collection, Harpers Ferry. Cited in *Wilderness in National Parks* (2009, University of Washington Press) by J. C. Miles.

Murtaugh, Paul A., and Michael G. Schlax. 2009. "Reproduction and the carbon legacies of individuals." *Global Environmental Change* 19, no. 1: 14–20.

Nassbaum, Martha. 2009. *Frontiers of Justice: Disability, Nationality, Species Membership*. Cambridge, MA: Harvard University Press.

Nelson, Michael Paul. 1998. "An amalgamation of wilderness preservation arguments." In *The Great New Wilderness Debate*, edited by J. Baird Callicott and Michael P. Nelson, 154–198. Athens: University of Georgia Press.

Nelson, Michael Paul. 2010. "Teaching holism in environmental ethics." *Environmental Ethics* 32, no. 1: 33–49.

Nelson, Michael Paul, and J. Baird Callicott, eds. 2008. *The Wilderness Debate Rages On: Continuing the Great New Wilderness Debate*. Athens: University of Georgia Press.

Nelson, Michael Paul, Michelle L. Lute, Chelsea Batavia, and Jeremy T. Bruskotter. 2016. "Should we preserve the wolves of Isle Royale? An empirical assessment of public input." Oregon State University, Corvallis, OR. https://www.researchgate.net/publication/305114465_Should_We_Preserve_the_Wolves_of_Isle_Royale_An_Empirical_Assessment_of_Public_Input.

Nelson, Michael Paul, and John A. Vucetich. 2009. "Abandon hope." *The Ecologist* 39, no. 2: 32–35.

Nicholson, Alexander John. 1933. "Supplement: The balance of animal populations." *The Journal of Animal Ecology* 2, no. 1: 131–178.

Nicholson, Alexander John. 1937. "The role of competition in determining animal populations." *Journal of the Council for Scientific and Industrial Research, Australia* 10: 101–106.

Nickas, George, and Kevin Proescholdt. 2005. "Keeping the wild in wilderness." *International Journal of Wilderness* 11, no. 3: 13–18.

Nussbaum, Martha C. 2009. *Frontiers of Justice: Disability, Nationality, Species Membership*. Cambridge, MA: Harvard University Press.

O'Connell-Rodwell, Caitlin E. 2007. "Keeping an 'ear' to the ground: Seismic communication in elephants." *Physiology* 22, no. 4: 287–294.

Odum, Eugene Pleasants. 1967. *Fundamentals of Ecology*. Philadelphia, PA: Saunders.

Odum, Eugene Pleasants, and Howard T. Odum. 1959. *Fundamentals of Ecology*. Philadelphia, PA: Saunders.

Oelfke, Jack G., Rolf O. Peterson, John A. Vucetich, and Leah M. Vucetich. 2000. "Wolf research in the Isle Royale Wilderness: Do the ends justify the means?" In *Wilderness Science in a Time of Change, Proceedings RMRS-P-15-VOL-3*, edited by Stephen F. McCool, David N. Cole, William T. Borrie, and Jennifer O'Loughlin, 246-51. Ogden, UT: USDA Forest Service, Rocky Mountain Research Station.

Olson, Sigurd. (1938) 1998. "Why wilderness?" Reprinted in *The Great New Wilderness Debate*, edited by J. Baird Callicott and Michael P. Nelson, 97-102. Athens: University of Georgia Press.

Paine, Robert T. 1980. "Food webs: Linkage, interaction strength and community infrastructure." *Journal of Animal Ecology* 49, no. 3: 667-685.

Panksepp, Jaak. 2004. *Affective Neuroscience: The Foundations of Human and Animal Emotions*. Oxford: Oxford University Press.

Pardo, Jose M., and Francisco J. Quintero. 2002. "Plants and sodium ions: Keeping company with the enemy." *Genome Biology* 3, no. 6 (2002): reviews1017-1.

Pearl, Judea. 2009. *Causality*. Cambridge: Cambridge University Press.

Peterson, Carolyn C. 2008. *A View from the Wolf's Eye*. Houghton, MI: Isle Royale Natural History Association.

Peterson, Rolf O. 1977. *Wolf Ecology and Prey Relationships on Isle Royale*. US National Park Service Monograph Series 11. Washington, DC.

Peterson, Rolf O. 1984. *Ecological Studies of Wolves on Isle Royale: Annual Report 1983-84*. Houghton: Michigan Technological University.

Peterson, Rolf O. 1988. *Ecological Studies of Wolves on Isle Royale: Annual Report 1987-88*. Houghton: Michigan Technological University.

Peterson, Rolf O. 1989. *Ecological Studies of Wolves on Isle Royale: Annual Report 1988-89*. Houghton: Michigan Technological University.

Peterson, Rolf O. 1991. *Ecological Studies of Wolves on Isle Royale: Annual Report 1990-91*. Houghton: Michigan Technological University.

Peterson, Rolf O. 1995a. *Ecological Studies of Wolves on Isle Royale: Annual Report 1994-95*. Houghton: Michigan Technological University.

Peterson, Rolf O. 1995b. *The Wolves of Isle Royale: A Broken Balance*. Minocqua, WI: Willow Creek.

Peterson, Rolf O., James M. Dietz, and Durward L. Allen. 1971 *Ecological Studies of Wolves on Isle Royale: First Annual Report, 1970-71*. Lafayette, IN: Purdue University.

Peterson, Rolf O., and Richard E. Page. 1988. "The rise and fall of Isle Royale wolves, 1975-1986." *Journal of Mammalogy* 69, no. 1: 89-99.

Peterson, Rolf O., Richard E. Page, and Philip W. Stephens. 1982. *Ecological Studies of Wolves on Isle Royale: Annual Report 1981-82*. Houghton: Michigan Technological University.

Peterson, Rolf O., and Philip W. Stephens. 1980. *Ecological Studies of Wolves on Isle Royale: Annual Report 1979-80*. Houghton: Michigan Technological University.

Peterson, Rolf O., and Philip W. Stephens. 1981. *Ecological Studies of Wolves on Isle Royale: Annual Report 1980-81*. Houghton: Michigan Technological University.

Peterson, Rolf O., Nancy J. Thomas, Joanne M. Thurber, John A. Vucetich, and Thomas A. Waite. 1998. "Population limitation and the wolves of Isle Royale." *Journal of Mammalogy* 79, no. 3: 828–841.

Peterson, Rolf O., and John A. Vucetich. 2002. *Ecological Studies of Wolves on Isle Royale: Annual Report 2001–02.* Houghton: Michigan Technological University.

Peterson, Rolf O., John A. Vucetich, Joseph M. Bump, and Douglas W. Smith. 2014. "Trophic cascades in a multicausal world: Isle Royale and Yellowstone." *Annual Review of Ecology, Evolution, and Systematics* 45: 325–345.

Peterson, Rolf O., John A. Vucetich, Gus Fenton, Thomas D. Drummer, and Clark Spencer Larsen. 2010. "Ecology of arthritis." *Ecology Letters* 13, no. 9: 1124–1128.

Pierce, John C., Nicholas P. Lovrich Jr., Taketsugu Tsurutani, and Takematsu Abe. 1987. "Culture, politics and mass publics: traditional and modern supporters of the new environmental paradigm in Japan and the United States." *The Journal of Politics* 49, no. 1: 54–79.

Pimm, Stuart L., Clinton N. Jenkins, Robin Abell, Thomas M. Brooks, John L. Gittleman, Lucas N. Joppa, Peter H. Raven, Callum M. Roberts, and Joseph O. Sexton. 2014. "The biodiversity of species and their rates of extinction, distribution, and protection." *Science* 344, no. 6187: 1246752.

Poirier, Jessica J., and Richard E. Taylor. 2007. *Images of America: Isle Royale.* Mount Pleasant, SC: Arcadia.

Popper, Deborah Epstein, Robert E. Lang, and Frank J. Popper. 2000. "From maps to myth: The census, Turner, and the idea of the frontier." *The Journal of American Culture* 23, no. 1: 91.

Pound, Roscoe, and Frederic E. Clements. 1900. *Phytogeography of Nebraska: I. General Survey.* Lincoln, Nebraska: Seminar.

Pradhan, Prajal, Luís Costa, Diego Rybski, Wolfgang Lucht, and Jürgen P. Kropp. 2017. "A systematic study of Sustainable Development Goal (SDG) interactions." *Earth's Future* 5, no. 11: 1169–1179.

Priest, Graham. 2014. "Beyond true and false." *Aeon*, May 5, 2014. https://aeon.co/essays/the-logic-of-buddhist-philosophy-goes-beyond-simple-truth.

Proulx, Gilbert, Shelley. M. Alexander, Hannah Barron, Marc Bekoff, Ryan Brook, Heather Bryan, Chris Darimont, et al. 2017. "Killing wolves and farming caribou benefit industry, not caribou: A response to Stan Boutin." *Nature Alberta* Spring 47: 4–11.

Public Broadcasting Service. 2009. "Adolph Murie," *The National Parks: America's Best Idea. Episode 6.* Accessed April 10, 2020. http://www.pbs.org/nationalparks/people/nps/2.

Purdy, Jedediah. 2015. "Environmentalism's racist history." *The New Yorker*, August 13, 2015. https://www.newyorker.com/news/news-desk/environmentalisms-racist-history.

Preston, Frank W. 1962a. "The canonical distribution of commonness and rarity: Part I." *Ecology* 43, no. 2: 185–215.

Preston, Frank W. 1962b. "The canonical distribution of commonness and rarity: Part II." *Ecology* 43, no. 3: 410–432.

Quammen, David. 1996. *The Song of the Dodo: Island Biogeography in an Age of Extinctions*. New York: Random House.

Quammen, David. 2012. *Spillover: Animal Infections and the Next Human Pandemic*. New York: W. W. Norton.

Räikkönen, Jannikke, John A. Vucetich, Rolf O. Peterson, and Michael P. Nelson. 2009. "Congenital bone deformities and the inbred wolves (*Canis lupus*) of Isle Royale." *Biological Conservation* 142, no. 5: 1025–1031.

Raley, Karen. 1998. "Maintaining balance: The religious world of the Cherokees." *Tar Heel Junior Historian* 37, no. 2: 2–5.

Risenhoover, Kenneth L., and Rolf O. Peterson. 1986. "Mineral licks as a sodium source for Isle Royale moose." *Oecologia* 71, no. 1: 121–126.

Rothstein, Stephen I. 2004. "Brown-headed cowbird: Villain or scapegoat?" *Birding* 36, 374–384.

Rotter, Michael C., and Alan J. Rebertus. 2015. "Plant community development of Isle Royale's moose-spruce savannas." *Botany* 93, no. 2: 75–90.

Royle, Nick J., Per T. Smiseth, and Mathias Kölliker, eds. 2012. *The Evolution of Parental Care*. Oxford: Oxford University Press.

Ruse, Michael. 2013. *The Gaia Hypothesis: Science on a Pagan Planet*. Chicago: University of Chicago.

Samuel, Bill. 2004. *White as a Ghost: Winter Ticks and Moose*. Edmonton, Alberta, CA: Nature Alberta.

Sand, Håkan, John A. Vucetich, Barbara Zimmermann, Petter Wabakken, Camilla Wikenros, Hans C. Pedersen, Rolf O. Peterson, and Olof Liberg. 2012. "Assessing the influence of prey–predator ratio, prey age structure and packs size on wolf kill rates." *Oikos* 121, no. 9: 1454–1463.

Sandel, Michael J. 2009. *Justice: What's the Right Thing to Do?* New York: Farrar, Straus, & Giroux.

Schmidt, William E. 1984. "Southern practice of eating dirt shows signs of waning." *New York Times*, February 13, 1984.

Schwartz, Charles C., and Lyle A. Renecker. 1998. "Nutrition and energetics." In *Ecology and Management of the North American Moose*, edited by Albert W. Franzmann and Charles C. Schwartz, 441–478. Washington, DC: Smithsonian Institution.

Shotwell, Alexis. 2016. *Against Purity: Living Ethically in Compromised Times*. Minneapolis: University of Minnesota Press.

Shrader-Frechette, Kristin. 1996. "Individualism, holism, and environmental ethics." In *The Environment in Anthropology*, edited by Nora Haenn and Richard Wilk, 336–348. New York: New York University Press.

Simberloff, Daniel. 2013. *Invasive Species: What Everyone Needs to Know*. Oxford: Oxford University Press.

Simberloff, Daniel. 2014. "The 'balance of nature'—evolution of a panchreston." *PLoS biology* 12, no. 10.

Solomon, Christopher. 2014. "Rethinking the wild: The Wilderness Act is facing a midlife crisis." *New York Times*, July 5, 2014.

Standing Bear, Chief Luther (1933) 1998. "Indian Wisdom." Reprinted in *The Great New Wilderness Debate*, edited by J. Baird Callicott and Michael P. Nelson, 201–206. Athens: University of Georgia Press.

Steinhart, Peter. 1995. *The Company of Wolves*. New York: Vintage.

Stoll, Albert., Jr. 1921. "Isle Royale." *Detroit News*, December 3, 1921.

Tabatabaie, Vafa, Gil Atzmon, Swapnil N. Rajpathak, Ruth Freeman, Nir Barzilai, and Jill Crandall. 2011. "Exceptional longevity is associated with decreased reproduction." *Aging (Albany NY)* 3, no. 12: 1202.

Treves, Adrian, and Kerry A. Martin. 2011. "Hunters as stewards of wolves in Wisconsin and the Northern Rocky Mountains, USA." *Society & Natural Resources* 24, no. 9: 984–994.

Triandis, Harry C. 2018. *Individualism and Collectivism*. New York: Routledge.

Trifonov, Edward N. 2011. "Vocabulary of definitions of life suggests a definition." *Journal of Biomolecular Structure and Dynamics* 29, no. 2: 259–266.

Turner, James M. 2002. "From woodcraft to 'leave no trace': Wilderness, consumerism, and environmentalism in twentieth-century America." *Environmental History* 7, 462–84.

Twain, Mark. 1866. "The great volcano Of Kilauea." *The Sacramento Daily Union*, November 16, 1866.

Tyson, Peter. 2012. "Dogs' dazzling sense of smell." NOVA website, October 4, 2012. https://www.pbs.org/wgbh/nova/nature/dogs-sense-of-smell.html.

US National Park Service. 1999. *Isle Royale: General Management Plan*. Houghton, MI: Isle Royale National Park.

US National Park Service. 2016. *Draft Environmental Impact Statement to Address the Presence of Wolves in Isle Royale National Park*. Houghton, MI: Isle Royale National Park.

Utida, Syunro. 1957. "Cyclic fluctuation of population density intrinsic to the host–parasite system." *Ecology* 38, no. 3: 442–449.

Vickery, Jim Dale. 1994. *Wilderness Visionaries*. Minocqua, WI: Northword Press.

Volterra, Vito. 1926. "Fluctuations in the abundance of a species considered mathematically." *Nature* 118, no. 2972: 558–560.

Vucetich, John A. 2016a. "Should humans intervene when climate change threatens an island's ecology?" *Natural History* 124, no. 7: 20–23.

Vucetich, John A. 2016b. "Witness Statement." In *The Status of the Federal Government's Management of Wolves*. Oversight Hearing, Committee on Natural Resources, United States House of Representatives. https://www.gpo.gov/fdsys/pkg/CHRG-114hhrg21616/pdf/CHRG-114hhrg21616.pdf.

Vucetich, John A., Jeremy T. Bruskotter, and Michael Paul Nelson. 2015. "Evaluating whether nature's intrinsic value is an axiom of or anathema to conservation." *Conservation Biology* 29, no. 2: 321–332.

Vucetich, John A., Jeremy T. Bruskotter, Michael Paul Nelson, Rolf O. Peterson, and Joseph K. Bump. 2017. "Evaluating the principles of wildlife conservation: A case study of wolf (*Canis lupus*) hunting in Michigan, United States." *Journal of Mammalogy* 98, no. 1: 53–64.

Vucetich, John A., Dawn Burnham, Ewan A. Macdonald, Jeremy T. Bruskotter, Silvio Marchini, Alexandra Zimmermann, and David W. Macdonald. 2018. "Just conservation: What is it and should we pursue it?" *Biological Conservation* 221: 23–33.

Vucetich, John A., Mark Hebblewhite, Douglas W. Smith, and Rolf O. Peterson. 2011. "Predicting prey population dynamics from kill rate, predation rate and predator–prey ratios in three wolf-ungulate systems." *Journal of Animal Ecology* 80, no. 6: 1236–1245.

Vucetich, John A., and David W. Macdonald. 2017. "Some essentials on coexisting with carnivores." *Open Access Government.* August Issue, 216–217. https://www.openaccessgovernment.org/essentials-coexisting-carnivores/35771.

Vucetich, John A., and Michael Paul Nelson. 2008. "Distinguishing experiential and physical conceptions of wilderness." In *The Wilderness Debate Rages On,* edited by Michael Paul Nelson and J. Baird Callicott, 611–632. Athens: University of Georgia Press.

Vucetich, John A., and Michael Paul Nelson. 2010. "Sustainability: Virtuous or vulgar?" *BioScience* 60, no. 7: 539–544.

Vucetich, John A., and Michael Paul Nelson. 2014. "Curation or conservation?" *New York Times,* August 20, 2014.

Vucetich, John A., and Michael Paul Nelson. 2018. "Acceptable risk of extinction in the context of endangered species policy." In *Philosophic Frames on Public Policy,* edited by Andrew I. Cohen, 81–104. Lanham, MD: Rowman & Littlefield.

Vucetich, John A., Michael P. Nelson, and Chelsea K. Batavia. 2015. "The Anthropocene: Disturbing name, limited insight." In *After Preservation: Saving American Nature in the Age of Humans,* edited by Ben A. Minteer and Stephen J. Pyne, 66–73. Chicago: University of Chicago Press.

Vucetich, John A., Michael Paul Nelson, and Jeremy T. Bruskotter. 2017. "Conservation triage falls short because conservation is not like emergency medicine." *Frontiers in Ecology and Evolution* 5: 45.

Vucetich, John A., Michael Paul Nelson, and Jeremy T. Bruskotter. 2020. "What Drives Declining Support for Long-Term Ecological Research?" *BioScience* 70, no. 2: 168–173.

Vucetich, John A., Michael Paul Nelson, and Rolf O. Peterson. 2013. "Predator and prey, a delicate dance." *New York Times,* May 8, 2013.

Vucetich, John A., Michael P. Nelson, and Michael K. Phillips. 2006. "The normative dimension and legal meaning of endangered and recovery in the US Endangered Species Act." *Conservation Biology* 20, no. 5: 1383–1390.

Vucetich, John A., and Rolf O. Peterson. 2004. "The influence of prey consumption and demographic stochasticity on population growth rate of Isle Royale wolves *Canis lupus*." *Oikos* 107, no. 2: 309–320.

Vucetich, John A., and Rolf O. Peterson. 2012. *Ecological Studies of Wolves on Isle Royale: Annual Report 2011-12*. Houghton: Michigan Technological University.

Vucetich, John A., and Rolf O. Peterson. 2014. *Ecological Studies of Wolves on Isle Royale: Annual Report 2013-14*. Houghton: Michigan Technological University.

Vucetich, John A., Rolf O. Peterson, and Michael Paul Nelson. 2013a. "Response to Gostomski." *The George Wright Forum* 30, no. 1: 101-102.

Vucetich, John A., Rolf O. Peterson, and Michael Paul Nelson. 2013b. "Should Isle Royale wolves be reintroduced? A case study on wilderness management in a changing world." *The George Wright Forum* 29, no. 1: 126-147.

Vucetich, John A., Rolf O. Peterson, and Carrie L. Schaefer. 2002. "The effect of prey and predator densities on wolf predation." *Ecology* 83, no. 11: 3003-3013.

Vucetich, John A., Rolf O. Peterson, and Thomas A. Waite. 2004. "Raven scavenging favours group foraging in wolves." *Animal Behaviour* 67, no. 6: 1117-1126.

Vucetich, John A., Leah M. Vucetich, and Rolf O. Peterson. 2012. "The causes and consequences of partial prey consumption by wolves preying on moose." *Behavioral Ecology and Sociobiology* 66, no. 2: 295-303.

Wagner, Frederic H. 2006. *Yellowstone's Destabilized Ecosystem: Elk Effects, Science, and Policy Conflict*. Oxford: Oxford University Press.

Wagner, Frederic H., and Joseph L. Sax. 1995. *Wildlife Policies in the U.S. National Parks*. Washington, DC: Island Press.

Wallace, Alfred R. (1853) 2010. *A Narrative of Travels on the Amazon and Rio Negro*. Reprinted, Cambridge: Cambridge University Press.

Wallace, Alfred R. (1858) 2003. "On the tendency of varieties to depart indefinitely from the original type." Reprinted in *Scientiae Studia* 1, no. 2: 231-243.

Wayne, Robert K., N. Lehman, D. Girman, P. J. P. Gogan, D. A. Gilbert, K. Hansen, R. O. Peterson, et al. 1991. "Conservation genetics of the endangered Isle Royale gray wolf." *Conservation Biology* 5, no. 1: 41-51.

Whiteley, Andrew R., Sarah W. Fitzpatrick, W. Chris Funk, and David A. Tallmon. 2015. "Genetic rescue to the rescue." *Trends in Ecology & Evolution* 30, no. 1: 42-49.

Wilderness Watch. 2014. "National Park Service Studying Isle Royale Wolf issue." *Wilderness Watcher* 25, no 2: 7.

Withey, Michael. 2016. *Mastering Logical Fallacies*. Berkeley, CA: Zephyros.

Wolf, Christopher, and William J. Ripple. 2017. "Range contractions of the world's large carnivores." *Royal Society Open Science* 4, no. 7: 170052.

Worster, Donald. 1993. *The Wealth of Nature: Environmental History and the Ecological Imagination*. Oxford: Oxford University Press.

Worster, Donald. 1994. *Nature's Economy: A History of Ecological Ideas*. Cambridge: Cambridge University Press.

Wright, George. 1932. Unpublished notes, May 23, 1932, Mammoth Hot Springs, Yellowstone National Park, Wyoming. Archived at University of California, Berkeley.

Young, Sera L., Paul W. Sherman, Julius B. Lucks, and Gretel H. Pelto. 2011. "Why on Earth? Evaluating hypotheses about the physiological functions of human geophagy." *The Quarterly Review of Biology* 86, no. 2: 97–120.

Yudell, Michael, Dorothy Roberts, Rob DeSalle, and Sarah Tishkoff. 2016. "Taking race out of human genetics." *Science* 351, no. 6273: 564–565.

Yurk, H., and A. W. Trites. 2000. "Experimental attempts to reduce predation by harbor seals on out-migrating juvenile salmonids." *Transactions of the American Fisheries Society* 129, no. 6 1360–1366.

Zuckerman, Laura. 2014. "Idaho to kill thousands of ravens to benefit imperiled bird species." *Scientific American*, March 19, 2014. https://www.scientific american.com/article/idaho-to-kill-thousands-of-ravens-to-benefit -imperiled-bird-species.

Adams, Jennifer, 187, 188
aerial observation, 80–81, 344n28; time spent in, 152–53; winter 2010, 196–219. *See also* bush pilots; moose counts; wolf counts
African wild dogs, 176–77, 318–19
African wildlife, 298, 301, 302, 318–19
Ahwahnechee, 282–83
Alexander, Henry, 250
algae, 288
Allen, Durward, 53, 92, 97, 121–22, 134–35, 151–52, 260, 272; balance-of-nature perspective, 125, 129–31; on captive-raised wolf release, 269–70; *Ecological Studies of the Wolf on Isle Royale, First Annual Report, 1970–1971*, 137–39; initiation of Isle Royale research, 6–7, 53–54, 67–70, 73; *Our Wildlife Heritage*, 68; at Purdue University, 69, 72, 73; retirement, 53, 144; *The Wolves of Minong*, 6–7, 129, 134, 341n3
alpha female wolves, 11, 167, 179, 180, 214
alpha male wolves, 11–12, 133, 135, 165–68, 173, 200; killed by other wolves, 179–80, 181, 221–23; Old Gray Guy, 165–68, 181, 187–91
alpha wolf pairs, 8, 9, 133, 147
alpha wolves, control of moose-kill access, 147
amensalism, 49
ancestry: common, 305–6, 356n37; of Isle Royale wolves, 189
Angleworm Lake, 180–81, 207, 214
Anning, Mary, 99
anthropomorphism, 12–14, 304, 305, 311, 335n5
Argumentum ad Naturam, 263, 269, 276, 297, 311

Aristotle, *Nicomachean Ethics*, 312
Arnett, G. Ray, 149–50, 343n20
arthritis, 19, 36, 161, 205–6, 223, 229
attack rate, 108, 110–11, 116–17
autographs, 288

bacteria, as protozoan prey, 112–18
balance of nature, 93–132; ancient Greek and Roman concepts, 95–97, 102, 131; biological control and, 118–22; commercial manipulation, 122–25; as cross-cultural concept, 94–97; Darwin's theory, 100–102, 129; Elton on, 123–25; environmental basis, 123–24; environmental ethics and, 305–23; exogenous forces, 142–44; Gauss's theory, 112–18, 121, 128; Huffaker's experiments, 117–21, 122; humans role in, 130–32; Odum on, 128; theological interpretations, 96, 97–98, 99–100, 125; Utida's experiments, 121–22; Volterra's theory, 106–12, 116, 117, 121, 122, 128, 129; Wallace's interpretation, 102–5. *See also* wolf-moose population dynamics, on Isle Royale
balsam firs: dendrochronology, 290–93, 354n8; moose browsing on, 161–62, 223, 291–94, 354n11
Bangsund cabin, 54
bears: black, 258; brown, 298; grizzly, 255
Beaver Lake, 196
beavers, 19, 71, 82, 86, 133, 137, 147, 165, 205, 213, 247, 261, 348n7 (chap. 8)
beetles: adzuki bean, 121–22; carrion, 186; flour, 124
Berry, Wendell, 245
beta wolves, 135
Big Jim (male wolf), 269–71, 351n16

"Big Pack," 133-34
Big Siskiwit Swamp, 86, 215
biodiversity crisis, 298-303, 316, 333;
    conflicting obligations in, 315-16;
    mitigation strategies, 302-3, 313-14,
    321-23, 331, 355n18; recovery time, 300;
    wolf predation and, 309
biological control, 117, 118-22
black (melanistic) wolves, 134-35, 191,
    269, 342n4, 342n6
blood analysis, of wolves, 153, 156-57
bot fly larvae, 43-46, 337n4, 337n8
Boy Scouts, 68, 279, 284
breeding population, of wolves,
    188-89
British Columbia, wolf-caribou popula-
    tion dynamics in, 315-17, 318, 319-20
browsing, by moose, 56, 161-62, 223,
    291-94, 328
Bryson, Bill, 239
Burgess, Jack, 76-77, 80-81
Burns, Robert, 10
bush pilots, 76-78, 144. See also Burgess,
    Jack; Glaser, Don E.; Murray, Don;
    Tomes, Arthur C.

Cahokia, 249, 283
Canada, wolf immigration from, 187-88,
    351n16
Canada yews, 42
canine parvovirus, 154-56, 157, 162-63,
    182, 285, 291, 344nn32-33
captive moose, release on Isle Royale,
    271-73
captive-raised wolves, release on Isle
    Royale, 260, 351n15
car camping, 253, 254, 278-79
carcass utilization index, 173-74, 175,
    345n17
caribou, 315-17, 330-31
carnivores: biodiversity and, 298-99,
    303; interactions with humans,
    297-99, 355n16; trophic cascade
    argument for, 296-97, 303
Cartier, Jacques, 247
Catoctin Mountain National Park,
    296
cats, 17, 46, 154-55

Cherokee people, 94, 131
chickadees, 173-74
Chickenbone Lake, 178, 180, 202, 203
child rearing, opportunity cost, 45
Chippewa Harbor, 166, 250
Chippewa Harbor Pack, 3-4, 196-97;
    alpha wolves, 200, 205, 211, 214, 217,
    223-24, 232; attack on lone wolf,
    200-201; competition with other
    packs, 178—81, 219-20, 222-23;
    disintegration, 224, 232; hunting
    behavior, 3-4, 7-8, 207; moose kills, 8,
    202-3, 206-7, 208, 212, 213, 215, 216;
    origin, 168; Romeo (male wolf) and,
    196, 200-205, 211-12, 218, 232; size, 219;
    territorialism, 195
Cicero, 95-96, 97
Cinderella (female wolf), 166-68, 178,
    180-82, 183
Clements, Frederic, 125-28, 129, 309,
    340n35
climate change, anthropogenic, 234, 237,
    238, 285, 308, 309, 313-14, 348n2
Cochran, Tim, 238
Cole, James, 72
commendation, 49
competition, 102; between humans and
    wolves, 6, 308; among wolves, 149,
    166-68, 178-81, 214
Con, Titus, 240
Conglomerate Bay, 250
consumerism, 284-85
Cook, James, 263
copper mining, 248-51, 252, 349n25,
    349n27
cottagers, 251, 252, 253, 257-58, 278-79,
    349n29
courtship and mating behaviors, 5, 80-81,
    168, 203-4, 210, 213, 215, 216, 347n5
coyotes, 255, 261
creation myths, 305, 356n36
Cumberland Point, 216

Daisy Farm, 162, 212, 214
Daisy Farm Pack, 148
Dall sheep, 255
D'Ancona, Umberto, 105-6
Darlington, Philip J., 264

Darwin, Charles, 47, 106, 111, 125, 341n40, 341n46; *On the Origin of Species*, 100–102, 104–5

Davidson Island, 87

dead moose, researcher's locating of, 162. *See also* death, of moose

death, of moose: moose's experience of, 31–32, 203, 315–16. *See also* moose kills

death, of wolves, 5, 9, 10, 133–34, 310–11; between 1981 and 1988, 152; adult mortality rate, 150; alpha males, 133; average annual number, 181–82; canine parvovirus–related, 154–56, 157, 162–63, 182, 285, 291; dating the time of, 70; by drowning, 223–24, 234, 268, 285; killed by other wolves, 149, 153, 221–23, 326; last of original Isle Royal wolves, 326; lone wolves, 146; mainland wolves, 153; by shooting, 153, 231–32, 310–11, 351n15; specimen #3529, 183; by starvation, 7, 9, 149, 153; wolf pairs, 146; wolf pups, 9, 181–82; during wolf restoration project, 325–26, 327

deer, 297; mule, 46; as tick hosts, 46, 47, 337n10; white-tailed, 257; wolf predation on, 259

deer mice, 40–46, 337n4, 337n6, 337n8

Deism, 99

Denali National Park, 255, 256, 259

De Natural Decorum (On the Nature of the Gods) (Cicero), 95–96

dendrochronology, 290–93, 354n11

Desbordes-Valmores, Marcelline, 107, 340n20

DNA analysis, 70, 156–57, 168, 186–88, 207

dogs, 9–10, 16, 21, 183–84

Douglas firs, 295

dreams, 16–20, 23–24

drownings, 223–24, 234, 268, 285

Durham, William, 97

eagles, 162, 287, 289

earth, as living organism, 309–11, 313, 314

East Pack: alpha wolves, 188, 195; competition with other packs, 166–68, 178–81; disintegration, 195; moose-kill sites, 163–64; social order, 165; territorialism, 147, 148, 195, 202

Ecological Society of America, 259

Ecological Studies of the Wolf on Isle Royal, First Annual Report, 1970–1971 (Peterson and Allen), 137–39

ecology, 96; quadrant method, 125–27, 340n35

Ecology, 121

ecosystems, 311; carnivore's role in, 297; energy and nutrient flows of, 128, 129; feedback within, 128–29; health, 296, 303–5, 311, 355n28, 355n32, 356n35; humans' relationships with, 303–7; individual's priority over, 315–16, 319–20; of Isle Royal, 233–34; as living organisms, 125–32, 309, 314; species-area relationships, 265–67; ungulate's role, 233, 236, 257

elephants, 45, 318

elk, 6, 68, 257, 295, 297, 350n10

Elton, Charles, 123–25

Emerson, Ralph Waldo, 277

emotions, 5, 6, 9–10, 19, 65, 335n1, 336n11

empathy, 14–15, 16, 336n11

energy flow, 49–51, 128, 129

entitlement, 307, 312, 356n39

environmental ethics, 305–23; conflicting obligations, 315–20; fair treatment, 306–7, 308–9, 310, 311–13, 314; restorative justice, 312–13

environmental impact statement (EIS), 325, 348n8 (chap. 8)

equality, 307, 312

Equilibrium Theory of Insular Zoogeography (MacArthur and Wilson), 265–67

equity, 307, 312, 356n42

Estes, James, 286, 289–90

ethical consistency, 306–7

European settlers, 250, 276, 278, 283, 307

evolution, 31, 99, 100, 103–4, 105, 299, 300, 339n11 (chap. 4), 341n46

experiences: mental, 19; wilderness as, 277, 279–80

extinction, 98–100, 111, 235–36, 298; on islands, 157, 261–62, 265–69, 294; local, 300, 301, 302; as natural process, 157, 235–36, 268–69, 299; predator-prey dynamics, 119–20; rate, 299–302 extinction, of wolves, 235–39; anthropogenic causes, 268, 285; National Park Service's position on, 157, 235–36, 270, 273–74, 294; as "natural" process, 261–62, 268–74; Peterson's view, 157–59; wolf reintroduction after, 157, 259–60, 270

F193 (female wolf), 326
fair treatment, 306–7, 308–9, 310, 311–13, 314
family, wolf packs as, 5, 8–9, 135; human-wolf comparison, 10–12, 335n7
Father of Animal Ecology (Charles Elton), 123
Father of Plant Ecology (Frederic Clements), 126
Feldtmann Lake, 29, 203, 229
feline distemper, 154–55
fishers, 69
Five Finger Point, 87
Flagship (airplane), 4, 7, 20, 27, 197, 198, 200, 201
food chains, 286–91, 289; wolf-moose-forest, 288, 290–96
food webs, 48–51, 127, 165, 289; diagrams, 289, 337–38n16
Forbes, Stephen, 127
Forester, Johann, 263–64, 350n5
forests, impact of moose browsing activities on, 151, 291–96
fossils and fossil record, 98–100, 102
foxes, 123, 133; arrival on Isle Royale, 261; as carcass scavengers, 39, 153, 162, 173, 221; red, 261; tracks, 70, 78–79; urine, as rabbit repellent, 286, 289, 354n4
Franklin, Benjamin, 250
Freedom of Information Act, 235, 325
frugivory, 49
fruit flies, 124, 156, 273
fungivory, 49

Gaia hypothesis, 310, 313, 314
Gang of Four, 148, 149
Gause, Georgy, 112–18, 121
gender roles, in wolf packs, 12
gene clones, 48
generations, duration, 116

genes, 48
genetic diversity: loss, 156–57, 158–59, 169; restoration through genetic rescue, 189–91, 234–35, 269–71, 293, 294, 324–25
genetic isolation, 157, 158–59, 163, 169
genetic profile, 187
genetic rescue, 189–91, 234–35, 269–71, 293, 294, 324–25
genetics, of families, 11
genetics, of wolves: of captive-raised wolves, 269–71; and lumbosacral transitional vertebrae deformity, 183–86. See also inbreeding
geocentricism, 311
geographic ranges, 300, 301, 302–3, 308, 333, 356–57n44
geophagy, 56–58. See also mud licks
ghost moose, 40
glaciation, 246–47, 248, 300, 349n24
Glacier National Park, 253
Glaser, Don E., 24, 30, 31, 163, 187, 197, 198, 200–201, 205, 206, 210, 215, 219; contributions to wolf-moose research, 144, 231; flying ability, 201
Golden Rule, 306–7
Gostomski, Ted, 237–38
Grace Creek, 29, 202, 203
Grace Harbor, 33, 34
Grace Island, 34
Grand Canyon, 239–40, 241
Grand Portage National Monument, 238
Grand Teton National Park, 257
granivory, 49
Great Smoky Mountains, 296
greenhouse gases, 308
Greenstone Creek, 203
Greenstone Ridge, 3, 202, 203
Greenstone Ridge Trail, 245
grief, 5, 9–10, 25, 335n1
grooming, 46–47, 337n10

habitat fragmentation, 266–68
hares, 123, 133; snowshoe, 71, 147, 149
Harvey Lake, 196, 197, 203, 205, 208
Harvey Lake Pack, 147, 148, 149, 152
Hatchet Lake, 202, 203, 205, 206, 213
Hawaii, colonization of, 273

Hawaii National Park, 240-41
Hay Bay, 216, 217, 221
Hedrick, Phil, 189, 190, 271
herbivores, 6, 297, 311
herbivory, 49, 268, 288, 331; assessment, 55; effect on balsam firs, 161-62, 223, 290-94; tick parasitism-related decrease, 50-51
homeostasis, 128-29, 309-10, 314, 341n40
Houghton Point, 25, 221, 231
howling, 55, 70, 80-81, 220
Hudson's Bay Company, 122-23, 126
Huffaker, Carl, 118-22
Hugginin Cove, 229
human-nature relationship, 4-5, 177-78, 305-6, 309; with carnivores, 297-99; conflicting obligations in, 315-20; culpability in, 317-18, 321; with ecosystems, 303-7, 311; fair treatment in, 306-7, 308-9, 310, 311-13, 314; humans' management of natural processes, 157-58; kinship, 6, 10-11, 306, 307, 309; reason's role in, 321-23; restorative justice in, 312-13
humans (Homo sapiens), 13; in balance of nature, 130-32; creation myths, 305, 356n36; differences and similarities to wolves, 5-6, 10-12, 13-14, 336nn9-10; geophagy in, 56; hatred of wolves, 149-50, 308; Paleolithic, 32; population growth, 314, 319; reproduction vs. longevity opportunity costs, 45
hunting, by humans: of carnivores, 297, 298; of wolves, 73, 76-77, 308
hunting strategies, of wolves, 80-81, 82-84; lone wolves, 201-3; moose calves, 3-4, 230; selection of target moose, 19-20, 21; for weakened moose, 27-31, 202, 203, 205. See also moose kills
hydatid cysts, 84, 134
hyponatremia, 56-57
hypothesis testing, 144

ice bridges, 67, 131, 134, 187, 191-93, 269, 342n6; effect of climate change on, 190, 193, 234, 285
identification, of wolves: DNA-based, 70; National Park Service policy, 358n4

imagination, 16
immigrants, effect on predator-prey dynamics, 117-20
immigrant wolves: arrival via ice bridges, 67, 131, 134, 187, 190, 191-93, 234, 269, 285, 330, 342n6, 348n11; captive-raised wolves as, 260; extinction risk effects, 267; genetic profile, 187-89; as genetic rescue, 189-91, 234-35, 269-71, 293, 294, 324-25; as packs, 134, 135; wolf pair, 157
inbreeding, 40, 52, 193, 232, 236, 332, 337n3; climate change and, 308, 309; depression, 189, 234, 346n31, 356n47; genetic deformities and, 184-85, 346n27; genetic diversity loss through, 156-57, 158-59, 169, 328, 329-30; genetic rescue, 189-91, 234-35, 269-71, 293, 294, 324-25, 348n9; incest, 14, 188, 190, 219; in island populations, 40-46
incest, 14, 188, 190, 219
Indiana Dunes National Park, 296
individualism, collectivism vs., 320, 321, 358n69
injuries, to wolves from moose, 18, 24, 29, 30, 152, 203, 205-6
insectivory, 49
insects, predatory, 118-19
Isabelle (female wolf), 224-28, 230-32
island biogeography, 238, 239, 260, 261-68. See also island populations; Isle Royale
"Island Complications: Should we Return Wolves to Isle Royale?" (Cochrane), 238, 239, 260, 261-63, 268, 269, 271, 272
Island Copper Company, 252
Island Mine, 250
island populations: extinction, 157, 261-62, 265-69, 294; inbreeding, 40-46
Isle Royale, 3-4; acquisition by US, 250; copper mining, 248-51, 252, 349n25, 349n27; ecosystem, 258; geologic formation, 241-44, 247, 349n21; humans' arrival, 247-50; island biogeography, 238, 239, 260, 261-63; logging, 252, 255, 257-58, 349n31; plant and animal colonization, 244, 246; topography, 243-45, 246-47, 349n25

Isle Royale National Park, 5–6; captive-raised wolves' release, 68–69, 260, 269–71, 351n15; creation, 237, 252–53, 254–55; first wolf-moose study, 255–56; last two wolves, 308–9; proposed introduction of wolves, 259–60; researchers' accommodations, 54–55, 73; shelter graffiti, 280–82, 353n43; wilderness character, 153, 158, 237, 238, 257–58, 259, 274–75; wilderness recreation in, 251–52; wolves' emigration from, 135; wolves' first immigration to (1948-1949), 67–68. *See also* wolf-moose population dynamics, on Isle Royale

James, George Wharton, 240
jaw necrosis, 19–20, 21, 28, 32, 84
Jefferson, Thomas, 98
Johnson, Wendel, 133
Jordan, Pete, 133
Journal of Animal Ecology, 123
junipers, 295
justice, 304–5; distributive, 312; punitive, 312, 317–18; restorative, 312–13, 357n57; social, 356n42

kelp forests, 286–87, 289–90
Keweenaw Peninsula, 242–43, 249, 349n25
Kierkegaard, Søren, 276–77
Kilauea volcano, 240–41
kill rate, 27, 80, 84, 88–89, 90, 152, 169, 229, 338n17, 338n20, 345n14, 346n34; carcass utilization index and, 175; definition, 170; estimation, 54, 170–71, 176, 336–37n16, 338n20, 345n15, 347n6; per capita, 170, 176–77, 195, 336n16, 342–43n13; predation rate and, 342–43n13, 345n12; seasonal variation, 338n17; Volterra's equations, 108, 109–12, 170; wolf population correlation, 171–72
kinship, 6, 10–11, 305–6, 307, 309, 345–46n22
Krefting, Laurits, 260

Lake Desor, 27, 29, 30, 183, 204, 228
Lake Mason, 25, 213, 214

Lake Minong, 247
Lake Richie, 180, 200, 203, 214
Lake Superior, 33, 73, 166, 167, 193, 196, 225–26, 241, 242–43, 336n15; formation, 247
land bridges, 48
laws of nature, 99, 106, 151, 265
Leopold, Aldo, 254, 259, 278, 279; *A Sand County Almanac*, 284; "Thinking Like a Mountain," 310–11
life / living organisms: Darwin's perspective, 125, 127; dignified life, 308, 357n46; ecosystems as, 125–32; hierarchical levels, 48–49, 50–51; varieties, 310
Lily Lake, 215
lion fish, 302
lions, 154, 176–77, 298, 301, 318–19, 355n16
Little Todd Harbor, 213–14, 217–18
live-capture, of wolves, 153, 156, 182, 187
live-trapping, of deer mice, 42–43
locus coerulus, 17
logging, 252, 255, 316–17, 349n31
logical fallacies, 262–63
lone wolves, 87, 88, 133–34, 146, 149, 197–98, 215; death, 146; hunting behavior, 201–3; wolf pack attacks on, 166–68, 212–13, 216, 224–28. *See also* Isabelle (female wolf); Romeo (male wolf)
longevity, reproduction *vs.*, 45–46
Long Point, 221, 226, 228, 349n31
Lovelock, James, 309, 310
lumbosacral transitional vertebrae (LSTV), 183–86, 346n28
Lyell, Charles, 99, 104
lynx, 123, 261, 298

M93. *See* Old Gray Guy (male wolf)
M183 (male wolf), death of, 326
macaques, 14–15, 336n11
MacArthur, Robert, 265–67
maggots, 36, 153, 162, 229. *See also* bot fly larvae
malnutrition, 147, 153, 160–63
Margulis, Lynn, 309, 310
Marshall, Bob, 278, 279–80
Mather, Stephen, 252–53, 254–55

mating season, 201–2. *See also* courtship and mating behaviors

McCargo Cove, 7, 87, 179, 197, 211, 229, 250

McLaren, Brian, 164–65, 290–91

meat eating, 32

Mech, David, 97, 121–22, 133, 151–52, 236–37, 272, 342n8; initial Isle Royale research, 69–70; moose counts, 85, 89–90; summer studies, 69–70, 71–72; Vucetich's relationship with, 342n8; winter studies, 73–81, 82–84, 87–90; wolf counts, 85–89; *The Wolves of Isle Royale*, 81, 82, 90, 92

melatonin, 17

memories, 19, 65

metaphor, 10, 335n6

mice, 133, 156. *See also* deer mice

Middle Pack, 24–25, 198, 224; alpha females, 167, 188, 205–6, 215, 218–19; alpha males, 165–68, 212, 213, 221–23; competition with other packs, 165–66, 166–68, 178–79, 195, 219–20; disintegration, 149, 222–23; genetics, 188; hunting behavior, 27–31, 204, 205, 213–14; interactions with wolf pair, 217–18; interaction with lone wolf, 212–13, 216; meat consumption, 206; moose kills, 170, 202, 217, 219; territorial boundaries, 147, 148, 195

Minnesota Forest Products Company, 252

Minong Ridge, 179, 207, 250–51

misanthropy, 304, 305, 311, 355–56n34

Mississippian culture, 283

mites, 119–20, 122

Mongolian people, 4

moose, 152; age, 122, 133, 152, 209, 229; antlers, 56, 59, 229; arrival on Isle Royale, 271–73, 352nn20–22; experience of death, 31–32, 203, 315–16; food chains, 288, 290–96; foot loadings, 140; hair loss, 39–40, 59–60, 63; heat stress, 22, 182, 346n25; individuality, 210–11; interactions with humans, 62; introduction to Isle Royale, 271–73, 342n20; as living beings, 64–65; migration to North America, 48; recording images of, 58–60, 61, 62,

65; responses to wolf attacks, 3–4, 18, 21, 22, 23, 24, 29, 30, 82–83, 152, 203, 205–6; size, 7

moose calves, 3–4, 52, 71, 138, 147, 161, 171, 212, 230; bone processing, 209; survival rate, 91; twins, 228

moose counts, 53, 89–92, 135, 208, 210–11, 229, 230; method, 152, 207–8

moose hunting, by humans, 237

moose-kill frequency. *See* kill rate

moose kills: annual number, 90; average per wolf per month, 53; carcass utilization index, 173–74, 175; failure, 22–23; frequency, 27, 52, 53, 54, 88–89; injuries to wolves during, 18, 24, 29, 30, 152, 203, 205–6; subordinate wolves' access to, 147; of weakened moose, 27–31, 32, 50, 84, 85, 160, 202, 203, 204, 205, 339n11 (chap. 3); by wolf pairs, 215–16; wolves' dreams of, 16–20. *See also* kill rate; *moose kills under specific wolf packs*

moose-kill scavengers, 173–74; foxes, 39, 153, 162, 173, 221; ravens, 20, 28, 82, 162, 173, 176, 177, 206, 211, 217

moose-kill sites, 20–21, 72, 80, 82, 197; age determination of moose, 133; guarding of, 202–3, 211, 212–13, 215; wolves' revisiting of, 30; wolves' scavenging on, 229. *See also* kill rate; *moose kills under specific wolf packs*

Moose of Isle Royale, The (Murie), 255

moose population: between 1930 and 1946, 255–56, 259, 260; in the 1960s, 90–91; between 1970 and 1981, 145–46, 148, 160–61; in 1990, 293–94; between 2004 and 2012, 182–83, 293–94; age structure, 52, 338n18; tick parasitism effects, 50–51, 66. *See also* wolf-moose population dynamics, on Isle Royale

moral dilemmas, 315, 318, 320, 322

Moskey Basin, 54, 196, 213, 221, 244–45

Mount Franklin, 201

Mount Ranier National Park, 257

Mount Siskiwit, 213, 215–16

Mud Lake, 217, 219

mud licks, 56–64

Muir, John, 136, 239-40, 254; "Our National Parks," 283
Murie, Adolph, 255-59, 272, 296
Murphy, Ty, 78
Murray, Don, 82, 85, 88; aerial survey technique, 80, 84, 89-90, 137; contributions to wolf-moose research, 77-78, 133; retirement, 144

National Geographic, 92
national parks, 350n32; anthropogenic threats to, 234, 237, 238-39, 348n2; extinctions in, 267; highway system, 253, 254; scientific mission, 157, 158; species-area relationships, 267
National Parks Association, 259
National Park Service (NPS), 67, 69, 73, 233-38, 252-55; captive wolf release decision, 269-70; competition with US Forest Service, 253-54; predator eradication policy, 257; wolf extinction position, 233-36, 237-38, 273-74, 324, 358n1; wolf-moose research position, 149-50; wolf restoration position, 313, 324-26, 348n5
National Rifle Association, 150, 343n22
National Science Foundation (NSF), 69, 136
Native Americans, 282-83. *See also* Cherokee people; Ojibwe people
natural, as good, 268-69, 271, 297
natural selection, 100-105, 175, 271
nature, fear of, 5-6. *See also* human-nature relationship; wilderness
necropsies, 8, 20-21, 24, 35-37, 54, 84, 162, 206, 345n2; bone processing, 209-10, 347n4; carcass utilization index, 173-74, 175; on wolves, 152-53, 222, 223
need, 307, 312
Nelson, Michael Paul, 285
Newmark, William, 267
Nicholson, Alexander, 123-25
North America, great rift zone, 241-45, 246, 349n24
Northeast Airways, 72, 73
Northeast Pack, 147, 148, 149

observation, 54, 58-59, 85, 125, 151-52; quantification in, 125-27, 129; transformation into insight, 141, 142-44. *See also* aerial observation
Odum, Eugene, 128-29, 340n39
Ojibwe people, 4, 238, 249-50, 305-6, 309, 311-12, 349n27
Old Gray Guy (male wolf), 166-68, 181, 234, 329, 346n34; genetic rescue by, 187-91, 293, 294, 351n16
Olson, Sigurd, 136, 137, 259, 279
Olympic National Park, 257
On the Origin of Species (Darwin), 100-102, 104-5
optimal foraging strategy, 175
organisms, 310; ecosystems as, 125-28; relationships to populations, 48, 49-50
otters, 261
"Our National Parks" (Muir), 283
Our Wildlife Heritage (Allen), 68
owls, 44, 55, 316

Paduka Pack, 196-97, 200-201, 207, 215; alpha wolves, 188, 202-3; territorial boundaries, 195
paleontology, 99
Palmisano, John, 286, 289
Paramecium caudatum, 112-18, 122, 128-29
parasite-host relationships: deer mice and bot fly larvae, 43-46; deer and ticks, 46, 47; moose and ticks, 37-40, 46-51, 65, 161
parasitism, 49, 56, 288
parent-offspring conflict, 10, 11
parent wolves, 8, 9, 345n22; relationship with older offspring, 10. *See also* alpha female wolves; alpha male wolves
partial prey consumption, 174-75
parvoviruses, 154-55. *See also* canine parvovirus
Passage Island, 41-42
pedigree, of wolves, 187
periodontal disease, 211
personality and personhood, 9, 12-13, 335n5
Peterson, Bill, 271-73
Peterson, Candy, 54

Peterson, Rolf O., 27, 30, 31, 34–35, 40–41, 53–54, 188, 219, 342n8; 1980s research, 146, 149–51, 153–54, 155–56; 2010 winter wolf aerial survey, 198, 200–201; Allen's recruitment of, 136; balsam fir research, 291, 293; *Ecological Studies of the Wolf on Isle Royale, First Annual Report, 1970–1971*, 137–39; first summer field season, 136–37; on genetic isolation of wolves, 163; interactions with National Park Service, 149–50, 157–58, 159; retirement, 336n14; scat analysis project, 71–72, 147, 149; snow conditions studies, 138–44; trophic cascade research, 290–91; Vucetich's relationship with, 342n8; on wolf-moose population dynamics, 151
Pickett Bay, 202
Pinchot, Gifford, 253–54
pine martens, 273
Pip (male wolf), 224, 225–26, 227, 228, 232, 348n7 (chap. 7)
piscivory, 49
planktivory, 49
plants: quantification, 125–27, 129; sodium content, 57; succession in, 126
Plato, 95, 96, 338n25
Plotinus, 96–97, 102, 111, 129
pollution, 93, 252, 302
poplars, 256
population dynamics, 48–49, 194–95. *See also* predator-prey dynamics; wolf-moose population dynamics, on Isle Royale
populations, relationships to organisms, 48–50
Pound, Roscoe, 125, 340n35
predation, 268; age of prey, 122; in balance of nature, 102; as creative force, 96–97, 129; food webs and, 48–49, 288; human rationalization, 32; rate, 52, 66, 344–45n1, 345n12, 349n10
predation loss, reproductive gains vs., 91–92
predation pressure, 223
predator extermination policies, 72–73, 256, 257, 295–96, 307–8, 313

predator-prey dynamics, 129–30, 150–51, 286, 288, 341n43, 342–43n13, 354n1; Gause's equations, 112–18, 122; Huffaker's experiments, 117–21, 122; Volterra's equations, 105–12, 116, 117, 170; wasps and beetles, 121–22
prefrontal cortex, 14–15
pregnancy, parasitism during, 56
Preston, Frank, 264, 265, 267
Prince Charming (male wolf), 178, 180, 181, 188
Proescholdt, Kevin, 274
protozoans, ciliated, 112–18, 122, 128–129, 178–79
Purdue University, 69, 72, 73

quadrant method, 125–27
Queenie (female wolf), 269, 270–71, 351n15

rabbits, 45, 286, 289
racism, 271, 283, 312, 352nn18–19, 353n47
radio-collared moose, 208
radio-collared wolves, 55, 180–81, 182, 196, 217; Chippewa Pack, 200, 201, 219, 220; Isabelle (female wolf), 224–28, 230–32; Middle Pack, 201; mortality mode, 221–22; Pip (male wolf), 224
radio-collaring, of wolves, 70, 153–54
radio telemetry, 210
Räikkönen, Jannikke, 183, 185
ravens: as carcass scavengers, 20, 28, 82, 153, 162, 176, 177, 206, 211, 217, 221, 228; scat ingestion, 186, 188
reason, 321–23
Red Oak Ridge, 203
religion, 32
REM sleep atonia, 17
reproduction, opportunity cost of, 45–46. *See also* courtship and mating behaviors
reproductive gains, predation losses vs., 91–92
reproductive opportunities, 195
Robinson Bay, 179, 214
Rock Harbor, 162, 250
Rocky Mountain National Park, 257, 296
Rodgers, Denise, 10

Romeo (male wolf), 196, 200–205, 207, 208, 210; as alpha male, 218–20, 223; death, 223–24; mate, 232; moose kills, 211–12; natal pack, 211, 213, 214, 215, 217, 218
Roosevelt, Franklin D., 255
Roosevelt, Theodore, 72, 279–80
rumen, 56, 228–29, 348n8 (chap. 7)

Sailing Alone Around the World (Slocum), 198
Sand County Almanac, A (Leopold), 284
sandhill cranes, 55
sanguivores/sanguivory, 49, 337n14
scat analysis, 55, 70–71, 147, 149, 186–88, 338n3
scientific knowledge, 178
sea otters, 286–87, 289–90
sea urchins, 286
selachians, 105–6
Semper, Karl, 127–28
sense of place, 245–46, 255
"The Sense of Place" (Stegner), 245–46
sexual behavior, 166–67. *See also* courtship and mating behaviors
Shelton, Phil, 133
siblings, 11, 70; all living beings as, 305–6, 356n38; of wolves, 17, 18, 22, 23, 24, 70, 187, 207, 213, 232, 329, 348n9
simile, 10
Siskiwit Bay, 221, 231
Siskiwit Lake, 22, 82, 188, 196, 203, 204, 205, 212, 213, 215, 219
smell, sense of, 21
Smith, Doug W., 149
Smits, Lee, 68–69, 260, 269
snow conditions, effect on wolf-moose interactions, 138–44, 161, 162, 206, 208, 216, 336n15
snow fleas, 216, 347n2
sociality/socialization, 5, 9, 80–81, 176, 177, 214
social order, 134–35, 165
sodium, 56–57
Southwest Pack, 147, 148, 149
species-area relationship, 263–67
spiders, 173–74
springtails, 124, 347n2
spruce trees, 161, 292–93

Standing Bear, Luther, 282
starvation, 7, 9, 44, 128–29; in moose, 32, 50, 84–85, 161–62; in wolves, 7, 9, 149, 153
Stebler, A. M., 260
St. John's wort, 118–19
Stoll, Albert, 252–53, 257–58
suffering, 276, 282, 315–16, 358n67
Superior National Forest, 254
sustainability, 303
swimming ability, 225, 272, 273, 352nn20–21
symbiosis, 49

teeth: moose, 209, 341n1; wolves, 23, 222, 231
territorialism, of wolf packs, 5, 54, 80–81, 146, 147, 148, 195, 202, 219–20; size of territories, 247
ticks: deer, 38; dog, 38. *See also* winter ticks *(Dermacentor albipictus)*
tigers, 298, 349n16
toads, 64
Todd Harbor, 87, 200–201, 250
Tomes, Arthur C., 72, 76, 77
tracks and tracking, 205, 347n3; of moose, 8; of wolves, 70, 75, 78–80, 85, 86–87, 227
trapping, 122–23, 126
Trio (wolf group), 224, 225–30, 232, 348n7 (chap. 7)
trophic cascade, 289–90; exogenous perturbations, 289–94
trophic interactions, 286–91
Turner, F. J., 278

United Nations, 319
University of Michigan, 255
US Fish and Wildlife Service, 67
US Forest Service, 197, 253–55
US National Park Service. *See* National Park Service
Utida, Syunro, 121–22

value, utility-based, 298–99, 300
vertebrates, 299, 314
vocalizations: howling, 55, 70, 80–81, 220; prior to hunting, 17–18; of wolf pups, 55

Volterra, Vito, 106–12, 116, 117, 121, 122, 143, 151, 340n21

Vucetich, Leah Cayo, 56–62, 41–43, 163, 187, 206

Wallace, Alfred Russel, 102–5, 125

Washington Creek, 28, 56, 202, 215

Washington Harbor, 33–34, 73, 78, 89, 197–98, 216, 217, 220, 227, 272, 293, 349n31

Washington Harbor Club, 272

wasps, parasitic, 121–22

Wayne, Robert, 156, 158

West Pack, 33–37, 165–166

white supremacists, 271, 352n18

"Why Wilderness?" (Olson), 279

wilderness: as abstraction, 275–77, 282; American character and, 276, 278–79; biblical concepts, 276–77; as deprivation of civilization, 276–78, 282; as experience, 277, 279–80; human conquest, 278, 279–82; legal designation, 274–75, 277; machismo appeal, 136, 279–80; philosophers of, 136

Wilderness Act (1964), 237, 259, 284

wilderness aviation, hazards of, 75–76. See also aerial observation

wilderness recreation, 251–55, 257–58, 277–79; "Leave No Trace" ethic, 283–85, 354n49

Wilderness Society, 259, 278

Wilderness Watch, 274, 303–4, 353n25

wildlife management, 157–58

Wilson, Edward O., 265–67

Wind Cave National Park, 296

Windigo station, 54–55, 212

winter ticks (Dermacentor albipictus), 36–40, 46–51, 65, 161; on dead moose, 36–37; as hair loss cause, 39–40, 59–60, 63; mooses' perception of, 46–48; as predation predisposing factor, 50–51, 66, 182

wolf-caribou population dynamics, 315–17, 318, 330–31

wolf counts, 53, 72, 85–89, 89–90

Wolfe, Michael, 133

Wolf Men, The (documentary), 136

wolf-moose population dynamics, on Isle Royale, 52–56, 85, 160, 341n3; in the 1950s, 53–54, 121; between 1959 and 2020, 135, 145–46, 147–51, 152; as balance of nature, 116; effect of canine parvovirus on, 162–63; first winter study, 73–81; following wolf restoration, 327–28; imbalance, 159, 233–34; rescinded cancellation of research, 149–50

wolf packs: balance of power in, 11–12; composition, 8–9; decrease in number, 145, 146, 195; distribution, 247; division of labor, 83–84; as family units, 5, 8–9, 10–12, 135, 329, 335n7; gender roles, 12; interpack competition, 149, 166–68, 178–81, 214; offspring, 8–9, 10; size, 9, 83–84, 87–88, 176–77, 335n3; social alliances within, 134–35; teamwork, 177. See also names of specific wolf packs

wolf pairs, 146; moose kills, 177, 211–12, 215–16; as wolf population founders, 156–57. See also alpha wolf pairs

wolf population, 232, 233–60; between 1958 and 2020, 52, 73, 87–88, 146; between 1970 and 1980, 146; in the 1980s and 1990s, 157, 158–59, 160, 165; between 2006 and 2010, 178, 182, 195; anthropogenic climate change and, 313–14; comparison with deer mouse population, 40–46; National Park Service's response to, 233–38, 324–25; public opinion about, 273–74; stable, 88–89, 90; of the US, 308, 356–57n44. See also wolf-moose population dynamics, on Isle Royale

wolf pups, 9, 147, 181–82, 224, 232, 335n4, 347n1; born to restored wolf population, 326; canine parvovirus in, 155; as percentage of wolf population, 150; personalities, 9

wolf restoration, 157, 324–33; biodiversity and, 303, 331; human culpability and, 317; "Island Complications: Should we Return Wolves to Isle Royale?" (Cochrane), 238, 239, 260, 261–63, 268, 269, 271, 272; National Park Service's policies, 294, 313; public opinion about,

wolf restoration (*cont.*)
324-25; trophic cascade argument for, 296-97, 303-5

wolves: age, 70, 168, 205, 206; diet, 6, 71, 147, 148, 149; differences and similarities to humans, 5-6, 10-12, 13-14, 336nn9-10; food chains, 288, 290-96; geographic range, 301; humans' hatred of, 149-50, 305; killed by other wolves, 179-81; physical appearance, 88; proposed introduction to Isle Royale, 259-60; Scandinavian, 184-85, 346n29; scavenging by, 162; stomach, 176; "unnaturalness" of, 261, 269-71, 273

Wolves of Isle Royale, The (Mech), 81, 82, 90, 92, 341n3

Wolves of Minong, The (Allen), 6-7, 129, 134

woodcraft, 279, 284

wounded moose, 27-31, 202, 203, 204, 205

Wright, George M., 256, 295-96

Yellowstone National Park, 235, 253, 256, 257, 282, 295-96

Yosemite National Park, 254, 282-83

Zahniser, Howard, 274